Science in the Spanish and Portuguese
Empires, 1500–1800

Science in the Spanish and Portuguese Empires, 1500–1800

Edited by

Daniela Bleichmar, Paula De Vos,
Kristin Huffine, and Kevin Sheehan

STANFORD UNIVERSITY PRESS

STANFORD, CALIFORNIA

Stanford University Press
Stanford, California

This book has been published with the assistance of the USC-Huntington Early
Modern Studies Institute; the USC College of Letters, Arts, and Sciences; and
the History Department, Graduate School, College of Arts and Sciences, and
Center for Latino and Latin American Studies at Northern Illinois University.

Printed and bound by CPI Group (UK) Ltd, Croydon, CR0 4YY

Library of Congress Cataloging-in-Publication Data
Science in the Spanish and Portuguese empires, 1500-1800 / edited by
Daniela Bleichmar ... [et al.].
p. cm.
Includes bibliographical references and index.
ISBN 978-0-8047-5358-6 (cloth : alk. paper)
1. Science--Latin America--History. 2. Science--Spain--Colonies--History.
3. Science--Portugal--Colonies--History. 4. Science--Latin America--Histori-
ography. 5. Latin America--History--To 1830. I. Bleichmar, Daniela, 1973-
Q127.L38S36 2009
509.03--dc22   2008041298

Typeset by Bruce Lundquist in 10/12 Sabon

Contents

Illustrations

Figures

Maps

Tables

Tables

Contributors

ONÉSIMO T. ALMEIDA is professor of Portuguese and Brazilian Studies at Brown University, which he chaired for twelve years. He is currently working on a book about the transformation of the Portuguese medieval mind as a result of the era's maritime explorations, the subject of a course he teaches titled "On the Dawn of Modernity." He has published several articles in both languages, including "Portugal and the Dawn of Modern Science" (in *Portugal—the Pathfinder: Journeys from the Medieval toward the Modern World—1300–1600*, edited by George D. Winius, 1995) and "R. Hooykaas and His 'Science in Manueline Style': The Place of the Works of D. João de Castro in the History of Science" (*Ibero-Americana Pragensia*, 1997).

ANTONIO BARRERA-OSORIO is associate professor of History at Colgate University, where he teaches the history of science in the sixteenth and seventeenth centuries, Atlantic World history (1500–1800), and early modern Spanish history. He is the author of *Experiencing Nature: The Spanish American Empire and the Early Scientific Revolution* (2006); "Local Herbs, Global Medicines: Commerce, Knowledge, and Commodities in Spanish America" (in *Merchants and Marvels: Commerce, Science, and Art in Early Modern Europe*, edited by Pamela Smith and Paula Findlen, 2001); and "Things from the New World: Commodities and Reports" (*Colonial Latin American Review*, 2006).

DANIELA BLEICHMAR is assistant professor in the departments of Art History and Spanish and Portuguese at the University of Southern California. Her research focuses on the visual culture of natural history in the Spanish empire, the history of collecting and display, and the history of print. She is the author of multiple articles on these topics, among them "Painting as Exploration: Visualizing Nature in Eighteenth-Century Colonial Science" (*Colonial Latin American Review*, 2006); "Training the Naturalist's Eye in the Eighteenth Century: Perfect

Global Visions and Local Blind Spots" (in *Skilled Visions: Between Apprenticeship and Standards*, edited by Cristina Grasseni, 2007); and "Books, Bodies, and Fields: Sixteenth-Century Transatlantic Encounters with New World *Materia Medica*" (in *Colonial Botany: Science, Commerce, and Politics in the Early Modern World*, edited by Londa Schiebinger and Claudia Swan, 2005). She is currently at work on a book about colonial botany and visual culture in the eighteenth-century Spanish world.

JORGE CAÑIZARES-ESGUERRA is the Alice Drysdale Sheffield Professor of History at the University of Texas at Austin. He is the author of the prize-winning *Puritan Conquistadors* (2006) and *How to Write the History of the New World* (2001). He has also written *Nature, Empire, and Nation: Explorations of the History of Science in the Iberian World* (2006) and has coedited (with Erik Seeman) *The Atlantic in Global History, 1500–2000* (2006).

FIONA CLARK is lecturer in Colonial Latin American Studies at Queen's University Belfast, Northern Ireland. Her early research focused on a socio-cultural and historical study of the *Gazeta de Literatura de México* and led to her current interest in Irish-Mexican networks in the history of medicine in the late colonial period. She is the author of articles published in *Dieciocho, Bulletin of Spanish Studies*, and *Studies on Voltaire in the Eighteenth Century*, and is currently coediting a collection of essays (with James Kelly) entitled *Ireland and Medicine in the Seventeenth and Eighteenth Centuries* for the Ashgate series History of Medicine in Context.

ALEXANDRA PARMA COOK has coedited (with Noble David Cook) *Good Faith and Truthful Ignorance: A Case of Transatlantic Bigamy* (1991) and *People of the Volcano: Andean Counterpoint in the Colca Valley, Peru* (2007). Their next book, *The Plague Files: Crisis Management in Late Sixteenth-Century Seville*, will appear in spring of 2009.

NOBLE DAVID COOK, professor of History at Florida International University in Miami, has authored *Demographic Collapse: Indian Peru, 1520–1620* (Cambridge University Press, 1981) and *Born to Die: Disease and New World Conquest, 1492–1650* (1998) and coedited (with W. George Lovell) *"Secret Judgments of God": Old World Disease in Colonial Spanish America* (1992).

PAULA DE VOS is associate professor of Latin American History at San Diego State University. She completed her dissertation, titled "The Art of Pharmacy in Seventeenth- and Eighteenth-Century Mexico," at the

University of California, Berkeley, in 2001. Her research interests lie in the history of science and medicine in Spain and Latin America in the colonial period. She is author of articles published in *Endeavour, Colonial Latin American Review, Journal of World History, Eighteenth-Century Studies,* and *Colonial Latin American Historical Review.* She is currently working on a book manuscript based on her dissertation, tentatively titled "The Craft of Medicine in Colonial Mexico: The Art and Science of the Apothecary."

JÚNIA FERREIRA FURTADO is full professor of Brazilian Colonial History and Modern History at Universidade Federal de Minas Gerais, Brazil. She was visiting fellow (2000) and visiting professor (2001) in the History Department at Princeton University. Her book *Chica da Silva e o Contratador dos Diamantes: O Outro Lado do Mito* (2003) received the Casa de las Américas 2004 prize. She is also a coauthor of *Cartografia da Conquista das Minas* (2004).

MARTHA FEW is associate professor of Latin American History at the University of Arizona. She is the author of *Women Who Live Evil Lives: Gender, Religion, and the Politics of Power in Colonial Guatemala, 1650–1750* (2002) and has published a number of journal articles, including "Chocolate, Sex, and Disorderly Women in Late-Seventeenth and Early-Eighteenth-Century Guatemala" (*Ethnohistory,* 2005), and "'That Monster of Nature': Gender, Sexuality, and the Medicalization of a 'Hermaphrodite' in Late Colonial Guatemala" (*Ethnohistory,* 2007). Professor Few has held research fellowships at the Newberry Library in Chicago, the John Carter Brown Library at Brown University, and the Huntington Library in Pasadena and is currently at work on her next book, provisionally titled "All of Humanity: Colonial Guatamala and New World Medical Cultures before the Smallpox Vaccine."

PALMIRA FONTES DA COSTA is assistant professor of History of Science at the New University of Lisbon, Portugal. Her research interests include the history of natural history and medicine in Portugal and England from the sixteenth to the eighteenth century and the history of the book and of visual representations. She is the author of multiple articles and the editor of *O Corpo Insólito: Dissertações sobre monstros no Portugal do Século XVIII* (2005). She is currently finishing the book *The Singular and the Making of Knowledge at the Royal Society of London in the Eighteenth Century.*

DAVID GOODMAN is professor of History and Technology at The Open University (Faculty of Arts), Milton Keynes, UK. His research concentrates on science, technology, and society in early modern Spain. His publications include *Power and Penury: Government, Technology and Science in Philip II's Spain* (1988) and *Spanish Naval Power, 1589–1665: Reconstruction and Defeat* (1997).

KRISTIN HUFFINE is currently an NEH fellow at the John Carter Brown Library, at Brown University. She is also assistant professor of Latin American history at Northern Illinois University and the Monticello College Foundation Fellow, 2008–2009 at the Newberry Library in Chicago. She has published articles on Jesuits, natural history, and the intersection of science, race, and empire in colonial Latin America and is currently preparing a book for publication, tentatively titled, "Producing Christians from Half-Men and Beasts: Jesuit Ethnography and Guaraní Response in the Colonial Paraguayan Missions, 1609–1790."

ANTONIO LAFUENTE is a senior researcher at the Instituto de Historia, CSIC (Madrid) and has been a visiting scholar at the University of California, Berkeley. His many publications include a critical edition of Voltaire's *Elements de la Philosophie de Newton* (1998), *Guía del Madrid Científico: Ciencia y Corte* (1998); a book about the exhibition *Imágenes de la Ciencia en la España Contemporánea* (1998); *Georges-Louis Leclerc, Conde de Buffon (1707–1788)* (1999); and *Los Mundos de la Ciencia en la Ilustración Española* (coauthored with Nuria Valverde, 2003). He also publishes the blog "Tecnocidanos": http://weblogs.madri-masd.org/Tecnocidanos/.

HENRIQUE LEITÃO is research assistant at the Center for the History of Science at the University of Lisbon, Portugal, where he teaches in the MA and PhD programs. His research interests include the history of exact sciences in Portugal from the fifteenth to the seventeenth century, and scientific activities in Portuguese Jesuit colleges. He is also interested in the history of scientific books and collaborates regularly with the Portuguese National Library, Lisbon. His publications include *O Comentário de Pedro Nunes à Navegação a Remos* (2002), *O Livro Científico Antigo dos Séculos XV e XVI: Ciências Físico-Matemáticas na Biblioteca Nacional* (editor, 2004), and *The Practice of Mathematics in Portugal* (coeditor, 2004).

ANNA MORE is assistant professor of Spanish and Portuguese at the University of California, Los Angeles. She specializes in colonial Latin American literature, particularly from Mexico and Brazil, as well as aesthetic and cultural theory. She is currently completing a book on the relationship between baroque aesthetics and issues of colonial sovereignty in the works of Carlos de Sigüenza y Góngora.

JUAN PIMENTEL is tenured scientist at the Instituto de Historia, CSIC (Madrid). He held a postdoctoral fellowship at the Université de la Sorbonne (Paris IV) and was a visiting scholar at the Department of History and Philosophy of Science at Cambridge University (1994–96). His publications include *La Física de la Monarquía: Ciencia y Política en el Pensamiento Colonial de Alejandro Malaspina, 1754–1810* (1998); *Testigos del Mundo: Ciencia, Literatura y Viajes en la Ilustración* (2003); and "The Iberian Vision: Science and Empire in the Framework of a Universal Monarchy, 1500–1800" (in *Nature and Empire: Science and the Colonial Enterprise*, edited by Roy MacLeod, 2001), among multiple articles.

MARÍA M. PORTUONDO is assistant professor of history at The Johns Hopkins University. Her principal interests are early modern geography, the transfer of scientific and technological knowledge, and the development of science in Latin America. Her book *Secret Science: Cosmography and the New World* (University of Chicago Press, 2009) explores the role of Spanish court cosmographers during the late sixteenth century and the assimilation of geographic information about the New World into European conceptual frameworks. She is also the author of "Plantation Factories: Science and Technology in Late-Eighteenth-Century Cuba" (*Technology and Culture*, 2003).

KEVIN SHEEHAN completed his licentiate in Missiology and Inter-Religious Dialogue at the Gregorian University, Rome, in 1995, and holds a Ph.D. in history from the University of California, Berkeley. His research interests focus on the dynastic union of Spain and Portugal, 1580–1640, and on the implications of this union for the relationship between the Portuguese and Spanish in the Asia-Pacific region.

NURIA VALVERDE is a researcher at the Instituto de Historia, CSIC (Madrid). Her book *Instrumentos Científicos, Opinión Pública y Economía Moral en la Ilustración Española* (2007) focuses on the construction of the values of precision within scientific networks in eighteenth-century Spain and public understandings of them. She is coauthor with Antonio Lafuente of *Los Mundos de la Ciencia en la Ilustración Española* (2003) and has collaborated on the exhibitions "Madrid, Ciencia y Corte" (directed by A. Lafuente, 1999) and "Monstruos y Seres Imaginarios en la Biblioteca Nacional" (directed by A. Lafuente and J. Moscoso, 2000).

TIMOTHY WALKER is assistant professor of history at the University of Massachusetts, Dartmouth, and a visiting professor at the Universidade Aberta in Lisbon, Portugal. His work focuses on the adoption of colonial indigenous medicines by European science during the Enlightenment, slave trading in the Atlantic and Indian Oceans during the seventeenth and eighteenth centuries, as well as commercial and cultural links between the Portuguese overseas colonies in Asia, Africa, and the Americas. He is the author of multiple articles on these topics and of the book *Doctors, Folk Medicine and the Inquisition: The Repression of Magical Healing in Portugal during the Enlightenment* (2005).

Preface

This volume represents the first English-language anthology of essays by multiple authors that is devoted exclusively to the practices of science in the Spanish and Portuguese empires. The collection covers a wide geographic and chronological range, from Iberia to Latin America to Asia and the Pacific Rim, and from the end of the fifteenth to the beginning of the nineteenth century. The book's main purpose is to show that the Spanish and Portuguese empires were active participants in the practices of science during the early modern period. In this way, it provides a much needed study of the Spanish and Portuguese contributions to the Scientific Revolution in Europe and abroad, a topic about which very little has been published in English, despite its importance for scholars and for students in the university classroom.[1] Similarly, although in the past few decades the complex relationship between science and empire has received enormous attention from specialists in a wide range of fields, the vast majority of studies have focused on the French, German, and British colonies, while the Iberian colonial experience has received remarkably little treatment in Anglophone literature.[2]

The idea for putting together this volume originated with two different panel presentations on Iberian imperial science on which several of the contributors participated, one at the History of Science Society's annual meeting in 2002 and the other at the Latin American Studies Association conference of 2003. The enthusiastic response we received for both presentations encouraged us to proceed with this volume. We were further convinced by the fact that over the last ten years, several writers surveying the existing literature on colonial and imperial science have decried the lack of publication on the Iberian cases and emphasized their crucial importance for redressing imbalances and promoting a more global understanding of the development of early modern science.[3] In addition, the proliferation of programs and course offerings in Latin American history and literature, world history, cultural studies, and sociology and philosophy of science has led to a growing body of scholars who are interested in these themes, and a deeply felt need for publications that can be used in their research and teaching. The impetus for this book thus arose out of these two preoccupations: a historiographical lacuna that needed to be addressed, and a desire to offer our own students, as well as other students and scholars, the opportunity to learn about the sciences in the Spanish and Portuguese empires through a series of articles, written by specialists, that would be helpful in the broadest possible range of disciplinary and topical settings.

The intended audience for this volume is a wide-ranging variety of scholars from different fields and disciplines. The volume should appeal to historians and literary scholars of colonial Latin America, early modern Spain and Portugal, and the Iberian Asia-Pacific; historians of early modern and Enlightenment science; and historians of medicine. It will also provide valuable comparative information and analysis for scholars of European imperialism and colonialism in general from the end of the fifteenth to beginning of the nineteenth century, but particularly scholars of the British empire. In its recognition of the social construction of science, the link between knowledge and power, and the role of knowledge in the formation of colonial identities, this volume, we believe, will also interest scholars from a variety of disciplines, including sociology and philosophy of science, cultural, postcolonial and subaltern studies, and cultural and medical anthropology. In addition to its appeal to specialists, the book is also intended for use in the classroom, for both upper-level undergraduate and graduate students in the fields identified above. Its chapters can be used to teach many different undergraduate and graduate courses, including courses in world history, Latin American and European history, history of science, technology, medicine, Latin American and European studies, and comparative colonialism.

ʊ

This book gathers original contributions from senior scholars, many of whom have produced key works in their field, and younger scholars, whose work illustrates the new directions in which this type of research is moving. The group is interdisciplinary, including authors from the fields of history, history of science, art history, and literature; it is also international, bringing together scholars based in the United States, Europe, and Latin America. We believe the heterogeneity of the group reflects the richness of the topic and of the scholarship currently being produced in a burgeoning field.

The volume is divided into four parts, designed to address significant themes and methodologies that have emerged in a variety of disciplines, including the history and philosophy of science; Latin American, Atlantic World, and Pacific Rim history; as well as literary, cultural, subaltern, and postcolonial studies. Chief among these themes, and related to each of the parts, is a general assessment of Iberian science in terms of its crucial yet undervalued influence on the development of the western scientific tradition in the early modern period. Thus Part I, "Reassessing the Role of Iberia in Early Modern Science," consists of historiographical contributions in which David Goodman, Palmira Fontes da Costa, and Henrique Leitão provide an overview of the current state of scholarship on early modern science in Spain and Portugal and their respective empires. These three authors emphasize the contribution of interdisciplinary studies in promoting better understanding of scientific activities in the Iberian world.

A second aim of the volume is to challenge traditional core/periphery models that place Europe at the center of scientific knowledge production by characterizing colonial science as essentially subordinate, derivative, or imitative. This theme is taken up in two separate but complementary sections of the volume. Part II, "New Worlds, New Sciences," challenges traditional "diffusionist" models based on the assumption that European scientific knowledge spread unidirectionally from a European "core" to a colonial "periphery." By focusing on the ways in which the natural world of the Americas served to complicate and alter European categories of knowledge and methods of determining truth, this section thus emphasizes the dialectical and multidimensional nature of knowledge production. In this vision, no single region is privileged at the expense of others in terms of the legitimacy or primacy of its scientific tradition. Part III, "Knowledge Production: Local Contexts, Global Empires," focuses on the importance of local scientific traditions that developed in the empire and the crucial issue of translating the meaning of this new knowledge between different cultures. It also points to the fact that "local" contexts do not necessarily mean "colonial": local knowledge was produced in Spain as well in response to information coming in from the Americas.

The final portion of the volume deals with the practicalities of empire. Part IV, "Commerce, Curiosities, and the Circulation of Knowledge," agrees with recent scholarship in breaking down the traditional dichotomy made between science for curiosity's sake and science carried out in the interest of economic gain.[4] The essays in this section address the commercial motivation for much of the scientific activity that took place in the Spanish empire, from the sixteenth to the eighteenth century. Yet they also examine the role of intellectual inquiry, exchange, and representation that was involved in this activity. In this way, they demonstrate that so-called "pure" and "applied" scientific activities are inherently intertwined and mutually reinforcing rather than mutually exclusive. The volume also includes an introduction by Jorge Cañizares-Esguerra and an afterword in response to the essays in this collection by Noble David Cook and Alexandra Parma Cook.

As a group, the essays have several underlying goals and assumptions that unite them. First, they aim to challenge traditional assumptions about the rise of early modern science by exploring and presenting new explanations concerning the nature of knowledge production, colonial hegemony, and the ways in which both power and knowledge move and transform in different contexts. These articles also recognize the political and economic motivations of the scientific enterprise in the colonial context. They acknowledge that scientific "truth" is not simply discovered but rather constructed within a complex network of legitimizing institutions and paradigms. In addition, they recognize the importance of indigenous participation and influence in the construction of colonial systems of knowledge, and the fact that so-called "peripheral" science is by no means a derivative handmaiden to that of the metropolis. Finally, this collection proposes alternative ways of studying the development of science, calling for a new emphasis on the specificity of local factors in the construction of knowledge.

Acknowledgments

A book that was as long in the making as this one cannot help but acquire numerous debts of gratitude to those who helped make it possible. First and foremost, the editors would like to thank Jorge Cañizares-Esguerra for his support of the project from its earliest stages. Not only did he agree to chair our panel on science in the Spanish empire at the 2003 Latin American Studies Association meeting in Dallas, Texas, but when we approached him with our idea to bring together scholars in the field for an anthology on Iberian colonial science, he was also very encouraging and agreed to write the introduction to the volume. In addition to his groundbreaking scholarship in Iberian science, Jorge has thus played an important role in mentoring a growing community of scholars in this field as well. We thank him for his advice and support.

We also received aid from others along the way. William Taylor, Paula Findlen, and Susan Deans-Smith generously agreed to review the proposal and provide valuable feedback, and they have been similarly encouraging of the project from the outset. Norris Pope at Stanford University Press has provided excellent editorial guidance and been responsive, communicative, and very supportive throughout each step of the publication process. Emily-Jane Cohen has been similarly efficient and conscientious. We could not have asked for a more positive experience in the publication process. We would also like to gratefully acknowledge the generous subvention provided to help defray publication costs by Peter Mancall, director of the Early Modern Studies Institute at the University of Southern California, and Kenton Clymer, Chair of the History Department at Northern Illinois University. And finally, we would like to thank all of the contributors to the volume. Not only have they provided here excellent examples of their scholarship, but they also have demonstrated collegiality and patience at every turn, and we have enjoyed the opportunity to collaborate with them.

D.B., P.D., K.H., and K.S.

Science in the Spanish and Portuguese
Empires, 1500–1800

Introduction

JORGE CAÑIZARES-ESGUERRA

After perusing this important collection of essays, readers will most likely reach the conclusion I have myself reached: "science" was the handmaiden of the Iberian empires. Cosmography and natural history were the backbone upon which the Portuguese and Spanish crowns built their mighty Christian monarchies. The systematic gathering of information, plants, curiosities, and indigenous knowledges was a trademark of both empires.[1]

Yet the public in general still harbors an image of these empires as built on the quicksand of superstition and greed. Students readily bring up the formula "glory, gold, and God" when prompted to describe the Iberian colonization of the New World. They have been educated and socialized into thinking of the "Spanish conquest" as an adventure led both by soldiers of fortune bent on pillaging and by a fanatic clergy obsessed with converting Indians. In such a narrative, there is no role for "science," for the latter belongs with the likes of Copernicus, Galileo, and Newton—not Cortés and Pizarro. Textbooks on the history of Western Civilization rarely pause to ponder the role played by astronomy and cosmography in the fifteenth-century Iberian southward Atlantic expansion.

The scientific revolution is often linked to the collapse of the Ptolemaic cosmos, as geocentricism and Aristotle's physics gave way to heliocentrism and a new physics of motion.[2] But just as significant was the collapse of Ptolemy's *Geography*: Portuguese and Spanish cosmographers and navigators literally redrew the map of the earth, as Onésimo T. Almeida and María M. Portuondo remind us in their essays. As the Portuguese inched their way around the Cape of Good Hope with the help of cross-staffs, astrolabes, and compasses, they helped rewrite

history on a global scale. The new maritime routes across the Atlantic
and the Indian Ocean helped create new global economies. Contrary
to common opinion, America did not yield gilded treasures but green
ones: naturalists, doctors, apothecaries, and merchants helped identify
new dyes, stimulants, pharmaceuticals, woods, and spices, creating new
fortunes and economies across the Atlantic, as well as new forced mi-
grations from Africa.[3] Iberian metallurgy also transformed the globe.
European merchants traded with their Asian peers using American silver, which was extracted largely with technologies first developed due
to the painstaking experimentation of Iberian metallurgists and alchemists in the Peruvian and Mexican highlands.[4]

But why do students and the general public think of the Iberian empires
in terms of ignorant, zealot friars and plundering conquistadors rather
than of savvy naturalists and learned cosmographers? The answer, to be
sure, lies in age-old religious battles harking back to the Reformation.
Under Philip II, Spain and Portugal became the leaders of the Catholic
Reformation, a movement that sought to stall the spread of Protestantism
in continental Europe and of the Ottomans in the Mediterranean. An
overstretched Spanish empire gradually bled itself white, and Protestant
printers decisively won a propaganda campaign that cast Iberia as a land
of murderous, rapacious conquerors, and benighted, zealous priests.[5] To
be fair, the propaganda campaign also managed to present Iberians as
soulful primitives—artistically gifted playwrights and painters.

Why were the Iberians unable to counter these claims? Surprisingly,
the answer lies in the nature of Spanish and Portuguese print culture. The
first to have suggested this was Alexander von Humboldt. His *Examen
critique de l'histoire de la géographie du noveau continent* (Critical Examination of the History of Geography of the New Continent; 1836–39)
was a massive, five-volume history of the origins of the early modern
Iberian expansion to the New World. His reconstruction of the history
of late medieval and early modern geography was possible, he remarked,
only because new archival sources had recently been made available by
the Spanish scholar Martín Fernández de Navarrete: it was in the archives where scholars should be looking for evidence of Iberian contributions to scientific knowledge. The sources Fernández de Navarrete
published between 1825 and 1837 had in fact been collected some fifty
years earlier by the Valencian Juan Bautista Muñoz. Like Humboldt,
Muñoz thought that the answer to a well-entrenched Enlightenment tradition that dismissed Spanish contributions to early modern knowledge
lay in the archives. Muñoz spent most of his adult life collecting evidence
of Spain's contributions to European knowledge, putting together ninety-
five folio and eighteen quarto volumes of materials he found in the ar-

chives. It is deeply ironic that Muñoz never saw any of these thousands of primary sources into print.[6]

Contributors to this collection have heeded the calls of Muñoz and Humboldt. Most essays are based on painstaking archival research. Timothy Walker, for example, reconstructs the ways Portuguese naturalists and physicians drew upon the indigenous knowledge of Africans, Asians, and Native Americans. His sources are six hitherto unpublished natural histories written in 1596, 1612, 1770, 1788, 1794, and 1799. Antonio Barrera-Osorio reconstructs the massive sixteenth-century culture of Spanish empirical trials, digging up countless sources in archives. Paula De Vos identifies 335 shipments of curiosities that naturalists and bureaucrats in the colonies sent to Madrid in the eighteenth century alone. And the list goes on.

It has taken the collective archival effort of generations of historians to be able to understand a few aspects of the role that science played in the Iberian colonial expansion. These efforts beg the question: Why would the Iberians let their collective efforts gather dust in archives? Anyone who has done work on the early modern Spanish empire has surely noticed that works often circulated in manuscript, not print. One answer to this puzzle lies in the bureaucratic culture of secrecy that the monarchy fostered, *arcana imperii* (state secrets), as María M. Portuondo in her dissertation has reminded us.[7] Knowledge was best kept in manuscripts and circulated only among trusted readers so as to deny any potential imperial rivals information about loosely held frontiers and territories. But in addition to a culture of bureaucratic secrecy, there seems to have been distrust in the media of print itself. As book historian Fernando Bouza has argued, early modern Spain was characterized by a lively scribal culture of manuscripts, the preferred means of circulating knowledge.[8] We know that this was a wider pattern of the Atlantic World. It took some time for the English elites to own up to the virtues of the printing press, originally preferring to circulate their wares in manuscripts, for print was thought to be a medium of charlatans, hackers, poseurs, and struggling lower-class "authors."[9] The circulation of manuscripts using pen names, for example, remained typical of the elite public sphere in eighteenth-century British America.[10] Clearly, Iberian intellectuals shared an even deeper ancien régime suspicion of authorship. Yet they also paradoxically realized the importance of print culture to gaining the propaganda wars for prestige. From the evidence, it is clear that the Iberians badly lost these wars.

The role of images in communication could also help explain the puzzling Iberian culture of print. In a nicely crafted essay in this volume, Daniela Bleichmar shows the sheer amount of botanical images that circulated in the eighteenth-century Spanish empire. The botanical expedition of Mutis alone completed 6,700 folio illustrations of plants. It has taken two centuries for some of these illustrations to be published. Curiously, of all the Enlightenment-age sciences, the Iberians excelled in one in particular: botany. It is true that the demands of empire put a premium on establishing monopolies of new agricultural commodities, thus promoting expeditions of all kinds.[11] Yet botanical images worked in the Catholic monarchy in strikingly similar ways, as did religious ones; namely, as mnemonic aids to help distant audiences experience the plant itself without traveling.[12] Within the empire, knowledge circulated widely through images. For both the learned and the masses, the path to memory, piety, and the emotions began with images, not print.[13] It is difficult for readers to grasp the all-pervasive presence and sheer monumentality of visual culture in the Catholic monarchy: large and complex urban economies of artisans and guilds sprouted from scratch. Their sole function was to keep up with the production of religious paintings and sculptures. The culture of the Iberian empires postulated that religious (and secular) knowledge was to be gained through the senses. Along with sounds, smells, and the choreographed motion of the theater, images and objects played key roles in this religious, empirical epistemology. It should therefore not surprise us that, along with botanical illustrations, there was also a brisk demand for natural objects to collect, keep, exchange, and study, as Paula De Vos has marvelously documented.

Perhaps one of the most important contributions of this volume is to remind readers of the particular religious culture in which the scientific practices of the Iberian empires flourished. Júnia Ferreira Furtado introduces the reader to José Rodrigues Abreu, an enlightened Portuguese medical reformer who, after visiting Minas Gerais in Brazil, was convinced that the original location of Paradise was in the American tropics. Rodrigues Abreu was concerned with the alchemical origins of gold and was deeply aware of the religious, millenarian overtones of alchemy and Paracelsianism.[14] Abreu belonged to a long tradition of learned, millenarian alchemists operating in Minas Gerais whose political impact is poorly understood. Adriana Romeiro for example, has described the plights and adventures of an Abreu counterpart: Pedro de Rates Henequin, who also thought that Paradise had originally been located in Minas Gerais. Henequin sought to persuade the younger brother of João V, Manuel, to lead an independent kingdom in Brazil, the much-awaited Fifth Monarchy of Daniel's prophecies. For all their contributions, we still do not know the alchemical foundations of the millenarianism of these naturalists.[15]

As Kevin Sheehan's fascinating study of the pilot-cosmographer Pedro Fernándes de Quirós suggests, varieties of this alchemical millenarianism flourished in the Spanish empire much earlier. With the help of powerful patrons in Rome—including the pope Clement VIII and a learned Spanish Neoplatonist, the Jesuit Juan Bautista Villalpando—Quirós devised a plan to discover the *quarta pars incognita*, an Antarctic continent expected to exist in the Antipodes to keep the planet in balance. With the help of several special apparatuses of his own design, including machines to measure longitude and to obtain fresh water out of seawater, Quirós eventually landed in 1606 in what today is New Guinea. He christened the new land "Austrialia" (*sic*) to honor the Austrian roots of the Spanish Habsburg Philip III. He also laid the groundwork for a new city in the wilderness, New Jerusalem. What is remarkable about this long-forgotten adventure is not so much the striking modernity of many of Quiros's devises, but the religious world that spawned the expedition. Villalpando wrote a Neoplatonic interpretation of Ezekiel's description of the Temple of Solomon, arguing that every detail of the temple was a prefiguration of the history of the Catholic church.[16] Villalpando also maintained that the Israelites, the temple, and King Solomon were prefigurations of the Iberians, the Escorial, and the Spanish Habsburgs, in that order. Quirós belonged to Villalpando's world of Catholic Neoplatonism and typological readings of history and the cosmos. It is not surprising, therefore, that Quiros envisioned Philip III as a new Solomon whose New Jerusalem in the New Ophir that was Australia was the fulfillment of biblical prophecies, including the arrival of a new millennium. The "modernity" of his mechanical devices belongs in the same world as his founding of New Jerusalem in Austrialia, part of the larger learned (Neoplatonic, typological), millenarian traditions of the Catholic monarchy.[17]

Juan Pimentel's subtle essay on Juan E. Nieremberg's natural history of American wonders is yet another example of the importance of baroque Catholic theology in the framing of interpretations of the natural world in the Iberian empires. Pimentel argues that seventeenth-century Spanish natural history went beyond utilitarian, commercial goals; it was ultimately religious. Nieremberg was a Neoplatonist, not unlike Kircher, who found in nature a language of religious signs to be decoded. Such messages pointed to occult sympathies, micro- and macrocosmic correspondences, and more important, narratives of human sin and Christian salvation. Iberian science, as Antonio Barrera-Osorio and María M. Portuondo suggest, was pragmatic and utilitarian. Yet it also belonged in a religious, cultural world that historians cannot afford to overlook. I am sure readers will enjoy the many learned contributions in this collection. It is my hope that they will also be drawn into mental worlds that still remain largely uncharted.

PART I

Reassessing the Role of Iberia in Early Modern Science

CHAPTER ONE

Science, Medicine, and Technology in Colonial Spanish America

New Interpretations, New Approaches

DAVID GOODMAN

"The history of Spain is one of the most attractive fields that lie open to the historical student. Its variety is infinite and the possibilities of new and important discoveries are unexhausted." So began the stirring preface to that pioneering and still-admired work of the early twentieth century, *The Rise of the Spanish Empire* by Roger Bigelow Merriman, professor of history at Harvard. And addressing himself to fellow American readers, he continued: "The principal interest of the subject will inevitably center around Spain's activities as a great conquering and colonizing power," a focus that demanded viewing "the history of Spain herself, which forms the background for the entire picture . . . from the standpoint of the great Empire which sprang from her."[1]

How different the portrayal of the same subject, three-quarters of a century later, in Henry Kamen's *Spain's Road to Empire!*[2] Here, the very idea of a Spanish conquest of the New World is rejected because the vastness of America rules it out. Rejected also is the traditional belief that Spain created an empire unaided. Instead, Kamen insists that Spain's empire existed only by collaboration with Italian, German, and Flemish military and technical experts, and through commercial connivance of English and Dutch competitors. Other foreigners, Chinese, sustained Spain's Far Eastern entrepôt of the Philippines by running Manila's economy, shipbuilding, and directing the trade primed by the annual arrival of the silver galleon from Mexico, across the Pacific. And Kamen highlights the importance of native Indians and African blacks for the retention of Spain's Caribbean possessions: they provided the labor for plantations and, as soldiers, played a key role in securing the strategic centers of Havana and Portobello.

Kamen's emphasis reflects the great alteration in approach in recent historical writing on all empires—Spanish, Portuguese, British, Dutch, and French. There is a new sensitivity to the perils of Eurocentrism and to unjustifiable relative neglect of the victims of colonization, sentiments encouraged by the now prevailing antipathy to imperialism. Avoidance of Eurocentrism and heightened interest in the contributions of Amerindians and Creoles (those born in Spanish America and descended from immigrants from peninsular Spain) are also conspicuous trends in recent research on science, medicine, and technology in Spain's American empire.

THE QUICKENING PACE OF RESEARCH

Over the last two decades there has been a surge of publication on science, medicine, and technology in colonial Spanish America.[3] In this considerable output, two quite distinct types of research are discernible. First, new interpretations have formed based on traditional empirical study of printed and—this noticeably increasing—archival sources. Second, new interpretations derived from radically new approaches to Ibero-American history are on the increase, inspired by disciplines outside of conventional history. My survey attempts to illustrate and assess these trends by critical discussion of a selection of the more important publications. Many of these carry the logo of an institution sponsoring the quincentenary commemoration of the discovery of America.

Of all the stimuli to recent research and publication in the field, nothing matched the motivating force of the quincentenary. In the United States, an academic could recall "the great wave of the quincentennial, from the time it began to swell in the late 1980s until it finally rolled to shore in 1992," bringing a multitude of "events, exhibitions, speeches, films, and writings."[4] In France in 1985, Tomás Gómez, professor at the University of Paris, could already detect a strong "revival of interest in the discovery of the New World" among French researchers, as the quincentenary approached. As a French Hispanicist doing research in Madrid, Gómez was ideally positioned to initiate collaboration between scholars mobilizing around similar research programs on both sides of the Pyrenees, programs in which he identified the common ground as "the stimulus to intellectual activity given by the discovery of America, especially to science."[5] In Spain, a new and ambitious research venture, with which Gómez sought to connect French scholars, was begun in 1984 by the Consejo Superior de Investigaciones Científicas (CSIC), Spain's Higher Council for Scientific Research. Here, the aim was to fund research on "the history of culture and science with reference to relations between the Old and New Worlds."[6] Six of the Consejo's research institutes were primed for this program and joined by academics from as many Spanish universities. José Luis Peset, appointed to direct the project, vividly recalled the hectic activity generated: "We could not lose sight of the celebration of 1992—it was almost obsessive at the time. . . . We wanted some good monographs to record the revival of interest amongst Spaniards and Spanish Americans in our common past."[7] Around one hundred researchers, engaged in ten projects, began work in 1985–87. Money was released for the urgent purchase of periodical sets, filling gaps in Spain's libraries. Professors, Spanish and non-Spanish, were hired to deliver courses on themes connected to the research and colloquia organized. And as soon as one project was completed (published as three volumes of eighty-one essays in well over 1,500 pages, ranging from life sciences to technology and society in colonial Spanish America), so another began, also to mark the quincentenary: "España y América: Ciencia y cultura entre dos mundos" (Spain and America: Science and Culture between Two Worlds).[8]

Any historian of science from the United Kingdom sampling those eighty-one essays is bound to be astonished by the inclusion of a contribution on political conflicts in Trinidad and another on urban violence in present-day Brazil—this one written by a member of the CSIC's Department of History of Science in Madrid. The explanation, of course, is that in the Hispanic world "science" (*ciencia*) has a much broader meaning, comparable in scope to usage in France and Germany. Of the various sections of the CSIC, one covered theological science, philosophical science, juridical science, and economic science; another covered historical science and philological science; others were devoted to natural and technological sciences. The CSIC's original religious drive may now have weakened, but the umbrella of meanings of "science" persists, in keeping with current usage in both peninsular Spain and Latin America. Awareness that this has potential for confusion provoked Mariano Peset, historian at the University of Valencia, to bring clarity from the outset to the proceedings of the Second International Congress on the History of Hispanic Universities, held in Valencia in 1995. In his prologue surveying developments in recent research on the history of universities, Peset remarked that historians of science had made important contributions, a comment that forced him to elaborate, because the history of science had "diverse senses": one sense signified the history of all university disciplines; the other sense, much more restricted, signified natural science. He needed to "establish this duality" because he wanted to identify the advances in understanding of the history of Hispanic universities made by scholars studying curricula as a whole and by historians of science (in the narrower sense) who had focused on the development of the precocious medical faculty of early modern Valencia.[9]

EMPIRICAL STUDIES

The first wave of scientific and technological excitement generated in peninsular Spain by the discovery of America had already appeared in the 1500s. The demands of transatlantic navigation had by then become acute: reliable charts, precise astronomical instruments to measure latitude and longitude, and the training of pilots. All of this began to be organized from 1508 at the Casa de la Contratación (House of Trade), recently founded by the Catholic monarchs in Seville, the official center monopolizing all trade and communication with the Indies. (See Portuondo's arguments in Chapter 3.) There is nothing outstanding in the way of recent research on Seville's *Casa*. But important work has appeared on later sixteenth-century efforts to achieve the same navigational goals.

Amid dissatisfaction with the quality of charts, instruments, and pilots produced in Seville, Philip II decided that urgent action was needed. In 1582 he appointed the Valencian cosmographer Jaime Juan to undertake a variety of tasks in voyages to New Spain and the Philippines. The duties included supervision of pilots to see how their practices could be improved, teaching them to use navigational instruments, making maps in the Indies, and determining the latitude and longitude (from lunar eclipses) of localities in New Spain. New light on Juan's observation of a lunar eclipse in New Spain in November 1584 to establish the longitude of Mexico City has come from archival research in Seville and Mexico by María Luisa Rodríguez Sala of the Universidad Autónoma de México (UNAM). She discovered not only the manuscript in which Juan recorded his astronomical observations and calculations, along with astrological predictions, but other new documents as well that show how the operation was performed. The technical skills of Cristóbal Gudiel, royal armorer and gunpowder manufacturer, were central. Gudiel made and installed the instrument used to observe the eclipse.[10]

At the same time that he employed Jaime Juan, Philip II created his mathematical academy in Madrid, sending experts from his newly con quered kingdom of Portugal to teach the mathematics that was indispensable for navigation and cosmography. (On the relationship between science and navigation, see the chapters by Almeida and Sheehan in this volume.) Philip's academy has long been a shadowy institution, but now less so as a result of the study by María Isabel Vicente Maroto and Mariano Esteban Piñeiro, who publish documents providing more information on the teaching personnel and the functioning of the institution which continued into the seventeenth century.[11]

The ambitious centralizing policy to achieve comprehensive information on every part of Spain's vast American possessions is a well-known feature of Philip II's administrative drive. Yet, as Jesús Bustamante shows, the origins of this inquiry are traceable to the previous reign, to Charles V's officials of the Council of the Indies in the later 1520s.[12] But the peak of this systematic inventorying that included much scientific data on natural resources was reached in the later sixteenth century with the fully developed *Relaciones* (geographical accounts). (For a study of the *Relaciones,* see Barrera-Osorio in Chapter 11.) These were responses to formal questionnaires devised by officials of the Council of the Indies, questionnaires seeking precise information on the population of a locality, the status of Indians (free or unfree), climate, disease, churches, fortresses, ports, plants, mines, and other matters of strategic and economic importance. By 1730 the questionnaire, sent to all parts of Spanish America, had grown to 435 questions. The replies depended on the knowledge of local Creoles and mestizos (persons of mixed Spanish-Indian parentage). It is now acknowledged that when this mass of information arrived in Madrid from across the Atlantic, it was immediately put in the state archives and so failed to become the basis for government action. Today, many of the *Relaciones* have been either lost or buried in archives, but many others have been conserved and some published in the nineteenth century, though in imperfect editions—hence the long-term project of Mexico City's University Institute of Anthropology to produce purer, critical editions.[13]

Historians have recently become increasingly aware of the richness of this source material for the colonial period, and analysis of the *Relaciones* has become a research project of CSIC's Department of the History of America in its Centro de Estudios Históricos in Madrid. How much can be gleaned from these sources on sixteenth-century natural history of colonial Spanish America is apparent from the careful and detailed study by Raquel Álvarez Peláez, historian of science at the CSIC.[14] Particularly impressive is the study on the questionnaires themselves. A CSIC edition now not only publishes for the first time the variety of questionnaires but also becomes the indispensable guide for reading this type of document.[15] An introductory essay in the same volume suggests that analysis of the questionnaires reveals changing ideas on health and disease in the colonial period and points to the recurrence of questions seeking information on medicines used by Indian healers, questions not readily answered due to the secrecy of Indian tradition.[16] (For a further examination of indigenous colonial medicine, see Few in Chapter 7.) Frustration over that lack of forthcoming information provoked this comment by an official drafting a *Relación* from replies to the questionnaire of 1604: "It is certain that the Spanish have not striven to discover them [the Indians' medicinal remedies] as much as they have to find gold." The author of this and the rest of a hitherto anonymous *Relación* has now been identified. Scholarly linguistic analysis of the manuscript, conserved at Madrid's Biblioteca Nacional, convincingly demonstrates that it was written by Pedro de Valencia, humanist and chronicler of the Indies. Valencia never crossed the Atlantic, but he was fired with intense curiosity about the natural history of the New World. Indeed, as this new study shows, his interest led him to undermine the questionnaire that had been devised by the Count of Lemos, president of the Council of Indies. Valencia's *Relación* did not follow the order of that questionnaire, demoting the priority of questions on silver mines by relegating the responses to a brief appendage. That reflected his belief that precious metals were a curse for Castile and misery for Indian mine workers. Instead, his *Relación* emphasized medicinal plants and geographical features.[17]

Pedro de Valencia had misrepresented Spanish endeavors to discover the secrets of Indian medicines. For Philip II had sent his court physician, Francisco de Hernández, to New Spain for that very purpose (see Barrera-Osorio in Chapter 11). The resulting monumental survey of the 1570s described some three thousand plants, almost the totality of the flora of New Spain, including descriptions of their medicinal virtues, revealed by Hernández's successful collaboration with Indian herbalist-healers. Much important new light on this climactic botanical expedition and previous accounts of New World flora has come from the highly productive team of historians of medicine and science at Valencia University. In one publication, José Pardo Tomás and María Luz López Terrada undertake a close textual study of the earliest reports by Europeans of the New World's exotic flora, a task intended to further understanding of how the Old World assimilated this astonishing extension of botanical knowledge.[18] There is judicious critical analysis of familiar and lesser-known texts, including the penetrating observations of Roman Paré, the Catalan friar who sailed on Columbus's second voyage to evangelize the Indians of Hispaniola and was soon describing some of the herbs they used in rituals and in healing. As for Columbus himself, from the journal of his first voyage to communications on his fourth and final voyage, his news included scattered details on American flora. But Columbus was no naturalist. As the authors show, his descriptions of species were imprecise and marred by false identification through his conviction that he was in the Far East. By contrast, the authors praise the quality of botanical descriptions in Gonzalo Fernández de Oviedo's *Sumario de la natural historia de las Indias* (Summary of the Natural History of the Indies; Toledo, 1526) and *Historia general y natural de las Indias* (General and Natural History of the Indies; Seville, 1535). These works are rightly seen as transcending mere chronicle by their concentrated focus on nature, the first true treatises of the natural history of the New World that include the earliest descriptions of guaiacum, macagua, and other medicinal plants. Fernández de Oviedo's insistence on the superiority of direct observation over accepting what the Ancients had written has long been recognized, but here Pardo Tomás and López Terrada reinforce that insistence by highlighting Oviedo's ability to overcome his own strong attachment to Pliny. Another fine point made is precocious recognition in the *Crónica del Perú* (Chronicle of Peru; Seville, 1553) of Pedro Cieza de León, soldier and chronicler, of an Andean zone of vegetation. In a valuable concluding section, historians Pardo Tomás and López Terrada present a catalogue that testifies to their progress in tackling the notorious difficulty of identifying American plants from their sixteenth-century descriptions.

Pardo Tomás has also fruitfully collaborated with the prolific José María López Piñero, professor of history of medicine at the University of Valencia and the then director of its Instituto de Estudios Documentales e Históricos sobre la Ciencia. The result is a book that persuasively reinterprets the European influence of Francisco Hernández's revelation of the botanical treasure of New Spain.[19] That influence was long discounted because of the destruction of Hernández's original manuscript in the seventeenth-century fire in the Escorial where Philip II had shelved it. Indeed, after careful examination of the Escorial library's current possessions of herbaria and botanical illustrations, Pardo Tomás and López Piñero found that not a scrap survives of Hernández's work. Yet they demonstrate his considerable influence. That was ensured by the existence of other texts elsewhere. They publish for the first time a seventeenth-century manuscript conserved in the library of the famous medical school of Montpellier and demonstrate that it is nothing less than a collection of copied texts from the destroyed original treatise by Hernández. The Montpellier manuscript is the key to reconstructing the content and structure of the lost original. In addition, the epitome of

Hernández's treatise was published in seventeenth-century Rome. The influence of Hernández on leading European botanists Robert Morison, Étienne Geoffroy, and John Ray is made clear.

The explanation by Carmen Benito-Vessels of the long delay in publishing Hernández's work has failed to convince. She cites censorship by Philip II to protect Catholic orthodoxy from Hernández's naming of plants (allegedly perceived by theologians as heretical, since only God could name and create living beings), Hernández's friendship with Judaizers, and his undue sympathy for heathen Indian medicine. Less convincing still is her argument for interpreting Hernández's posting to New Spain as exile.[20]

After the death of Philip II, the later Spanish Habsburgs presided over a steady decline in Spain and its empire, culminating in the nadir of Charles II's reign (1665–1700). Then Spain's fortunes revived with the enlightened reforms of a new dynasty, the eighteenth-century Bourbons: Philip V (1700–24; 1724–46), Ferdinand VI (1746–59), Charles III (1759–88), and Charles IV (1788–1808). So runs the still-influential conventional history. This view then began to change through a challenging reinterpretation on two fronts in the 1980s. First, Henry Kamen showed that the reign of the last Spanish Habsburg was not nearly as black a period as had generally been supposed: in fact, impressive economic, political, and scientific developments had occurred.[21] Then, in a judicious revisionist survey, John Lynch argued that it was questionable whether the Bourbons were an improvement on the Habsburgs. Charles III was "overrated" by historians, he was "not enlightened" and "not an intellectual innovator."[22] With very few exceptions, Bourbon ministers were anything but enlightened radical reformers. José de Gálvez, Charles III's minister of the Indies, is judged an "anti-Enlightenment bigot."[23] Yet, as Lynch acknowledges, Charles III's government sought to promote technical skills and practical knowledge, above all as a means to strengthen the state, which meant a more determined imperial policy. Lynch's portrayal convincingly demonstrates that a changed policy to America was one of the most distinctive features of the Bourbon monarchs. Vigorous reassertion of imperial control resulted in a "second colonization" of America,[24] dedicated to greater control of America's resources and resolute defense against foreign, especially British, rivalry for empire. In America, taxation became oppressive, Indian laborers became more exploited, and concerted efforts were made to remove Creoles from the positions they occupied in the church and bureaucracy, replacing them by peninsular Spaniards deemed more reliable. The role of the Indies was to produce for metropolitan Spain. By the 1790s, Jorge Escobedo, intendant of Lima and councillor of the

Indies, could declare that the colonies were "countries from which we seek to squeeze the juice."[25]

Behind all these policies lay the Bourbons' priority to develop a powerful navy, and that, above all, meant modern scientific and technical training for officers. That policy would have repercussions in colonial Spanish America. In an influential interpretation, Antonio Lafuente (of the CSIC's Centro de Estudios Históricos in Madrid) and Manuel Selles (of UNED, Spain's distance-teaching university) argue that for lack of civilian scientific institutions, Spain's armed forces became the main institutional cultivators of modern science in the peninsula. (For Lafuente's arguments concerning the control of geographical space and the extension of colonial hegemony, see Valverde and Lafuente in Chapter 10.) They study the creation (in 1717) and fluctuating progress of the Academia de Guardias Marinas at Cadiz with its eventual introduction, in 1783, of a course of advanced studies, followed by sister institutions at Cartagena and Ferrol. The courses included trigonometry and astronomy, but there was internal resistance to this by one director of the Cadiz academy who judged inappropriate the use of advanced textbooks that sought to raise naval cadets to the "sublime heights of geometry and physics."[26]

In a well-researched study based on state and private archives, the scientific career of one graduate of the naval academies of Cadiz and Ferrol is traced by Dolores González-Ripoll Navarro, historian of science at the CSIC in Madrid. The cadet's training brought posting to the strategically important Strait of Magellan to complete a cartographic survey. On return he joined the team at the Cadiz naval observatory and then, in 1792, was sent to command an expedition to prepare a maritime atlas of North America, resulting in twenty-six maps today housed in the Museo Naval, Madrid. The inventory of his library shows his possession of numerous French astronomical and mathematical treatises.[27]

Lafuente and others have similarly made a strong case for the importance of Spain's eighteenth-century military academies as prime centers for teaching modern scientific and technical knowledge, a development characterized as "an enormous interest and heavy investment in science" with expenditure on scientific books, instruments, and professors' salaries that put Spain's universities into the shade.[28] The creation of these academies in Barcelona, Cadiz, Madrid, and Oran; the numbers of student cadets; and details of the curriculum and manuals in use have now been researched.[29] In funding these academies, the Bourbons' intention was to supply proficient military engineers to strengthen the monarchy's defenses, but not only that. In 1768–1800, 211 of these trained experts were at work throughout Spanish America, from Chile to Florida. Along with defense works, they were engaged in a multitude of civil engineering activities: water supply, tobacco works, and the building of hospitals, prisons, palaces, bull-rings, and "churches large and small all over America."[30] Miguel Constanzó, trained in a peninsular military academy, was posted to New Spain in 1764, remaining there until 1814. This lifelong service included participation in the strategic operations to secure the northwest extremity of New Spain from Russian incursions, a task requiring improved mapping through astronomical observations. Similarly he surveyed the zone around Vera Cruz, planning defense in case the British occupied that port. But he was also busy with urban improvement in Mexico City, directing the relocation of the central market, restructuring the Plaza Mayor, improving street paving, drainage, and canals.[31]

The most striking manifestation of scientific activity in colonial Spanish America under the Bourbons is the remarkable cluster of expeditions sent to the Indies. Between 1735 and 1805, as many as fifty-four arrived in America (another nine expeditions sailed beyond America to the Philippines). A very large bibliography on the topic had already accumulated when, in the 1980s and '90s, a torrent of new publications expanded it considerably: monographs and a series of glossy, commemorative works (to mark the quincentenary)—some multivolume—publishing for the first time numerous documents and plates generated by particular expeditions and including an edited collection of brief essays of varying quality. The best of this output—some outstanding scholarship here—was achieved by deeper archival research (assisted by some superb new inventories produced by archivists) and more profound interpretation.

A first step toward supplying an overall view of this burst of expeditions was made in a very helpful pioneering essay by Ángel Guirao de Vierna of the Department of the History of America attached to the CSIC's Centro de Estudios Históricos in Madrid. By means of bar charts, graphs, and analytical tables, the number and frequency of expeditions, their main objectives, destinations, and ministries organizing them are all clarified. The results show peaks and troughs. The sharp decline in the number of expeditions dispatched after 1774—not one was sent during 1776—is here attributed to ministerial changes. But from 1777 on, through the greater preoccupation with America and energetic efforts of José Moñino y Redondo, the Count of Floridablanca (secretary of state) and José de Gálvez y Gallardo (secretary of the Indies), a decade-long surge in expeditions occurred, some of them on an enormous scale. The greatest peak of expeditionary activity was reached in 1792, in Charles IV's reign, but as Guirao indicates, some of these had already been authorized by Charles III. The analysis shows that responsibility for most of the expeditions (70 percent) belonged to the Secretary for the Navy and the Indies and the two secretariats that subsequently resulted from administrative division.[32]

A start has also been made on assessing the costs of these expeditions. This task is made difficult by the lack of complete data on expenditure for even individual expeditions, also because funds came not just from the royal treasury but from private individuals and groups as well. At a time of economic crisis, the crown was forced to look to the reserves of Madrid's powerful guilds, the Cinco Gremios Mayores, and this financial source, suggests María Luisa Martínez de Salinas Alonso (historian of the University of Valladolid), may be "the key to the preparation of these expeditions." She emphasizes the mounting costs even when an expedition had ended: archival manuscripts record the costs of transport of specimens collected, of construction of special containers, the years of continuing payment of salaries of scientific personnel returning to Spain to study results, and the expense of publication.[33] In a paper based on research of documents conserved at Madrid's Museo Nacional de Ciencias Naturales, Rodríguez Nadal and González Bueno found that Charles IV, with a treasury emptied by spiraling costs of war, sent a circular to viceroys, prelates, municipalities, and universities in Spanish America and the Philippines seeking voluntary donations to fund the expensive publication of an illustrated flora Americana. The resulting 43,000 pesos, including a "derisory" 6,000 from prosperous New Spain, is judged by the authors completely inadequate.[34]

What was the motivation for launching all these expeditions? Is it justifiable to continue to call them "scientific" expeditions, as they have commonly been designated? In the early nineteenth century, Alexander von Humboldt, intrepid Prussian explorer of the natural history of the New World, pronounced his famous declaration that no European government had sacrificed sums of money as great as Spain had to encourage knowledge of plants. Interpretation has changed in the last two decades. The most conspicuous trend has been to reinterpret the expeditions as geostrategic missions that combined scientific activity, thus deeming them politically motivated and better justifying the expenditure. Jean-Pierre Clément, Hispanicist of the University of Poitiers, rightly warns historians to be skeptical of the philanthropic rhetoric of ministers like José de Gálvez, who announced that the expedition he had just organized and sent to Peru in 1777 was to collect plants that promised to be "of great benefit . . . for the good of Humanity." His statement concealed the reality of high political and material stakes in the fierce rivalry between competing imperial powers. Accordingly, botanists on the expedition avoided shipping descriptions of new plants to Spain lest they were captured by the British navy. And the French member of the expedition,

Joseph Dombey, after discovering saltpeter on Peru's southern coast, made sure that his account of this basic component of gunpowder was published in a French journal.[35]

In their penetrating analysis of the earlier geodetic mission to Peru, Antonio Lafuente and Antonio Mazuecos recognize that the French and Spanish governments would not have spent so much just to settle the prevailing scientific dispute on the shape of the earth. Instead, imperial politics was the driving force, at least for the governments. The two Spanish members assigned to this expedition, Jorge Juan and Antonio de Ulloa, were young naval officers who had just graduated from the Cadiz academy. The authors show how, in the course of the expedition, they matured into men of science. Juan was soon given recognition by the Académie des Sciences. Ulloa's separate return journey was more dramatic: he was captured by the British—he threw into the sea sensitive political information on the Indies—and taken to England as a prisoner of war. Not long afterwards he was elected fellow of the Royal Society.[36] In another publication, based on archival research on both sides of the Pyrenees, Lafuente reveals the secret maneuvers of Maurepas, Louis XV's minister of the navy: instructing Champeaux, ambassador in Madrid, to allay Spanish suspicions that the proposed French expedition to Peru was motivated by commercial ambitions in Spain's Indies, while simultaneously dispatching a coded letter to the military governor of the French possession of Petit Gôave to exploit the opportunities offered by the geodetic expedition for contraband trade with the Spanish colonies. This paper also reveals the consequences of inadequate French funding. Some members of the expedition survived by practicing medicine, all participated in illegal traffic in precious metals, and Louis Godin sought salvation from poverty by marrying into a wealthy family of French settlers in Quito—he also secured appointment to the chair of mathematics at the university of Lima.[37]

The mixture of geopolitical and scientific motives—more, their inseparable combination—in the Malaspina expedition (1789–94), the climax in scale and ambition of the eighteenth-century expeditions, has been persuasively argued in Juan Pimentel's brilliant study (see also Pimentel's essay in Chapter 5).[38] Two recent developments laid the foundations for the success of this powerful essay in intellectual history. First, a detailed inventory was published of over four thousand documents generated by the expedition (records of scientific observations, maps, sketches, diaries, and correspondence) and housed in Madrid's Museo Naval.[39] Pimentel found this mass of information indispensable for "entering the labyrinth." Second, a manuscript of great importance was discovered in 1989 in the Archivo Histórico Nacional, Bogotá, by Manuel

Lucena Giraldo. Soon studying this with Lucena, Pimentel recognized that he had found the key to the questions that were at the heart of his research project: What were the political motives of Malaspina's expedition? And where was the connection with science? Now the very title of the unearthed manuscript pointed to the answers. Written by Malaspina himself, the document carried the title *Axiomas Políticos sobre la América* (Political Axioms about America), words that immediately conjured up the application of Newtonian natural philosophy, with its structure of axioms, to politics. This was, therefore, another example of the Enlightenment genre of seeking to use the methods of Newtonian science to explain the workings of society. Malaspina was attempting to elucidate the laws governing the machine that was the Spanish empire. Written before the expedition—an expedition that was conceived not by the crown but by Malaspina himself—Pimentel came to see that expedition as an experimental test of the axioms Malaspina had postulated in the manuscript.

Pimentel portrays Malaspina's scientific and technical education, strongly permeated with Newtonian influence, from his origins in Bourbon Sicily to his studies at an Augustinian college in Rome and then as a naval cadet at Cadiz's Academia de Guardias Marinas. By 1788 he sent Antonio de Valdés, minister for the navy, a proposal soon approved by the crown, for a comprehensive voyage to investigate the political state of Spanish America, map remote regions, collect plants, investigate metallurgical practice, and promote trade.

Publication of the rich results is only now being achieved as a multivolume set.[40] A separate publication reveals the activities of Antonio Pineda, selected by Malaspina for the expedition to head a small international team of naturalists that included the Czech botanist Thaddeus Haenke. Pineda's contribution, it now emerges from this archival research, was to describe a host of little-known species of American fauna.[41]

Like Malaspina's expedition, the botanical expedition to New Granada was not originated by the crown. In fact it took twenty years for its proposer, the physician José Celestino Mutis, to secure royal authorization. Now we have a superb critical study, by Marcelo Frías Nuñez, solidly based on research in archives in Spain and Colombia, that corrects numerous errors of earlier works and provides new perspectives.[42] Like Pimentel's study, this was a doctoral thesis supervised by José Luis Peset as part of the research program at the CSIC's Department of History of Science in Madrid. This is particularly good on the gestation of Mutis's plans and his motives. Already as a student at the College of Surgery in Cadiz, Mutis's interest in medicinal plants was awakened by the college's botanical garden. By the time he sailed to America to serve as physician to the viceroy of New Granada, Mutis was intent on investigating the plants of the New World, to go beyond what Francisco Hernández had achieved in Philip II's reign. Years later, another viceroy, Archbishop Caballero y Góngora, responded with such enthusiasm to Mutis's aims that he activated the botanical survey even before receiving approval for the recommendation that he had sent to Madrid. Frías shows how Mutis functioned—for twenty-five years—as the indefatigable, paternalistic director until his death. Ninety-three persons working with the expedition are identified: plant collectors, Creole artists, cooks, carpenters, and slaves.[43] Some new data on expenditure is presented. The results were a herbarium of twenty thousand plant specimens and a multitude of botanical illustrations. But as for texts on this flora, doubts remain over their survival or even if they ever existed; the origin of manuscript texts conserved in the archive of Madrid's Botanic Garden has yet to be elucidated. What is clear from Frías's study is the stimulus of Mutis and his expedition to the scientific awakening of New Granada—not surprising considering that Mutis's energy and enthusiasm bubbled in the viceroyalty during his residence of almost half a century. (On various aspects of the concepts of collection, utility, and visual representation, see the chapters by De Vos and Bleichmar in this volume.)

The Nordenflicht mining mission—it is usually called a "mission" and not counted as an expedition—was sent to the Andes in 1788. By that time, the silver mines of New Spain had long overtaken those of Peru to become the mainstay of Bourbon revenue. The boom in New Spain, peaking in the 1770s, has been attributed by Brading (University of Cambridge) not to technological innovation but rather to the crown's tax exemptions and entrepreneurial skills of local merchant-capitalists.[44] But Peru's mines were then seen in Madrid as requiring an injection of European technology, and to provide it Charles III's minister, Gálvez, sent two Spanish mineralogists to recruit experts from Saxony; as many as twenty-nine were contracted. The mission, historians agree, was a failure, but they have been disagreeing over the causes. For Marie Helmer, arrogance of the Germans and inflexible Peruvian miners prevented success. For John Fisher (University of Liverpool), the inexplicable expulsion of the only member of the team "who could have understood the language and mentality of the Creole miners," a Peruvian working in Spain, was decisive—a conclusion endorsed by the findings of the Council of the Indies. But Carlos Contreras (Instituto de Estudios Peruanos) and Guillermo Mira (University of Salamanca) put the arriving mission in the context of a prevailing local power struggle between miners and government officials, and also blamed the lack of a system of liquid capital for investment in technology.[45]

New archival research has thrown light on Bourbon administrators of Huancavelica, Peru's mercury mine that supplied the material for refining silver by the amalgamation process. The enterprising governorship of the marquis de Casa Concha is revealed and the reputation of a later incumbent, Antonio de Ulloa, rehabilitated.[46] And fascinating documents are uncovered by Carlos Sempat Assadourian (Colegio de México) on early eighteenth-century initiatives for the technological transfer of Newcomen engines from Britain to drain the mines of New Spain.[47]

For the Habsburg period, growing emphasis has been put on the importance of Amerindian skills in prospecting, assaying, and refining in the first decades of Spanish rule.[48] As for the notorious *mita*, the coercive system of forced Indian labor that added to the black legend of Spanish cruelty, revisionist interpretation of the evidence has been more positive. Peter Bakewell (University of New Mexico) portrays a more human side to Viceroy Toledo, who implemented the *mita*, and the coexistence of voluntary Indian labor in the mines.[49] And Jeffrey Cole, using manuscripts in Bolivia, argues that many Indians escaped the *mita* by migration, testimony against the image of Indians as passive victims of conquest.[50]

Recent concern with environmental pollution has inspired research on the health of mine workers in colonial Spanish America. Kendall Brown (Brigham Young University) combines study of archival documents in Seville with the latest medical literature on mercury poisoning to conclude that Huancavelica was an "environmental tragedy" that deserved its reputation as the "mine of death."[51] But José Sala Catala also reveals the sensitivity of viceroys to the risks to health.[52]

The trends in history of medicine in colonial Spanish America have been more intense archival research combined with widening perspectives. Epidemics, for example, continue to attract strongly, but increasingly in a fuller social, economic, and political context and increasingly focused on the less-studied peripheral regions of the Spanish empire. That is apparent in the impressive inquiry by Linda Newson (professor of geography, King's College, London University) into the mortality rates from disease in sixteenth-century Ecuador.[53] She shows how the introduction of Old World diseases struck Indian communities with varying severity, depending on the diverse environmental and cultural conditions existing in the different regions of Ecuador as well as the degree of intensity of Spanish settlement. Similarly, Suzanne Austin Alchon (University of Delaware) argues for the neglected study of the biological resilience of Amerindians, a development inseparable from the social and political experience of Indians under colonial rule. She discovers a unique seventeenth-century population recovery in Ecuador.[54] In addition, the continuing intractability of identifying epidemics and establishing their origin, and the growing interdisciplinary character of this study, squarely focused on the Indians under colonial rule, all emerge from the papers presented at the forty-sixth International Congress of Americanists.[55]

A series of local case studies has elucidated the responses of viceroys and municipal authorities to outbreaks of smallpox and yellow fever.[56] The studies also show the effective activity of clergy against epidemics: campaigning against the tradition of burials within churches, exhorting parishioners to vaccinate, and even, where physicians were lacking, themselves performing vaccinations. The precarious financial resources of hospitals in eighteenth-century Lima has been demonstrated by David Cahill (Macquarie University, Sydney).[57] And pioneering statistical analysis on the health of soldiers in the Bourbon "Army of America" has been undertaken at Seville's Escuela de Estudios Hispano-Americanos. (For a further examination of colonial medicine, see Few in Chapter 7.)[58]

A QUESTION OF IDENTITY

Latin American historians of science, these days, do not take kindly to assumptions and assertions that science first came to their countries in the colonial era by simple transmission from Europe, from cultural center to periphery. Joined by a growing number of sympathizers from other parts of the world, they now resent and reject such images of passive reception and intellectual subordination. Some, in reaction, even respond with evidence for scientific colonization in reverse: an Ecuadorian (Pedro Dávila) who became director of the Royal Cabinet of Natural History in Madrid; a physician from New Spain (Pablo de la Llave) who, in the metropolis, officiated as president of the Real Academia de Medicina; and Francisco Zea, a botanist from New Granada who rose to director of Madrid's Royal Botanic Garden.[59] Others complained of "Occidentalism" and urged that, with the then approaching quincentenary of the discovery of America, it was time to get rid of "old patterns of thought" and develop "a new explanation" for national scientific development in Latin America.[60] And repeated protests resound against the prejudice of Eurocentrism in the editorial introduction (by Juan José Saldaña, director of research in History of Science at the Universidad Autónoma de México) to a work that is presented as a "local" history of science in Latin America.[61] This insistence on the importance of local conditions in scientific development was a basic source of disagreement at an international conference on colonial science that forced the editors of the published proceedings to acknowledge a division of the contributors into two camps: half of the papers argued that science had a global character, everywhere the same; the rest perceived science taking on a different character in different colonial localities.[62]

Local conditions here meant a distinctive environment in every sense (physical, political, cultural) and distinctive attitudes of the inhabitants—indeed the two are inseparable, as has now become clear: mentality was strongly conditioned by New World environment. Historians have recently established that a Creole identity already existed in seventeenth-century Mexico and Peru. Its development is explained in part as a reaction against the contempt with which Creoles were treated by the officials of metropolitan Spain, a reflection of that wider eighteenth-century European intellectual belief, propagated by Buffon and others, in the degeneracy of the New World and its inhabitants, both human and animal. According to this deterministic and disparaging theory, constellations and climate dictated that all born in the New World would have despicable characteristics. The Creole response was to develop an identity that expressed pride and superiority. Central here was pride in a heroic Indian past formed by intermarriage of conquistadores with Indian nobility. And there was pride also in the Creoles' portrayal of the Mexica people as an advanced civilization (specifically, the past indigenous civilization rather than the despised present-day Indians) possessing remarkable scientific skills, including a calendar so impressive as to persuade Humboldt to deduce contacts with China.[63] Debate continues on the scientific and technological achievement of Amerindians, ranging from the origin of humoral medicine in Latin America to the puzzling construction of the *puquios* of Nasca, an impressive system of irrigation by filtration-galleries found in Peru.[64]

How was Creole identity nurtured? Were books an important means? The history of the book in Latin America is taking off. For Carlos González Sánchez (University of Seville), knowing what books were read by Spaniards in the New World "can be the best way to understand their behavior and way of life." He therefore analyzed the official documentary records of the *bienes de difuntos* (property of the deceased who died in the New World without heirs). He found inventories listing books—mostly religious but also medical works. Unfortunately, as he concedes, there is no indication that these books were ever read nor can this source show books once possessed but no longer owned at the time of death. Similarly, in his research on private libraries in colonial Peru, motivated by inquiry into the roots of Creole identity, Teodoro Hampe Martínez (professor at the Universidad Católica of Peru) finds a "profusion of scientific and technological books" in the great library of Francisco de Ávila, cleric of Cuzco, but admits it is "difficult to say" whether the books were really read.[65] But Mutis's famous scientific library was for collective use of New Granada's intellectual elite—they even shared correspondence, evidence of group identity.[66]

Is the evidence any clearer for the influence of universities on Creole identity? In Portuguese Brazil there were no universities, the consequence of government policy in Lisbon that forced the colonial elite to study in Portugal and so retain intellectual dependence on the mother country. But in colonial Spanish America, between 1538 and 1810, thirty universities were founded. Their historical study has now become fully contextual, including conferences dedicated to tracing the role of these universities in the formation of Creole elites.[67]

It is eminently plausible that attendance at university should have nurtured Creole identity. Yet the precise influences are difficult to establish. Indeed, the curriculum at colonial universities was no different from peninsular Spain: overwhelmingly scholastic and resistant to the entry of modern natural philosophy—not a promising characteristic for generating distinctive Creole identity. The results of current research on this are mixed. Celina Lértora Mendoza (CSIC, Buenos Aires) believes she can detect some modernizing light in the philosophy lectures in the 1770s at the university of Cordoba, a tendency to substitute modern physics for Aristotelianism, "a glimpse of praise of science" as fundamental for progress, a view that she alleges must have affected the way of thinking of those who attended the lectures.[68] But in the university of Mexico at this time, Enrique González González (University of Alcalá de Henares) finds the governing body had not the slightest interest in reform, closing the door to a chair in anatomy and other modern science. In a persuasive study, González interprets official pronouncements by the university authorities testifying to the institution's openness to learning as empty propaganda that can mislead the unwary historian.[69]

Frustrated at their universities' lack of provision of useful and modern scientific knowledge, appetite for which was a central characteristic of Creole elites, they turned to patriotic societies and academic institutions outside of the traditional universities. In this they were no different from the intellectual elites of peninsular Spain. But recent research shows distinctive traits. The patricians of Havana, rejecting the "useless" education offered by the university, sent their sons instead to Baltimore and Philadelphia.[70] In New Granada, the Creole elite preferred what Renán Silva (Universidad del Valle, Cali-Colombia) calls "that other university," imbibing the scientific education of the Botanical Expedition at work there.[71] And Jean-Pierre Clément has made an excellent analytical study of the *Mercurio Peruano*, a periodical published in Lima in the early 1790s with the dual aim of propagating useful scientific knowledge and inculcating love of Peru—its editorial team declared "we are more interested in knowledge of what is happening in our nation than what concerns Canadians, Lapps, or Muslims."[72] Thus, no cosmopolitanism is apparent here. Instead, there was determination to exploit local resources and promote prosperity with the aid of applied science. Hence the periodical's concentration on natural history, chemistry (Lavoisier's new nomenclature was reproduced in full), and practical medicine. The *Mercurio Peruano* marveled at the achievements of Inca science and technology. Clément refers to a recurring emblem in the issues of the periodical that he sees as a signature of creolism. It depicts Peruvian plants and a condor, together asserting American character, but also a dog and ruins of classical antiquity, signifying surviving European character. Did the editors of *Mercurio Peruano* thus see themselves as a nation distinct from Spain? Clément's answer: "Probably not."[73]

The formation of identity does not necessarily produce fighters for independence. That is apparent from the most important result of José Luis Peset's inquiry into the connection between eighteenth-century Creole scientists and the wars of independence from Spain. He found that José Antonio de Alzate, impassioned campaigner for a patriotic science in New Spain, exalting traditional Indian botany and rejecting the European classification system of Linnaeus, was no revolutionary. Contrary to the usual prevailing portrayal, Peset presented Alzate as loyal to the crown and an advocate of gradual reform.[74] Yet he recognizes that Alzate and Mutis—another nonrevolutionary—by their scientific activity contributed to the rise of independent nations.

The influence of ideology on political revolutions is a notoriously difficult subject. The role of the Enlightenment in the outbreak of the French Revolution remains controversial and elusive. In the case of independence in Latin America, the trend in recent interpretation is to reject or qualify previous explanations in terms of a direct influence of Enlightenment thought; instead, a largely homegrown Creole cultural nationalism is emphasized.[75] Creole identity thus remains connected in some way with the independence movement, as does Creole science. Two notable further explorations of that connection have appeared. Thomas Glick (Boston University) connects Creole scientific awakening by the botanical expeditions to New Granada, Peru, and New Spain to political independence movements in those regions. The links for him were Creole scientists' resentment over scientific dependence on Europe and perceptions of scientific isolation, both emotions driving them into the independence movements.[76] This fits some of the Creole scientists but not Alzate, who stayed loyal to the crown. In another interpretation, Rafael Sogrado Baeza (Catholic University of Chile) similarly focuses on the importance of the scientific expeditions to the Indies, seeing the "sociability by contact" between arriving European scientists and Creoles as the cause of development of "critical mentality" among Creoles that "stimulated the independence movement."[77] The argument is clear but tends toward a monocausal explanation of the complex independence movement. Were all Creole fighters for independence motivated by science?

Was there such a thing as Indian identity in colonial Spanish America? Not only has this begun to be alleged but a close connection has also been made with science, just as has been done with Creole identity. Traditional Indian medicine is perceived as the vehicle for identity formation. According to Suzanne Austin Alchon, the spate of Indian uprisings in the highlands of Ecuador from 1730 reflects the stronger survival of native traditions. For this site was an outlying region that had escaped the full force of Spanish efforts, which were felt in Peru, to extirpate Indian customs and beliefs. In Ecuador, Alchon argues, Indians preserved "unity and identity" by "continued reliance on pre-conquest beliefs and ceremonies including healing rituals." And she reinforces this by drawing on recent historical studies by Frank Salomon, an anthropologist, that show how healers and shamans became political leaders of this Indian resistance.[78]

Recognizable Indian identity is the conclusion reached in a brilliant essay by Carlos Viesca Triviño (head of the Department of History and Philosophy of Medicine, UNAM) on the "cultural confrontation" of medicine at the time of the Spanish conquest of Mexico. This is a masterly analysis of the *Codex Cruz-Badiano*, a mid-sixteenth-century medical manuscript discovered in the Vatican in 1929 and now returned to Mexico. The author, Martín de la Cruz, was an Indian noble and physician at the Colegio de Santa Cruz at Tlateloco, a Franciscan institution founded to teach theology and philosophy to Indians. Probably originally written in Náhuatl, the Codex was translated into Latin by Juan Badiano, Indian lecturer at the Colegio. Viesca considers the likely purpose of the *Codex* was to demonstrate the rationality of Indians, then being contested. Viesca shows how the *Codex* is structured by chapters on the various parts of the body affected by disease. And, revealingly, the body is correlated with different regions of the universe in accordance with pre-Columbian cosmology. While European influence is apparent in the names of diseases like epilepsy and melancholy, traditional Indian medicine is the therapy adopted: herbal remedies based on balancing the hot or cold nature of supposed cosmological agents causing disease. This leads Viesca to see the *Codex* not just as testimony to the strength of Indian culture in adverse conditions of subjugation, but even as a document that shows the Mexican Indian "affirming and displaying his cultural identity, daring to express it in Latin in a bold invasion of the reserved territory of the conqueror's science."[79]

Current concentrated interest in colonial identity is attributable to the influence of postmodernism—a sprawling cultural movement that arose in the 1970s, dedicated to the elimination of "modernism"—seen by its proponents as the pernicious, disingenuous rational and scientific view of the world that dominated Western intellectual life since the Enlightenment. Its attractions usually explained in terms of filling the ideological void left by the demise of Marxism, postmodernism has made its presence felt in all branches of history. For its fiercest critics, postmodernism threatens the very existence of the history discipline, while its supporters praise its wonderful broadening horizons that liberate conventional history from confining boundaries: postmodernism is a "fresh breeze for some, acid rain for others."[80]

Postmodernism sprang from multiple origins, which include cultural anthropology with its focus on ceremonies and rituals, interpreted by a whole range of meanings seen as equally valid. But literary critical theory has been the greatest influence, to the regret of conventional historians who blame it for invading their discipline with utter relativism and attitudes devoid of any sense of period. Indeed, for Jacques Derrida, French literary critic, nothing mattered beyond a given text—the only task was to "deconstruct" it, to decode and penetrate the text's surface in order to perceive its various meanings. This "discourse theory"—the prioritizing of language over all other human activity—motivated Haydn White, an intellectual historian, to become the leading campaigner in the 1970s and '80s for a postmodern reinterpretation of the practice of history: that history writing was literary production, and no real boundary delineated fact and fiction—that historians "invented" the past. Also in the 1970s, French philosopher Michel Foucault provided the additional, highly influential view of presenting language and discourse as forms of power that could oppress. Such views inspired Edward Said's widely read *Orientalism* (1978), which deconstructed European authors' texts on Orientals, revealing prejudices and demeaning stereotypes designed to justify Western colonial domination. And soon, in the early 1980s, the same postmodernist deconstruction of texts was applied to the history of the British Raj, inverting the traditional focus by replacing colonial rulers with underprivileged Indians, in a series of volumes titled *Subaltern Studies*.

A distinguished historian of the British empire, A. G. Hopkins (Smuts Professor of Commonwealth History, University of Cambridge), has identified postmodernism as "the most important single new influence on Imperial history" during the 1990s.[81] While acknowledging that it "has helped to open up new lines of enquiry and to inspire work of considerable merit," he also sees postmodernism as "promoting a distinctive cult of the obscure" and that the whole approach may in the future be seen as "an indulgence."[82]

Historians of Latin America also include critics of postmodernism. Alan Knight (University of Oxford), historian of the Mexican revolution, warns that "an overdose" of deconstructing texts "can lead to a surreal detachment from reality . . . morbid imagination triumphing over solid good sense."[83] Renán Silva complains, "we now suffer the worst excesses, fortunately temporary, of culturalists and the 'linguistic turn.'"[84] And John Lynch (professor emeritus of Latin American History, University of London), doyen of historical studies on the end of colonial Spanish America, felt the need to comment on contemporary methodological trends in a 1990 speech at the University of Seville, on the occasion of receiving an honorary doctorate. Recalling that "the methods that I learnt and followed were strongly empirical," he regretted that recent "theoretical concepts and models, far from clarifying history, distort it. They deform reality by pressing it into a mold created prior to the evidence." And, criticizing the dominant trend, he offered correction: "In an age of postmodernism it is not superfluous to affirm that history is a process of discovery, that truth is a matter to be ascertained, not invented, discovered rather than constructed."[85]

On the other side of the fence, supporters of postmodernism see real gains for understanding colonial Spanish America. Lyman Johnson (professor of History, University of North Carolina) and Susan Socolow (professor of Latin American History, Emory University) predict that, since the most familiar source material on first contacts of Spaniards with the New World consists of "self-promoting" accounts, "it is highly likely that the next generation of studies of the early period will draw heavily on the methods of contemporary literary criticism" to read Bernal Díaz and other conquistadors as "discourses to be decoded rather than as objective or disinterested relations of events."[86] And Elizabeth Hill Boone (Tulane University) recorded the remarkable switch in popular perception, attributable to postmodernist influence, as the quincentenary approached: an initial heroizing of Columbus faded into another rhetoric of a "two-sided Encounter," commemorative rather than celebratory, "given the ultimate destructiveness of the encounter for American cultures." Accordingly, acknowledging postmodernist inspiration, the quincentennial contribution of a Dumbarton Oaks symposium would be to focus on the resilience of indigenous cultures after the Spanish conquest.[87]

Strong postmodernist influence is discernible or made explicit in recent studies on the history of science in colonial Spanish America. The results are uneven. When in 1982, the Sociedad Latinoamericano de Historia de las Ciencias y la Tecnología (Latin American Society of the History of Science and Technology) was founded in Puebla, Mexico, it was soon resolved that its journal (first published in 1984) should have the Amerindian title: *Quipu*, the name of the knotted strings used by the Incas for calculations in censuses, numbers of taxpayers, and contents of warehouses. At the same time, Miguel León-Portilla, undertaking a study of Spanish hospitals built for Indians in New Spain, declared that "unlike previous discussions of this, I will focus on the Indian communities' attitudes to these hospitals, not the Spanish," which brought the new finding that Indians came to welcome an institution previously unknown in Mesoamerica.[88] Interpretations of the eighteenth-century scientific expeditions have been colored by postmodernism. The very title of Lafuente and Mazuecos's book on the Franco-Spanish geodetic expedition to Peru—*Los Caballeros del Punto Fijo* (The Knights of the Exact Point)—shows this, as it refers to the local Indian perception of European surveyors as magicians using strange instruments to sight distant points, even the stars. "We tried to put ourselves in the position of the Andean Indian," Mazuecos said in explaining the puzzling title that brought the Indians to center stage.[89] Pimentel's monograph on the Malaspina expedition acknowledges the postmodernist influence of Hayden White's *Metahistory*,[90] though it is not easy to see in what way White's work inspired this excellent study. And a cultural anthropologist, Fermín del Pino (Department of the History of America, Centro de Estudios Históricos, CSIC, Madrid) has endeavored to shed light from his discipline on the motivation behind the expeditions. Why did the Bourbons and other European monarchs spend lavishly on these expeditions? Potlatch is the answer. This ritual of Indians on the northwestern coast of America consists of gift giving to invited chiefs of other tribes, giving away wealth in a public display designed to establish superior status over rivals. In exactly the same spirit, eighteenth-century monarchs engaged in imperial rivalry, seeking a superior image by investing in grandiose expeditions.[91] The question here is whether potlatch really adds anything to the understanding we already have of prestige and rivalry.

Rather more light from ritual comes from Batia Siebzehner (Hebrew University of Jerusalem) in a fascinating discussion of the ceremony honoring the visit of Viceroy Iturrigaray to the University of Mexico in 1803.

A banquet was held featuring a dessert shaped into a temple of immortality, representing knowledge, with a hierarchy of statuettes that vividly portrayed the place of modern science in the colonial Spanish university curriculum. Statuettes representing theology and canon and civil law dominated the edible display. Next to these central figures came Aristotelian philosophy, represented by a statuette of a young woman holding a golden key. At the sidelines of this official culinary creation were figures representing medicine, botany, chemistry, agriculture, and mechanics, interpreted as potentially of value provided they did not obstruct the traditional Scholasticism.[92]

Jorge Cañizares-Esguerra (SUNY-Buffalo) is the latest historian of science to draw on postmodernism (see also his introduction to this volume). A migrant from Ecuador to the USA, and from medicine to history of science and then to historiography, his prizewinning book is dedicated to discovering "all sorts of submerged voices that dwell in the body and margins of texts" by the techniques pioneered by postmodern literary criticism. He argues, questionably, that discourse theory is already detectable in eighteenth-century Spanish American patriots, such as Alzate and Clavigero, in their defense of New World nature under attack from Buffon. According to this interpretation, Buffon's error stemmed from linguistic ignorance. In its revisionist account of the origins of European hostility to nature in tropical America, the entire monograph takes an assertive literary stance, from the outset to the closing words: "as far as audiences in the United States are only offered stories of violence . . . instability, and corruption in Latin America (a narrative conceit . . .), there are going to be storytellers like myself to recreate alternative worlds."[93]

No area of study of colonial Spanish American science has benefited more from postmodernist inspiration than cartography. In his latest book presenting a history of maps, Jeremy Black (Professor of History, University of Exeter) tells readers that thirty years ago, any such account would have concentrated on the European development, but today due weight is given "to other mapping traditions which are no longer regarded as inherently weaker than those in Europe." The "spatial values and practices of a great variety of cultures . . . have a different language and this needs to be understood."[94] Perhaps more than anyone else, Brian Harley (Professor of Geography, University of Wisconsin) has brought about this remarkable interpretative switch. The very titles of his influential papers of the late 1980s betray adherence to postmodernism (indeed, the acceptance of the doctrinal lead of both Derrida and Foucault is declared): "Texts and Contexts in the Interpretation of Early Maps," "Maps, Knowledge and Power," "Silences and Secrecy: The Hidden

Agenda of Cartography in Early Modern Europe," "Deconstructing the Map."[95] Breaking with "our Western culture" in which "cartography has been defined as a factual science . . . and inaccuracy . . . a cartographic crime," Harley instead insisted that "maps are never neutral or value-free or ever completely scientific."[96] A map "operates behind a mask of seemingly neutral science,"[97] and the role of the historian is to uncover the mask, to "deconstruct" the map. Cartographers may continue to disguise their products in "a technical specification—scale, survey instruments, design,"[98] but "the cartographer has never been an independent craftsman," rather a pawn in the "deliberate distortion of maps for political purposes."[99] Consequently, we need to abandon the tendency to look for achievement in maps measured by scientific accuracy and instead view maps as literary texts, as "controlled fiction."[100] The aim then becomes "to explore the discourse of maps in the context of political power."[101] Only then can we perceive the Eurocentric imperial propaganda that, in maps, justified territorial appropriation;[102] for example, those maps of the New World that portrayed vast stretches of land as empty space or relegated the Indian "to the status of a naked cannibal" in the maps' "marginal decoration" (see Valverde and Lafuente in Chapter 10).[103]

Harley's final, posthumously published paper modified this view of colonial maps as sheer Eurocentric propaganda. In it, he argued that "a good case can be made for asserting that most European maps disguise a hidden stratum of Indian geographical knowledge" and that, in their own mapmaking, "Indians adopted European uses for maps to reappropriate them as tools of resistance in a colonial struggle."[104] This cultural interaction in mapping, in the case of New Spain, has been effectively investigated in a pioneering study by Barbara Mundy, art historian, in which she dismisses the argument of historians of science that Europeans introduced into the New World technological advances in the form of more accurate maps, because "accuracy is in the eye of the beholder."[105] Mundy concentrates on those maps produced around 1580 in response to Philip II's endeavor for full information about the Indies. Officials in Madrid awaited receipt of local maps constructed in the established European form of grids, geometrical projection, and perspective. But what they were sent was very different and dashed expectations of compiling a comprehensive atlas of New Spain. For the majority of these maps were filled with the symbols and pictographic toponyms characteristic of Indian representations. The Indian artists had painted them with "double consciousness" for their local community and for a "shadowy, powerful Spanish patron."[106] Their maps therefore incorporated pre-Hispanic historical and social-structural features. Madrid was thereby presented with a bewildering "archipelago of individual and separate communities each with a unique sense of identity" expressed in maps.[107] This map production by Indian artists "is testament to the resilience of indigenous self-conceptions," which Mundy found still alive in Oaxaca.[108] A less culturally competitive, more collaborative image of mapmaking in New Spain has come from Duccio Sacchi (Department of Political Studies, University of Turin). He studied maps generated in villages by lawsuits over pastures and water rights. These maps were instruments for negotiating territorial settlements with the colonial rulers and show "a complex recombination of traditional components (European and Mesoamerican) which changes both of these cultures."[109] Commissioned, recognized, and authorized by the viceregal administration, these bicultural maps served for "inter-ethnic territorial agreements."[110]

The last two decades of intensive study of science, medicine, and technology in colonial Spanish America have yielded profound advances in understanding. Is there anything missing in this rich and varied output? A comparative approach is needed that considers Ibero-American and British American colonies together. Alone in touching on this, Thomas Glick sketched the potential of comparing the availability of scientific instruments in Spanish and British America.[111] The full program is yet to be taken up.

CHAPTER TWO

Portuguese Imperial Science, 1450–1800

A Historiographical Review

PALMIRA FONTES DA COSTA AND HENRIQUE LEITÃO

Even a cursory perusal of a map immediately shows the most salient and remarkable fact of the Portuguese maritime expansion of the fifteenth and sixteenth centuries: its staggering magnitude. Starting in the first decades of the fifteenth century with a territorial expansion to North Africa, in the next decades Portuguese seamen rapidly advanced all through the Atlantic, along the western coast of Africa and then the coast of Brazil. In the first decades of the sixteenth century, this expansion intensified and spread from the east coast of Africa and the Indian Ocean, to Southeast Asia and the seas of China. By the mid-sixteenth century, Portuguese trading posts were scattered all through the Atlantic coasts, from the eastern coast of Africa to the Indian subcontinent, and as far as Japan. Ships were regularly sailing from Lisbon along routes to places as diverse as the Azores islands, Brazil, or Goa. A reasonably effective military control of the seas where the trading routes operated had been achieved, a network of administrative control was being implanted, and missionary work was beginning in earnest.[1]

The vastness of the Portuguese empire posed enormous logistical problems. Besides the establishment of settlements along the coast, the fate of the trading routes depended also on gaining control of the seas. The problems involved and the solutions attempted are much too complex to be dealt with here, but it is important to keep two fundamental aspects in mind.

First, the Portuguese established trading posts and engaged in diplomatic relations or warfare with civilizations and cultures of the most diverse types. Differently from the Spanish empire, which, despite its dimension was mostly concentrated in South America and the Philippines, the Portuguese empire was above all characterized by its enormous geographical dispersion. The Portuguese came into contact with such wide-ranging cultures as the indigenous tribes of Brazil and Africa, the Arab merchants of North Africa and the Indian Ocean, the various kingdoms of India, myriad peoples in Southeast Asia, and the highly developed civilizations of China and Japan. Needless to say, it would be futile to try to find a common pattern or strategy of dialogue when so much diversity is at play.

Secondly, the Portuguese overseas enterprise was always affected by an acute demographic problem. By the mid-sixteenth century, the population of Portugal was roughly 1 million to 1.5 million inhabitants. Whatever the issue one focuses on—be it administrative, military, commercial, or religious—the imbalance between the task to be performed and available human resources is always enormous.

These two facts—the gigantic geographical dispersal of Portuguese presence around the world and the extreme scarcity of human resources—make the designation of "empire" a complex one to use when applied to Portugal's huge network of settlements, scarcely populated and loosely controlled by a central administration, a fact that historians have reminded us of frequently. Yet the surprising longevity of Portuguese presence around the globe signals the existence of durable, if informal, modes of relation with local populations and the sagacity of some strategies adopted.

The Portuguese empire was mostly a sea empire and navigation was crucial to maintain and foster it. (For a further examination of this question, see Almeida's contribution in Chapter 4.) The long-term success of the maritime expansion required the creation of state-supported institutions dedicated to the training of nautical personnel, certification of instruments and charts, and regulation of all procedures related to nautical activities. But the Portuguese empire was also, in the apt designation of Russell-Wood, "a world on the move," permanently traversed by merchants, travelers, soldiers, missionaries, and all sorts of voyagers who in many informal ways collected and exchanged information about nature.

Throughout the centuries, the Portuguese empire experienced great changes, and although even a summary description of these events is impossible to make, the principal lines of this evolution must be recalled. Needless to say, the chronological changes of the Portuguese empire will have direct implications on scientific activities. Up to the early seventeenth century, the Portuguese expansionist movement proceeded almost unhindered by other European competition. Disputes with Spain being mostly settled due to the celebrated Treaty of Tordesillas, the Portuguese were challenged above all by non-European powers. At the turn of the century this pattern began to change, mostly due to the arrival of the British and the Dutch in East Asia. The contraction of the Portuguese Asian empire led to strategic focusing in other regions, namely Brazil. All through the eighteenth century and until the first decades of the nineteenth century, when it became independent, Brazil was of vital importance to Portugal. A significant colonial presence in Africa is a phenomenon of the later nineteenth and twentieth centuries.

The history of the Portuguese maritime expansion is the most important topic in Portuguese historiography. The study of the creation, building, and ultimate fate of the Portuguese empire is the theme most investigated by Portuguese historians.[2] Scholarly production in these topics greatly exceeds any other aspect of Portuguese history. Yet, despite the massive erudite production on these topics, there is nothing that can aptly be called a study of Portuguese imperial science. This state of affairs, which is due, most of all, to the fragile nature of studies in the history of science in Portugal, has only slowly been changing in the past few years, but it has not yet been consolidated into an autonomous historical discipline.[3] Historians interested in the study of Portuguese imperial science must therefore collect their information from the myriad of studies that have inspected partial aspects of this story. In other words, one can only hope that sometime in the near future, from the variety of approaches to history of the Portuguese empire, a picture of Portuguese imperial science will emerge.

SCIENCE, TECHNOLOGY, AND THE MARITIME EXPANSION

For more than one century, the designation "Portuguese imperial science" simply meant "Portuguese nautical science." Either pressed by external demands or pursuing the discovery of a specific Portuguese contribution to science, historians have turned to the period of maritime expansion and concentrated on the technical and scientific innovations that gave Portuguese sailors advantage over their European (and non-European) competitors. Thus, historians have analyzed many different topics that include navigation techniques (manuals, teaching, and instruments), ship-building techniques, gunnery, cartography, literature of voyages and geographical descriptions, and more. The field is thus very broad and has frequently been shaped by the desire to ascertain

Portuguese priority or novelty in scientific and technical matters. Needless to say, it has been deeply influenced by the vagaries of the politics of each time.

This historiographic trend can be discerned from the early nineteenth century and has been maintained until today, making it one of the most constant and durable topics of investigation in Portugal.[4] In fact, for a country that has been slow to engage in studies of the history of science, the history of Portuguese nautical science is a happy exception. Although earlier attempts existed, the origins of this historiographic tradition may be associated with works that were published in the mid-twentieth century, such as the Viscount of Santarém's (1791–1856) foundational three-volume work, *Essai sur l'histoire de la cosmographie et de la cartographie pendant le Moyen Age et sur les progrés de la géographie après les grandes découvertes* (Essay on the History of Cosmography and Cartography during the Middle Ages and on the Progress of Geography after the Great Discoveries; 1849–1852). A few decades later, Luciano Cordeiro (1844–1900) inaugurated the study of Portuguese navigation techniques with his work, *De como navegavam os portugueses no começo do século XVI* (How the Portuguese Navigated at the Dawn of the Sixteenth Century; 1883). These works were soon followed by a series of other contributions, among which two exceptional books are notable: Henrique Lopes de Mendonça, *Estudos sobre navios portugueses dos séculos XV e XVI* (Studies on Fifteenth- and Sixteenth-Century Portuguese Ships), and Sousa Viterbo, *Trabalhos náuticos dos portugueses nos séculos XVI e XVII* (Nautical Works of the Portuguese in the Sixteenth and Seventeenth Centuries).[5]

The polemic between Joaquim Bensaúde (1859–1952) and some German scholars who questioned Portuguese priority led to the publication of a number of important works. In a series of seven volumes under the title of *Histoire de la science nautique portugaise* (The History of Portuguese Nautical Science), Bensaúde published a collection of essential documents and proceeded to demolish the German claims in a number of separate works, of which the most important was *Les legendes allemandes sur l'histoire des découvertes portugaises* (German Legends of the History of Portuguese Discovery; 1917–20).

A certain romanticism permeates these works and they are sometimes influenced by a nationalistic bias. Scientific accomplishments became a weapon that was used to define and increase national politics. It is no surprise that a connotation of "glorious" period or "golden age" was coined to describe the scientific achievements of this period and that historical facts were sometimes greatly exaggerated in order to justify a Portuguese excellence in nautical matters.[6] But despite its failings, this historiographical tradition revealed a considerable amount of facts that showed a deep connection between the needs of the empire and the practice of science.

The next generation of scholars introduced higher levels of precision and technical command. Authors such as António Barbosa (1892–1946), Luciano Pereira da Silva (1864–1926), Abel Fontoura da Costa (1869–1940), and Quirino da Fonseca (1868–1939) combined a full technical competence of the matters addressed with a keen sense of historical problems. These works laid the foundations for a modern and scholarly study of the Portuguese contributions to science.

This approach would reach its high point with the works of Avelino Teixeira da Mota (1920–1982), Armando Cortesão (1891–1976), and above all with the prolific contributions of Luís de Albuquerque (1917–1992). These authors brought the field to its maturity and produced scholarly works of the highest level such as, for example, the many critical editions of nautical guides published by Albuquerque or the magnificent *Portugaliae monumenta cartographica* (Cartographical Monuments of the Portuguese), by Cortesão and Teixeira da Mota.[7]

It was from within this tradition of studies that the question of Portugal's role (if any) in the Scientific Revolution was framed. This question interested not only some Portuguese historians but also distinguished foreign specialists, in particular the Dutch historian Reijer Hooykaas.[8] Two aspects in particular have been highlighted: the fact that, on the one hand, as a direct consequence of the maritime expansion, from a very early date in Portugal the authority of the old masters (Ptolemy, Pliny, Pomponius Mela, Strabon, and so on) was severely questioned and, on the other hand, that it is possible to identify in some Portuguese authors, in particular the celebrated D. João de Castro, an "experimental" and "modern" approach to the study of nature. (See also Almeida in Chapter 4.)

Yet, for all their many accomplishments, these historians and the intellectual tradition they established were restricted to a cognitive approach that omits cultural practices, social considerations, more institutional aspects, and material culture from the field of inspection.

Much broader perspectives are required, for example, in the case of institutions. The needs created by the Portuguese empire led to the creation of new institutions, but the story of these institutions and their relation to scientific practice is still poorly known. Until now, historians have focused only on one such institutional setting, the position of *cosmógrafo-mor* (chief cosmographer).[9] The state responded to the need for personnel trained in nautical matters by establishing the position of chief cosmographer, whose nature (duties and privileges) is defined in the

"Regimentos do Cosmógrafo-Mor." While this legislation is known to have existed from at least 1559, the first extant document is from 1592. According to this "Regimento," the chief cosmographer should examine makers of nautical instruments and cartographers; it was also his duty to authenticate all nautical charts, globes, and instruments; and finally, he was in charge of training future nautical pilots in mathematics, cosmography, and astronomy. Creation of the position of chief cosmographer signals the attempt to centralize the training and certification of pilots. The position was kept in essentially the same form until 1779, when it was completely reformed.

Much less is known about the relation between the university and the technical achievements and scientific practice of the maritime expansion. The university has traditionally been described as a very conservative institution, unaffected by the novelties brought from overseas by sailors and merchants. But there are reasons to suspect this to be a simplistic description of events.

꧁

A different historical tendency has focused attention in the informal, noninstitutional mechanisms of collecting and exchanging information. The many different groups that constituted the Portuguese empire—sailors, missionaries, soldiers, merchants, and the like—engaged in massive collections of information about the regions they were living in or traveling through. (See also Walker's arguments in Chapter 13 concerning the circulation of this knowledge in the Portuguese empire.) This information is preserved in an abundant literature—such as travel books, geographical descriptions, *roteiros* (mariners' logbooks), and missionary correspondence—most of which has been examined only partially and never integrated in a larger narrative of imperial science.

BUILDING NETWORKS FOR
SCIENTIFIC PRACTICE

The somewhat loose channels of communication and practice of science in the Portuguese empire underwent profound changes with the arrival of the Society of Jesus.

The examination of the Jesuit contributions to science is a recent phenomenon in Portugal. While it is also relatively new as a field of study in historiography per se, its emergence in Portugal seems to have been more delayed than elsewhere. This is surely a consequence of a persistent anti-

Jesuit and, more generally, anticlerical bias that has afflicted much of Portuguese historiography, especially in the history of culture and education. Two good examples are the immensely influential, four-volume history of the university by Teófilo Braga, *História da Universidade de Coimbra* (Lisbon, 1892–1902), whose virulent critiques of Jesuit education have become canonical, and Francisco Gomes Teixeira's, *História das matemáticas em Portugal* (Lisbon, 1934), whose main theses were adopted by the majority of the historians of science.[10]

॰॰॰

Recently, however, under the pressure of an international historiography that has directed a great deal of attention to "Jesuit Science," Portuguese historians have looked into these questions with more objective and more dispassionate eyes. These new developments are crucial to the understanding of Portuguese imperial science since, whatever judgment one may make, it is beyond doubt that the Jesuits radically altered the educational and scientific scenario in Portugal and that their action was done on a planetary scale.

From 1540, date of the arrival of the first members of the Society of Jesus to Portugal, to 1759, the year of their expulsion (which anticipated the suppression of the Society by the pope, in 1773), the Jesuits created a system of colleges of which the main characteristics, from the point of view of the history of science, can be summarized as follows. The Jesuits established the first organized and stable educational network at the "secondary" and "preuniversity" level and inserted educational institutions located in Portugal in a large, supranational context. Furthermore, they established the first truly regular teaching of mathematical disciplines in Portugal and sponsored scientific activities on a scale never before experienced there.

The complexity of the Jesuit educational enterprise in Portugal cannot be addressed here, but the sheer numbers are impressive, at least for Portugal. It is estimated that, by 1759, when the Jesuits were expelled and their educational network disrupted, approximately twenty thousand students were attending Jesuit institutions. Comparable numbers of youngsters attending schools in Portugal were only reached again more than a century later.[11] Moreover, this network extended beyond the European borders of Portugal, covering most regions where Jesuits were engaged in missionary work. Education in Jesuit colleges was planned along the rules of the *Ratio studiorum* (Method of Study), but this document gave only general guidelines that left ample room for local variations and even for the initiative of isolated teachers.[12]

Furthermore, these guidelines were not always put into practice. Thus, although in a broad picture one can say that all Jesuits had a similar training, a more microscopic analysis shows that regional variations could be very marked.

In 1574, following a demand made by the king, the Jesuits created their first mathematical class in Portugal—the Aula da Esfera (Course on the Sphere)—at the Colégio de Santo Antão, in Lisbon. Differently from mathematical studies in Jesuit colleges in other countries, this class was established not exactly to comply with the Jesuits' own plan of mathematical studies but rather with the objective of providing mathematical and technical training to personnel engaged in the Portuguese maritime expansion, that is, with the needs of the empire in view. The mathematical curriculum at this college was therefore substantially different from that of other Jesuit colleges in Europe. Although part of a supranational context, Jesuit scientific teaching and practice in Portugal reflected the specific conditions prevailing locally. After the pioneering studies by Luís de Albuquerque and especially after the careful investigation by Ugo Baldini, the Aula da Esfera of the College of Santo Antão is today recognized as the most important center for scientific activities in the period from the end of the sixteenth century to the expulsion of Jesuits in Portugal.[13]

From the point of view of the history of science, the network of educational institutions that the Jesuits established in Portugal and its empire is of great interest and corresponds to a new situation. Unlike any other educational system ever implanted in Portugal, the Jesuit system was organically connected to other institutions in Europe and to the missionary activities outside Europe. It was a stable and reasonably well organized long-range network of communication within which different scientific activities were practiced.

℧

Recent works have overcome the ideological constraints imposed by older historiography and greatly improved on our knowledge of the practice of science within the Jesuit network, both on a global scale and on a detailed level.

The work of Ugo Baldini on the scientific practice in the Iberian Assistancies and, in particular, on the Lisbon college of Santo Antão provides both fundamental elements and the theoretical framework to understand the nature and peculiarities of Jesuit science in the Portuguese empire.[14] Historians have been progressively addressing the many issues related to the history of Jesuit science in the Portuguese empire. Some important

Portuguese Jesuit scientists such as Inácio Monteiro were studied;[15] the relations between the Jesuit mathematicians in Beijing and the Academy of Sciences of St. Petersburg were analyzed;[16] and specific cases of transfer of scientific instruments via the Jesuit network were studied.[17] In confirmation of the increased interest of these matters, three international conferences were set up to study the scientific activities of Portuguese Jesuits in Portugal and the Portuguese empire.[18]

Brazil is obviously a special case within the Portuguese empire, and it would require a separate study. Some studies have already analyzed the practice of science in Brazil mediated by the Jesuits. Authors have studied aspects of natural history and also astronomical observations of Jesuits, most of all by the Bohemian missionary Valentin Stancel.[19] A topic that has interested historians for several decades and continues to attract some fine scholarship has been the study of cartographic missions in Brazil.[20] The role of the Jesuits in studying and diffusing natural historical knowledge has also attracted some attention.[21] However, much is still to be accomplished. The famous *Flora cochinchinensis* (Flora of Cochinchina), by the Jesuit João de Loureiro, is still to be studied in detail and to be incorporated in the broader narrative of imperial science.[22] The same can be said of other important works on this subject, such as the observations by the Jesuits Manuel da Nobrega, José de Anchieta, and Fernão Cardim on the natural world of colonial Brazil.[23] The need for further studies on the contribution of the Jesuit Luís de Almeida to the introduction of European medical practices in Japan should also be stressed.[24] A comprehensive study of Jesuit scientific practice in Portuguese India is still a desideratum. Some older works have inspected isolated episodes,[25] and the number of studies on natural history and medical activities seems to be growing, but there is still much to be done.[26] A comparative study between the different Jesuit missions has not been attempted. In general, the immense quantities of information collected by missionaries on indigenous scientific practice and non-European nature are still awaiting a historical analysis.

MEDICINE, NATURAL HISTORY, AND THE VISION OF THE EMPIRE

Most of the historiography on the Portuguese discoveries praises the contribution of Garcia de Orta's *Colóquios dos simples e drogas da India* (Colloquies on the Simples and Drugs of India) (1563) to the development of modern science. In fact, this was the first work to systematize the knowledge and medical applications of some of the new plants encountered by the Portuguese in Asia.[27] However, it is rarely mentioned that the work become known in Europe due to the efforts of a foreigner. Carolus Clusius acknowledged that, on his visit to Lisbon in 1564, he found de Orta's *Colloquies* by chance, and since they pleased him so greatly and were written in a language that was understood by so few, he decided to publish an abridgement of the original in Latin with additional commentaries.[28] This work was printed in Antwerp in 1567 and enjoyed several editions.[29] In contrast, a second Portuguese edition of de Orta's work was published only in the late nineteenth century.[30]

The majority of sources with original and valuable information on medicine and natural history in the context of the Portuguese colonial enterprise did not have the same fortune as de Orta's *Colloquies*. Several of them were never or only recently published. Others were destroyed by human agents or by natural disasters such as the 1755 Lisbon earthquake. A significant number remain uncatalogued in the Archives. One of the recurrent and crucial problems in the study of this subject has been the insufficient diffusion of sources and of scholarly work in the Portuguese-speaking countries as well as abroad. Already in the First Congress of the History of Portuguese World Expansion, organized in July 1937, Luis de Pina alerted his colleagues to the crucial need to publish bibliographical summaries of relevant Portuguese documents in a language more accessible than Portuguese.[31] Unfortunately, the suggestion was not followed, and only very recently can we trace a few individual efforts to disseminate studies on Portuguese imperial science in other languages.

It is during the first half of the twentieth century that the beginnings of an interest in the history of Portuguese colonial medicine and natural history can be seen. (For a brief historiography of Portuguese colonial medicine, see Walker in Chapter 13.) The authors of works produced during this period usually shared a scientific background and a fascination for the pioneering role of Portuguese authors in accurate description of new species and diseases. It is with indignation and perplexity that some of them complain about insufficient acknowledgment abroad of the Portuguese role in the development of modern science. In an article from 1926, Portuguese physician Carlos França emphasized his surprise at the absence of Portuguese names in the history of the natural sciences. Knowing the scientific orientation of the Portuguese discoveries, he felt a great repugnance in being forced to admit that what is known of natural history "in the lands we first stood [on] and colonized" is due only to the efforts of foreigners.[32]

It is therefore not surprising that one of the common themes of this historiography was the search for Portuguese predecessors in the development of medicine and natural history. Scholars from this period vindicated not only the pioneering role of the Portuguese in the discovery of plants, animals, and diseases previously unknown in the Old World, but also the Portuguese development on a new basis for a series of disciplines. Pina, the most prolific author on the subject, even declared that botany, zoology, mineralogy, and tropical medicine "were born in Portugal during the age of the discoveries."[33] Such exaggeration of Portuguese achievements should be understood within the political and historiographical situation of the period. Several of the contributions were made in a context of commemorative celebrations in which the political agenda of the regime sought political legitimization. It is, therefore, not surprising that this generation of historians has focused mainly on the sixteenth-century apogee of the empire. At the same time, we should not forget that the Portuguese contribution to the development of modern science has been and continues to be largely excluded in Anglo-American, French, and German scholarship.[34]

It is true that this historiography is, in great part, patriotically biased, but we should not forget its valuable contribution to the field, since it involved the first efforts to uncover some of the primary sources associated with the history of Portuguese colonial medicine and natural history. One of the features of the works produced by this generation of scholars is its descriptive nature, together with the inclusion of long quotations from original documents. It may, consequently, be said that it still has considerable informative value. At the same time, some of the arguments running through this body of work still need to be challenged and reassessed by new historiographic approaches. It should also be noted that some of the authors from this period were the first to point, however concisely, to the diversity of literature, agents, and places involved in the Portuguese colonial experience. At least until the eighteenth century, the information relevant to the study of this subject was not to be found in medical and natural historical treatises, but in chronicles, travel literature, and letters written by authors who were involved in the Portuguese efforts to conquer and trade or to preach Christianity. The works of the physicians Amato Lusitano and Garcia de Orta are an exception to this trend.[35] A few studies also make reference to the reciprocal influences and multidirectional flows of animals, plants, and medical knowledge among the various dominions of the Portuguese empire as well as in Europe.[36] In addition, some studies reveal an initial interest in institutions associated with medicine that were first founded by the Portuguese in Asia and Africa, such as hospitals and charitable institutions.[37]

THE STATE AND THE APPROPRIATION OF THE
COLONIAL NATURAL WORLD

The historiography of the subject has taken a new turn in the last two decades. The importance of scientists writing the history from which had flowed a Whig narrative of progress decreased considerably, and they were replaced by professional historians sharing some of the concerns and aims of the new historiography of imperial science.

The study of scientific expeditions in the Portuguese overseas territories during the second half of the eighteenth century has been the main topic of the new scholarly approaches to the field. William Simon's pioneering and wide-ranging study on this subject and several other recent works have pointed to an effective program, as broad as the Portuguese empire of the time, that led to a systematic inventory of Portuguese Asia, Africa, and America.[38] This was a program with political, economic, scientific, and cultural dimensions. On the one hand, the Portuguese colonies were crucial to the economic development of the metropolis and were of primordial importance from a geographical and strategic point of view. On the other hand, it is during this period that, following the Marquis of Pombal's rise to power during the reign of D. José I (1750–1777), a series of reforms were implemented in the Portuguese economic and educational system. These reforms were considered to be crucial for the Portuguese state to maintain hold over its empire.

One of the significant results of the project was the reformation of the University of Coimbra.[39] For the first time, it became possible to educate at the university level naturalists who were later employed in the Portuguese overseas expeditions. Consequently, and in contrast to the previous period, several of the authors of works on natural history had a specialized education in this subject. It should also be remarked that economic and educational reforms were closely linked, as can be seen in the work of Domingos Vandelli, professor of Chemistry and Natural History at the University of Coimbra and one of the main advocates of philosophical voyages in continental Portugal and in the overseas territories in order to define a strategy for the optimal allocation of available resources.[40] Vandelli also had a very important role in the 1768 creation of the Ajuda Palace Museum and Gardens in Lisbon, institutions that were a focal point for Portuguese overseas collecting activity.[41] The reforms were further reinforced during the reign of D. Maria I (1777–1816), and it was during this period that the Royal Academy of Sciences in Lisbon was founded in 1779. This institution was of paramount importance in guiding scientific research conducted overseas, as well as in promoting discussion and the publication of numerous memoirs on the subject.[42]

In the context of the educational reforms, it should also be stressed that when the Society of Jesus was expelled from Portugal (1759), military institutions became more and more influential as centers for the training and practice of mathematical sciences. Nearly all the late eighteenth- and nineteenth-century Portuguese mathematicians of some distinction had been trained in or were professors at military institutions, and these institutions were fundamental in the education of cartographers who participated in overseas expeditions. The importance of the military schools and of mathematicians affiliated with military institutions is well recognized today. However, the studies available are only exploratory attempts, and much work remains to be done.[43] One would like to have a comparative analysis of the organizational structure of military schools and universities; a study of the ways in which the needs of the empire affected or even shaped the establishment of military scientific activities.

Due first of all to its significant sugar production and then as a profitable source of precious metals, Brazil become of paramount importance to the economic balance of Portugal during the eighteenth century. The great importance given to Brazil is reflected in the historiography of the history of medicine and natural history. The increasing contribution of professional Brazilian historians of science to the subject should also be stressed.

The scholarship on the nine-year expedition of Alexandre Rodrigues Ferreira to the Amazon region from 1782 to 1792 is vast.[44] In his work on scientific expeditions in the Portuguese overseas territories, Simon provided an account of the aims and various stages of this trip together with information on Ferreira's description of several animal and plant specimens and their shipment to the Ajuda Palace Gardens and Museum in Lisbon. Information is also presented on the reports and memoirs of the naturalist as well as on the wide scope of Ferreira's collection and its fate during the Napoleonic invasion of Portugal. A significant part of it was expropriated by Étienne Geoffroy Saint-Hilaire to the Paris Muséum d'Histoire Naturelle. Ângela Domingues has shown that the expedition cannot be dissociated from the activities of territorial demarcation of the Portuguese and Spanish governments in the region.[45] Moreover, she has emphasized that political, geographical, astronomical, natural historical, anthropological, and ethnographical aspects of the expedition were profoundly intertwined. She has also pointed out that the Portuguese project of localizing, measuring, describing, and representing—which was conducive to the appropriation of the Amazon territory—involved a multitude of actors with various forms of expertise as well as persons with no scientific background and indigenous people. In addition, she has stressed that the main efforts associated with the expedition were frustrated, since the issue of demarcation did not have a logical conclusion and the scientific results of the mission were limited to omission for a very long period of time.

Recent literature on the subject has emphasized that naturalists were not the only actors crucial to the venture of describing, collecting, and understanding the nature of the Portuguese empire. Governors and administrative employees, as well as draughtsmen, were also crucial to the success of this enterprise. In fact, governors of Portuguese America would often give instructions to alter voyage routes as well as halt or help investigations to suit their interests. The contribution of the draughtsman José Joaquim Freire to the natural knowledge of the Portuguese overseas territories has been the subject of a recent, valuable, and beautifully illustrated study by Miguel Faria.[46] The book examines the importance of drawing for natural history and military engineering during the eighteenth century, giving special relevance to the drawings produced by Freire in the context of the Ferreira expedition to the Amazon region. Faria's work is especially insightful in the affiliation between military and natural historical drawings during this period and on the complementarity of draughtsmen and naturalists in the description and study of nature. He also provides useful considerations on the relationship between text and image and on the general importance of drawings in the appropriation of Portuguese overseas territories. The gaps in the historiography of the illustration of nature in relation to the Portuguese empire are still considerable, and this is a very profitable area for future studies. (For a study of the importance of illustration for the Spanish colonial science, see Bleichmar's Chapter 15).[47]

Mining knowledge in colonial Brazil is the other topic that has attracted increased attention. Silvia Figueiroa and Clarete da Silva have been at the forefront of the scholarship on this subject. Their excellent study on eighteenth-century mineralogists provides an introduction to the new approaches to mining in this period, together with a brief analysis of the work of two Luso-Brazilian mineralogists, Manuel da Câmara and José Vieira Couto.[48] From roughly 1700 until 1775, gold and diamonds from Brazil brought Portugal great wealth. Later, the production of these two minerals declined significantly. The Portuguese government set out to revitalize the industry by selectively introducing modern scientific theory and technology. Several students, most of them born in Brazil, were sent to the newly restructured University of Coimbra in order to learn the new scientific methods of metallurgy. A few of them were even subsidized by the Portuguese state to attend courses in mineralogy, chemistry, metallurgy, and other mining arts at the universities and technical schools of Freiberg and Paris. The education received on the subject introduced these students to the growing importance of other minerals such as coal and iron and to new methods of extraction, as well as to the importance of mining administration. It was considered to be crucial for the Portuguese government to know that the scientific methods of metallurgy might lead to more efficient production, not only of gold but also of iron, coal, and other minerals that had a growing importance in Europe.

After Câmara's studies in Portugal and in Europe, he was commissioned to supervise important mines in Brazil. However, the reforms he desired were never implemented. Couto also studied at the University of Coimbra and was afterwards commissioned to undertake a mineralogical survey in the region of Minas Gerais in Brazil. One of the results of his activities were the mineralogical memoirs he published in the journal of the Royal Academy of Sciences in Lisbon. He also produced descriptions of the geography, climate, and population of the territory. Figueiroa and Silva have stressed that the study of this case reveals that Couto absorbed an ideology of integrated development and that, for him, the interests of the Portuguese colony and the crown were the same. They concluded that within the Portuguese empire there was great continuity of policy between the metropolis and colonial scientists, but with certain distinctions among the colonies. Silva's book on Couto based on a contextual analysis of his memoirs further enlightens the relevance of Couto's work.[49] The book's main aim is to understand Couto's vision of science, his scientific activity in relation to that practiced by scientists working in the same area, his position in mineralogical and geological controversies of the period, and the various difficulties that Couto had to face during the course of his work.

A relevant essay was published recently on the work of mineralogist João da Silva Feijó.[50] The authors acknowledge how historiography has literally condemned this naturalist to obscurity and aim to illustrate his contribution to the development of modern science. Between 1783 and 1797, Feijó had his first assignment in the islands of Cabo Verde and then in Ceará, Brazil. In Cabo Verde, he studied and clarified his ideas about volcanoes and listed the products that could be exploited commercially in these islands, such as sulphur, salt, and sulphates. In Ceará, he studied the characteristics for the identification of mineral deposits, and he investigated saltpeter, following explicit orders from the Portuguese government due to the mineral's military importance. He also conducted botanical and paleontological investigations of the region. The authors make it clear that mining remained at the center of the Portuguese Enlightenment political project and that it relied on co-opting groups of Portuguese in America. They demonstrate that the life and work of Feijó illustrates practices, negotiations, and modes of publication that are typical of the process of globalization of science.

Great expeditions and scientific journeys collected natural historical material and data. As Janet Browne has emphasized, for a long period the material collected for medicine followed the same collecting institutions as natural history.[51] At the same time, natural historical activities were often carried out by physicians. Recent studies on the history of medicine have also been done in the context of the Brazilian Portuguese empire. Lycurgo Santos Filho has provided a general view of the history of Brazilian medicine.[52] It covers not only the impact of Iberian medicine in Portuguese America but also indigenous notions of illnesses and their treatment, African influences on the concept and treatment of diseases, and the role of the Jesuits in diffusing medical knowledge. Santos Filho relies on a vast number of sources but sometimes suffers from anachronism in his classification of a "prescientific" and a "scientific" medicine, associating the former with the medical practices of indigenous peoples and of Africans as well as with Jesuit priests and European healers without a medical degree.

In a more recent book, Márcia Ribeiro offers a concise but more appealing and complex view of the history of medicine in Brazil in the eighteenth century.[53] It provides insight into the rich mix of traditions that molded colonial science and into the relationship of interdependence between ruler and ruled, expert and general population. Ribeiro deals with the impact of some of the diseases introduced into Brazil by the Europeans and Africans during the period of colonization. In her view, colonial medicine should be seen as a scheme of knowledge that was the result of frequent exchanges between European, indigenous, and African cultures. She also makes clear that the great limitations in receiving medicines from Lisbon in sufficient number and in good condition enhanced the importance of the healing practices of non-European cultures. At the same time, throughout its colonial period, Brazil suffered from a chronic lack of surgeons and physicians. In Ribeiro's view, it should therefore not be surprising that surgeons with little practice, amateur healers, midwives, wizards, and mere charlatans had such an important role in medical practices. The author stresses as well how the importance of the supernatural in the treatment of disease was shared by common people and the higher ranks of society, including many physicians. Considerable attention is given precisely to the issue of the thin boundary between medicine and magical practices.

Vera Marques's book on medicine and pharmaceutics in eighteenth-century colonial Brazil is another valuable contribution to the subject.[54] It shares some of the concerns of the previous work in terms of revaluating the importance of the various actors involved in the history of healing in this territory with special relevance to the indigenous people. In contrast to Ribeiro, Marques does not attribute significant relevance to the diminutive number of surgeons, physicians, and materia medica that reached the colony from the metropolis. In her view, the diversity of cultural backgrounds of the Brazilian residents would be in any case sufficient to explain the persistence of plural healing practices. The narrative reveals processes in which the experience of colonizers and colonized intersected, showing how some of the plants of the territory were gradually incorporated into the Portuguese pharmacopoeia in a process that involved exchange of information, description, and classification. This was a process of appropriation that, during the eighteenth century and also before, was partly carried out in the colony and partly in the metropolis. Lisbon received from Brazil several specimens of medical interest, some of which were later exported back to the colony as accredited remedies. The author also calls attention to the role of the Medico-Chirurgical, Botanical and Pharmaceutical Academy founded in Rio de Janeiro in 1772.[55] This was to become a space of sociability for various men interested in medical knowledge and practices, which was guided by Enlightenment principles of ordering and classification in an attempt to control, and profit from, tropical nature. The author is also concerned with issues of secrecy motivated by the economic interests of the Portuguese government or by personal profit and, more important, by the fact that magic, religion, and science were, for a long time, intertwined in the territory. The two aforementioned works show how this is a fertile area for further studies.

The formation of networks of information and the processes of diffusing knowledge were crucial to the development of Portuguese colonial medicine and natural history. As was already mentioned, until the second half of the eighteenth century the Jesuits played a pivotal role in the circulation of European and indigenous information relevant to medical and natural historical practices. After this period, the agents involved in the exchange of information become more diversified and more closely linked to the state. In her 2001 essay, Ângela Domingues presented a clear and useful overview of the actors and institutions that were involved during the eighteenth century in the development of an information network that focused on the economic potential of the Portuguese colonies.[56] She has pointed out that information about the natural world flowed from all parts of the empire; it was propagated by agents with various backgrounds, aims and professional positions, and she has stressed that the main recipient was the Portuguese state. In her view, medical and natural historical knowledge gained from the colonies had an eminently practical nature, which should be associated with the intended achievement of scientific, political, economic, and social goals. Domingues is, however, very doubtful of the success of the enterprise and, in particular, the ways the Portuguese state controlled and managed the available information, the ways the various kinds of information were integrated into a global knowledge of the Portuguese empire, and to what extent the information was used to improve the well-being of the population or was just lost in various archives.[57]

Undoubtedly, one of the crucial problems running though the history as well as through the historiography of medicine and natural history in the context of the Portuguese colonial experience has been the insufficient dissemination of all the information gathered in the overseas territories, and this problem has manifested itself not only abroad but also at the very center of an empire whose high aspirations, purposes, and efforts were often annihilated by a dismissive view of the importance of the diffusion of knowledge.

CONCLUSION

From its fifteenth-century origins to the end of the period covered in this essay, the scientifico-technical dimension formed an integral element of Portuguese colonialism. Nautical, mathematical, and astronomical studies as well as natural historical and medical practices were crucial in defining and establishing the Portuguese colonial enterprise. The historiographical survey presented here shows that a significant amount of research has already been carried out on nautical science and on the Jesuit contributions to the development and diffusion of science. At the same time, the study of colonial Brazil during the second half of the eighteenth century has recently emerged as a fertile and even fashionable area of study. In addition, serious attempts have already been made to address complex and more theoretical questions such as, for example, the role of a specific Portuguese contribution to the "Scientific Revolution."

However, huge gaps still remain in our knowledge of the field in terms of period, place, and subject. New approaches to the sixteenth-century apogee of the empire are still absent, and the seventeenth century is almost completely avoided by historians. Very few works have been produced on material culture and visual representations of nature, or on environmental history. The history of the book and of the various processes of circulation of knowledge still need much further research. What knowledge was acquired and what was discarded, what knowledge was assimilated or transmitted because of European and colonial notions of gender is also an area open to future explorations.

The various studies covered in this essay suggest that an interest in the area of science and colonial studies is emerging in the Portuguese-speaking countries, but it is still difficult to speak of a discipline as such. The specific nature of the Portuguese empire and especially the gigantic geographical dispersal of the Portuguese presence around the world make it difficult to obtain a comprehensive view of the subject, but at the same time this will be all the more appealing and rewarding in terms of comparative studies with the history of other empires.

PART II

New Worlds, New Sciences

Cosmography at the *Casa*, *Consejo*, and *Corte* During the Century of Discovery

MARÍA M. PORTUONDO

How do you describe a new world? Europeans who attempted to describe America, from the sailor to the most learned scholar, faced this monumental challenge during the century after its discovery. What did this world contain? Where did these lands lie in relation to Europe? Who lived there? Were they like us? One group of scientific practitioners—cosmographers—recognized in these questions one of the greatest challenges of its time and made it their mission to explain these new lands and people. Nowhere were the efforts to answer these questions more steadfast than among Spanish royal cosmographers who worked during the sixteenth century for an imperial administration driven by an insatiable need to know the New World. This essay explores how the discovery forced Spanish royal cosmographers to address some epistemological and methodological problems at the very heart of scientific practice and history of science, and develop novel cosmographical practices in response.

Renaissance cosmography, defined broadly as it was at the time, encompassed aspects of the modern disciplines of geography, cartography, ethnography, natural history, history, and certain elements of astronomy. Although cosmographical production occupied some university scholars,[1] by the late sixteenth century cosmographical production in Spain shifted to the Casa de la Contratación (House of Trade, established in 1503 to regulate all commerce and navigation to the Indies) and the Consejo de Indias (Council of Indies, established in 1524 to govern the Spanish territories in the New World), as well as the *Real Corte* (royal court of the Habsburg monarchy). Most of the cosmographers working in these institutions had royal appointments and enjoyed the prestige—if not always the monetary rewards—that came with being in the king's service. They were under obligation to comply with the responsibilities associated with their posts as specified by their appointment order or by guidelines set by each institution's governing body. Private individuals, perhaps in expectation of royal favors, also submitted contributions to the empire's cosmographical storehouse, some in the form of *Relaciones* (geographical accounts), while others wrote ambitious geographical and historical works.

Each institution—*Casa*, *Consejo*, and *Corte*—cultivated distinctive cosmographical styles molded in large part by its administrative duties. As Antonio Barrera-Osorio points out in Chapter 11, an increasing valorization of empirical practices at these sites of knowledge production became the most trusted way of addressing scientific and technical questions.[2] Given this focus, cosmographers first had to address epistemological problems that arose from collecting information about the New World. At issue was the fundamental question of what constituted a cosmographical fact. For example, was a pilot's eyewitness description of a coast as valuable as that of a cartographer who had never visited the area but estimated its latitude mathematically? Consequently, does scientific theory trump experience? This was one of the key issues discussed between pilots and cosmographers at the Casa de la Contratación. Cosmographers also had to address methodological issues. One pressing problem for the Spanish monarchy was how to effectively legislate on and administer possessions, the closest of which lay thousands of miles from Madrid. How then to collect and compile ethnographical information about hundreds of different indigenous groups? How can this information be represented so that it can be used to legislate effectively? Does an encyclopedic descriptive narrative suffice? Cosmographers also had to address the issue of personal agency in science. For example, could a person without training in natural history adequately describe the flora and fauna of the New World? Or does a trained eyewitness investigate natural history best? The discovery and colonization of the New World forced practitioners to reconceptualize what constituted the very discipline of cosmography in one of the clearest examples of how western science changed because of the events of 1492. This essay introduces the work of selected royal cosmographers at the *Casa*, *Consejo*, and *Corte*, those whose career and work illustrate best how cosmographers tackled the tension that the discovery of the New World caused in the discipline of cosmography.

RENAISSANCE COSMOGRAPHY

The goal of every Renaissance cosmography, as Ptolemy stated in his *Geography*, was "to describe the whole of the known world."[3] A product of Renaissance humanism, cosmography combined the mathematical tools and epistemology of Ptolemaic geography with classical descriptive geography, modeled after Strabo and and Pomponius Mela, while resting firmly upon an Aristotelian understanding of nature. Renaissance cosmography therefore possessed the conceptual tools to answer many questions about the New World, and it had the necessary taxonomic categories to effectively address a number of administrative, financial, military, religious, and political concerns of the expanding Spanish empire. Geographic descriptions answered the obvious questions about the land, the empire's extent, and its borders, while Ptolemaic cartography promised to represent this geography with mathematical precision. Hydrographic reconnaissance judged the navigability of coastal waters and rivers, the existence of suitable ports, and the most efficient sea routes to the new lands. Descriptive narratives summarized ethnographic information about the native peoples and their customs and religion. Chronicles recorded the history of Spanish conquest and discovery in the new lands, while natural history yielded inventories of natural resources. Cosmography was, in short, the science that explained the earthly sphere by locating it within a mathematical grid bounding space and time. Natural phenomena and human actions that defied mathematization were described; words were the tools that took the phenomena from the realm of the unknown into the known. The relatively young discipline of cosmography would strain under the avalanche of new knowledge that resulted from the discovery of the New World.

The cosmographer's first task was to incorporate the New World into a new "universal" cosmography. In continental Europe, Gemma Frisius, Sebastian Münster, and André Thevet were among the cosmographers who took on this task. Their ambitious cosmographical projects required an exhaustive exercise in compilation and erudition which placed the territory described within the web of symbols and classical correspondences that defined the European understanding of the world.[4] The New World challenged what these armchair cosmographers could achieve, trained as they were in the humanistic tradition to extract information from books and to turn to classical sources for references. There were no classical sources describing the Incas, the Aztecs, or the Arawaks. Furthermore, Spain denied other countries access to the new lands and, during the second half of the sixteenth century, took steps to safeguard cosmographical information. Thus, French, German, and other continental cosmographers had to rely on spotty and, at best, third-hand information about most of the New World to compose their works. Spanish cosmographers worked under an additional pressure; not only did the cosmographical information have to incorporate continual new discoveries into a comprehensive world picture, it also had to answer a number of specific questions posed by the cosmographer's institutional patrons.

They increasingly required that cosmographical knowledge be useful. This utilitarian mandate implied a body of cosmographical knowledge that had to be current, timely, and accurate.

THE CASA DE LA CONTRATACIÓN AND THE SEA

Chartered in 1503 by Ferdinand and Isabella to manage private commerce and royal monopolies with the newly discovered lands, the Casa de la Contratación and its host city, Seville, became the locus of information exchange about the New World. Since its founding, officials at the Casa de la Contratación recognized the strategic value of their access to eyewitness knowledge about the New World, and put in place the institutional structure to secure and effectively leverage this information. Chartered with overseeing all navigation to the Indies, in 1508 the *Casa* named its first pilot major, responsible for establishing—after interviewing pilots and others arriving from the Indies—the officially sanctioned *carrera de Indias* (navigation routes to the Indies). The resulting nautical charts and rutters were known as the *padrón real* (royal rutter). In addition, the pilot major also determined the navigation techniques that constituted best practice, and although he was initially responsible for training pilots, by midcentury his duties were limited to examining and licensing them. Amerigo Vespucci was the first pilot major, serving from 1508–1512. After a series of interim appointments, Sebastian Cabot served from 1518 until he left for England in 1548.[5]

As early as 1519, the Casa de la Contratación had specialists in instrument and mapmaking, and in 1523 named its first cosmographer to a post that eventually became that of cosmographer major. One of his duties consisted in examining and licensing all nautical charts and instruments made by independent cartographers and instrument makers. A number of other individuals also received titular appointments as cosmographers of the *Casa*, among the most prominent, Hernando Colón (Christopher Columbus's son and erudite bibliophile), Alonso de Santa Cruz, and Pedro de Medina. These appointments gradually increased the profile of the cosmographical arts in the navigation program of the Casa de la Contratación.

The mathematical underpinnings of Renaissance cosmography, and particularly Ptolemaic cartography, were perceived as tools that could clarify geographical questions with heretofore unprecedented certainty. Geodesic measurements became politically important after Pope Alexander VI decreed in the 1493 papal bull *Inter Caetera* that lands discovered west of a line of demarcation encircling the globe would belong to

Spain, while those lying east of the line would belong to Portugal. The subsequent Treaty of Tordesillas between Spain and Portugal left the language associated with the line's location open to interpretation. In the prevailing political climate, it was a strategic necessity to have in royal service cosmographers who could convincingly debate such questions and attempt to determine longitudinal location with exactitude. Spanish cosmographers enthusiastically tackled the challenge, convinced that Renaissance cosmography was a science firmly rooted in mathematical rationalism and possessing the tools necessary to represent geography accurately. In various juntas called to discuss the "Tordesillas Question," cosmographers compared methodologies, shared skills, and developed ways of argumentation to support cosmographical facts, while in the process coalescing as a community and establishing a professional identity as specialized scientific practitioners.[6]

As the pace of exploration and settlement of the New World increased after the conquest of Mexico in 1521, so did the demand for pilots capable of guiding a ship safely across the Atlantic, and later in the century, across the Pacific Ocean. During the initial years of the *carrera de Indias*, pilot apprentices learned their crafts at sea, through careful observation of the sea and sky and with the compass and portolan charts as technological tools.[7] Since the thirteenth century, however, the astronomical and mathematical principles behind finding a ship's position at sea were well understood in academic circles but had found limited acceptance among experienced sailors who were more interested in reaching their final destination safely than knowing their exact position while at sea. Astronomical navigation began per se with the introduction in the mid-fifteenth century of two nautical instruments that simplified the astronomical and mathematical computations necessary for determining latitude while at sea: the Jacob's staff for measuring the elevation of the North Star, and the astrolabe for determining the height of the sun at noon (see Figure 3.1). The use of these nautical instruments, carefully conceived and designed to yield theoretically coherent results, appealed to academically trained cosmographers. Using the instruments on board a rocking ship, however, as pilots rightfully pointed out, quickly dispelled any confidence about the values obtained.

The utility of cosmographical knowledge in navigation was far from uncontested during the first half of the sixteenth century. Conflicts raged during the 1530s and 1540s between experienced pilots (who rejected astronomical navigation) and a group described as "theory proponents," which included many cosmographers who advocated the use of Ptolemaic cartography and astronomical navigation.[8] (On the Portuguese aspects of these developments, see Almeida in Chapter 4.) Pilot training

FIGURE 3.1. *Determining the sun's elevation using an astrolabe, 1563*
SOURCE: Pedro de Medina, *Regimiento de navegación* (Seville, 1563),
fol. 15v. Nettie Lee Benson Library, The University of Texas at Austin.

became the flash point between the two camps. Should licensed pilots be
expected to master the principles of astronomical navigation, or did it
suffice that they have the expertise necessary to guide a ship safely across
the ocean? Another issue debated among pilots in the *carrera* was the use
of navigation charts drawn with two scales, one with coordinates based
on true north and the other with a scale adjusted to correct for the effect
of magnetic variation on the compass needle. These debates resulted in
a series of lawsuits between cosmographers Alonso de Chaves, Pedro
de Medina, and Pedro Mexía, on one side, and Diego Gutiérrez, master
instrument maker at the Casa de la Contratación, who had the support
of then Pilot Major Sebastian Cabot, on the other. The bottom line with
the lawsuits was often competition over the lucrative instrument and
cartographic trade in Seville and the monopoly that Gutiérrez, backed
by Cabot, maintained to the exclusion of the other cosmographers. The
battlefield, however, concerned the supremacy of science or whether

mathematically based theoretical knowledge had intellectual authority over practical experience.

By the time Philip II ascended to the Spanish throne in 1556, royal authority had sided with cosmographers and established them as de facto experts on all cartographic, geographic, and navigational matters at the *Casa*. Cosmographers earned a significant victory when a royal decree in 1552 created a professorship of navigation attached to the Casa de la Contratación to train pilots in the art of astronomical navigation. The first chair was a university-trained cosmographer, Jerónimo de Chaves, whose father Alonso de Chaves was serving as pilot major (1552–1587). The institutional directors of the Casa de la Contratación saw in classroom instruction—and astronomical navigation—a way of perhaps shortening the years of apprenticeship necessary to learn the craft of piloting. In addition to educational duties, the *catedrático del arte de navegar y cosmografía* (professor of navigational arts and cosmography) also advised the Casa de la Contratación and the king on cosmographical matters.[9]

A new literary genre, the navigation manual, grew out of this climate of debate and proved a capable advocate of astronomical navigation. The navigation manual sought to establish the art of navigation on a theoretical footing consistent with established principles of natural philosophy. Authors drew from among the principles of the sphere as synthesized by Sacrobosco, the fundamentals of Ptolemaic cartography, and in varying measure, from practical nautical experience to produce this very popular genre. These books often also included brief geographical descriptions, often emphasizing the Indies and the Orient.

The first navigation manual printed in Spain was the *Summa de geographia que . . . trata largamente del arte de marear* (Compendium of Geography that Addresses the Art of Sailing; Seville, 1519), by Martín Fernández de Enciso. The book, written in Castilian, was one of the first works to meld the different theoretical strands that made up the discipline of cosmography with practical aspects of navigation. Enciso's text established the format followed by subsequent navigation manuals. It begins by explaining the theoretical underpinning of astronomical navigation, the theory of the sphere. That is, how projecting the celestial poles and equator onto the terrestrial sphere results in a grid of tropics, meridians, and parallels that forms the basis for latitude and longitude coordinates. He also explains various astronomical ways of determining latitude at sea. Bowing to the cosmographical tradition that preceded him, Enciso follows the chapters with a geographic description of the world along the stylistic lines of Pomponius Mela. Yet, he sets aside the classic division of the earth into the continents of

Europe, Asia, and Africa, instead dividing the world into two halves: the Orient and the Occident.

Toward the mid-1500s, the focus of the navigation manual shifted to a format that discusses exclusively nautical topics and only those cosmographical principles that pertained to navigation. The genre retained the theoretical section that introduced Ptolemaic concepts such as the poles, the equator, latitude and longitude, winds, climates, and basic cartography, but was now followed by practical instruction. A reader could expect to learn how to use maritime charts to set a course at sea, how to determine a ship's latitude using quadrants or astrolabes, and how to apply solar declination tables to adjust values. Following the successful models of Peter Apian's *Cosmographia* (1524) and of Gemma Frisius (1508–1555), Spanish authors used these manuals to introduce new nautical instruments. The new medium also served as a forum for discussing solutions to navigation problems. One such problem was the deviation of the compass needle from north caused by magnetic variation. In long, transoceanic voyages, the phenomenon added a new dimension of difficulty in charting a course at sea, since the deviation from north seemed to vary according to one's meridian. Navigation manuals became a forum for advocating different ways of coping with the problem, either by designing instruments that "corrected" the compass reading or trying to use the needle's deviation as an indicator of longitude.[10]

In Spain, as in the rest of Europe, the two most widely read navigation manuals were Pedro de Medina's *Arte de navegar* (The Art of Navigation; 1545) and Martín Cortés's *Breve compendio de la sphera y de la arte de navegar* (Brief Compendium on the Sphere and the Art of Navigation; 1551). Between 1561 and 1630, Cortés's book had at least six English translations with full attribution, while Medina's book enjoyed a wider diffusion, with twenty, mostly French, editions.[11] Both claimed to have been motivated to write the manuals out of frustration with pilots' incompetence and pilots' reluctance to learn the principles of astronomical navigation, something the authors believed would turn this dangerous craft into a more certain science. Cortés, in particular, cautioned against the use of navigation charts based on erroneous geographical information provided by pilots, going as far as to urge the king to send, "learned cosmographers and experts in the art of navigation to verify the latitude of the ports, capes, islands, and maritime ports, and that likewise the coasts be described truthfully, specially those pertaining to the navigation of the East Indies or New World."[12] Despite the authors' assertion that their books were for pilots, it is hard to imagine that these books were within the average pilot's reach. For although pilots held a position of utmost responsibility on board, often being the final authority on all matters pertaining to navigation and the technical aspects of operating a ship at sea, most were illiterate and notoriously undisciplined.[13] Navigation manuals of the midcentury were written indeed to formalize the principles of astronomical navigation, but hardly with the hopes of instructing practicing pilots in the nautical sciences.

The navigation manual and the school for pilots of the Casa de la Contratación were a reaction to a mounting crisis overtaking Spain's ventures overseas. The increasing number of transatlantic expeditions created a critical shortage of pilots with sufficient experience in oceanic navigation. When the shortage threatened the imperial program and the efficient exploitation of the newfound resources, elites of the time resorted to the familiar methods of their personal education—the classroom and the book—to try to resolve the problem. Cosmographers provided the didactic material and backed the program with the intellectual and administrative authority that their service to the crown had earned them. Therefore, the emphasis on teaching pilots astronomical navigation came about not because of any demonstrated technological superiority of astronomical navigation over traditional methods, but in response to a critical shortage of pilots in the Spanish empire.

By the 1570s, cosmographical activity at the Casa de la Contratación coalesced around the figure of Rodrigo Zamorano (1542–1620), a university-trained cosmographer. During his almost forty years at the *Casa*, he occupied at one point or another all of its cosmographical posts, as well as that of pilot major.[14] Zamorano cultivated American plants in his garden in Seville and exchanged botanical specimens with Dutch naturalist Charles de L'Écluse (1526–1609).[15] Although he never published on the subject of natural history, he wrote on astrology, mathematics, and navigation and published repeated editions of a calendar and chronology. Unfortunately, as is the case with most of the cartographic production at the Casa de la Contratación, few examples survived of Zamorano's cartography, which he practiced both privately and in his official capacity.[16]

In 1575 Zamorano was appointed for a period of five years as professor of cosmography at the Casa de la Contratación.[17] The following year, he published a Spanish translation of the first six books of Euclid's thirteen-volume *Elements* (Venice, 1492), which earned him consideration for the professorship of mathematics and astrology at the University of Salamanca. Zamorano was not content with simply teaching navigation, and he actively sought involvement in a number of cosmographical projects organized by the Council of Indies. In a 1582 petition letter to the king, Zamorano presented himself as a man of science capable of carrying out the necessary inquiries to advance the cosmographical art.[18]

As was customary in a communication aimed at gaining the monarch's attention, Zamorano was careful to select from among his many activities those he thought would increase his reputation before the king. He points out that he collaborated in a number of conferences aimed at updating the *Casa*'s cartography, including observing a number of lunar eclipses for the purpose of determining longitude and for which he personally built several instruments (see Figure 3.2). He also reminded the king of his recently published navigation manual, the *Compendio de la arte de navegar* (Compendium on the Art of Navigation; Seville, 1581), with much-needed updated astronomical tables.[19] In the *Compendio*, Zamorano explains that he has carried out careful astronomical observations to correct solar values he considered contributed to imprecise latitude calculations by pilots.[20]

Zamorano's career illustrates the pragmatism that guided the labors of cosmographical practitioners at the Casa de la Contratación. Theirs

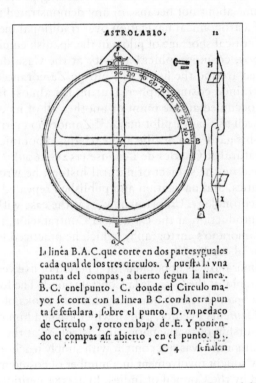

ASTROLABIO. 12

La línea B.A.C. que corte en dos partes yguales cada qual de los tres circulos. Y puesta la vna punta del compas, abierto según la línea. B.C. enel punto. C. donde el Circulo mayor se corta con la línea B C. con la otra punta se señalara, sobre el punto. D. vn pedaço de Circulo, y otro en bajo de .E. Y poniendo el compas así abierto, en el punto. B.
C 4   señalen

FIGURE 3.2. *Instructions for making a nautical astrolabe, 1588*

SOURCE: Rodrigo Zamorano, *Compendio del arte de navegar* (Seville, 1588), fol. 13. Reproduced by permission of The Huntington Library, San Marino, California.

was a cosmography stripped of its humanistic roots and that emphasized solely the mathematical products of the discipline: cartography and astronomical navigation. It was perhaps this focus that fed the interest and admiration other European nations had for the *Casa*'s cosmographical work. England, France, and later the Dutch Republic recognized the valuable storehouse of strategic geographical and nautical information kept—more or less—securely at the *Casa*, which made it an important intelligence target.

THE COUNCIL OF INDIES AND THE LAND

As the royal council responsible for the administration of Spain's American and Pacific colonies, the Council of Indies recognized not long after it was chartered in 1524 that its ability to carry out its duties depended on receiving timely and accurate information from the empire's domains. Religious, military, and government officials routinely reported on the situation overseas and occasionally included descriptions about the people, geography, and natural resources of the New World in their reports. Sporadic and written by nonspecialists, these reports did not form a coherent body of knowledge that the Council could easily utilize. Therefore, during the first half of the century, the Council relied on the work of cosmographers and historians loosely associated with the Council to compile and compose historical, geographical, and natural historical reference works. This was the case with two early contributors to the Council of Indies, Gonzalo Fernández de Oviedo (1478–1557) and Alonso de Santa Cruz (1505–1567).

Renaissance cosmography was well suited to meet the Council's need since it had retained, in imitation of its classical models, the historical narrative as part of a geographical reckoning of a territory. In Spain, these kinds of works generally separated the text into two distinct parts, a *historia natural* and a *historia moral*. Together, they drew from the Aristotelian hierarchical view of nature in which the natural world sustains man's moral actions. Man's moral soul and freewill placed him in a category apart from other creatures, at the apex of the natural hierarchy and yet distinct from it. Therefore, a *historia natural y moral* was typically divided into two sections, one describing the land and its geographic features as well as plants, animals, and natural resources; and one detailing the actions of man, both native and European, for that given space. This genre allowed for a mostly expository and intensively descriptive narrative in the sections that dealt with the natural elements and relegated interpretation to the "moral"—or what we would recognize today as the historical—section of the text. The most comprehensive of these works was Gonzalo Fernández de Oviedo's extraordinary *Historia general y natural de las Indias Occidentales* (General and Natural History of the West Indies; Seville, 1535). In 1532, the Council of Indies accepted his offer to write the history and natural history of the Indies and agreed to modestly underwrite the endeavor as long as the work was submitted to censorship by the Council prior to publication. The work set the standard for what constituted a complete—meaning moral and natural—history of the New World. It was also influential in establishing the epistemological criteria used in subsequent New World natural histories, one based on empiricism and eyewitness testimony.[21]

Alonso de Santa Cruz's association with the Council was at first tangential. He began his professional career as a cosmographer at the Casa de la Contratación, having earned the reputation of being an innovative instrument maker.[22] Along with his appointment, Santa Cruz also obtained a series of royal orders backing his emergent cosmographical interests. Although Santa Cruz garnered his salary from the *Casa*, he seems to have operated outside the administrative structure of the institution, preferring instead to use his connections at court (he was a *contino*, or king's attendant) to set in motion his cosmographical projects. During the years he was at court, he discussed astrology and cosmography with Emperor Charles V and tutored the future Philip II on the sciences.[23] His work was essentially a personal pursuit, encouraged by an admiring monarch and rewarded by royal favors.

Several times in his life, the monarch asked Santa Cruz to advise the Council of Indies on cosmographical matters and to serve in cosmographical conferences. In 1554 the Council requested Santa Cruz to prepare a report on different methods of calculating longitude at sea in order to evaluate an instrument made by German Cosmographer Peter Apian that promised to solve the problem of determining longitude. In response, Santa Cruz wrote the *Libro de longitudes*, a survey of all the known methods of calculating longitude, along with a commentary and précis of Ptolemy's *Geography*.[24] Santa Cruz never published his extensive cosmographical work, probably as a result of secrecy regulations that restricted the publication of what the Habsburg monarchy considered sensitive information. He referred several times to having spent years working on—but never felt he had finished—a descriptive geography and atlas of the known world, his now-lost *General geografía e historia* (General Geography and History).[25] Notable among his surviving works are the *Islario general de todas las islas del mundo* (General Map of All the Islands in the World), *El astronómico real* (Royal Astronomical Study), and an extraordinary map of the world.[26]

(For late-sixteenth-century and early-seventeenth-century developments in nautical instruments, see Sheehan's essay in Chapter 12.)

Santa Cruz's association with the Council was often rocky. In a petition made to the king in 1557, he complained to the future Philip II that men of science were needed yet not welcomed at the Council of Indies. The letrados (jurists) that served in the Council, he lamented, "can not stand any man practicing a science other than their own."[27] Santa Cruz's observation was not wasted on the future king. In 1571, as the result of a series of reforms initiated by Juan de Ovando, who later became the president of the Council, the Council instituted the office of Cosmographer-Chronicler Major of the Council of Indies. Ovando's reforms institutionalized cosmography as the preferred way of collecting and organizing geographic, hydrographic, ethnographic, and natural historical information about the Indies. Many of the practices associated with Renaissance cosmography became the government-sanctioned way of producing a body of knowledge that accurately described the sea, the land, and the peoples of the Indies.

Ovando's reforms consisted of a series of ordinances and instructions that included the duties of cosmographer-chronicler. These regulations also specified, in at times excruciating detail, the cosmographical methodology and mode of representation to be employed by the cosmographer-chronicler.[28] The law sought from the cosmographer some very specific products: cosmographical descriptions, maps of the Indies, and a current historical chronicle. These products were to be kept up to date and readily available to members of the Council in a "book of descriptions" that was securely guarded in the Council's archivo de los secretos (archive of secrets). The cosmographer had to compile the book from material continuously collected from overseas, following mechanisms the law provided and that were woven into the administrative fabric of the empire. The instructions implicitly recognized that neither encyclopedic natural history along the lines of Oviedo's nor all-encompassing universal cosmography such as the one Santa Cruz had spent a lifetime writing satisfied the Council's need for up-to-date information about an ever-expanding empire. With the book of descriptions, the Council of Indies no longer had to wait for a cosmographer to finish a universal cosmography. After implementing the reforms, the Council expected to have access to timely reports and descriptions, catalogued by jurisdiction and summarized yearly by the cosmographer-chronicler. If the legal statute was interpreted broadly, the instructions provided an efficient mechanism for collecting and cataloguing cosmographical information in a timely manner; yet if interpreted narrowly, it had the potential of simply functioning as a bureaucrat's to-do list.

Juan López de Velasco (ca. 1530–1598) was the first Cosmographer-Chronicler Major of the Council of Indies. A former legal assistant to Juan de Ovando, López de Velasco was not a university-trained cosmographer. His personal interests lay in education and linguistics, yet he served diligently as cosmographer-chronicler for twenty years. Throughout his career, he resorted to different methods in an effort to comply with the directives regulating his post. During the first four years of his tenure, he dedicated himself to writing a cosmography along the stylistic lines of the Renaissance cosmography; its style, however, was influenced by its sole intended audience: government officials. The result was the *Geografía y descripción universal de las Indias* (Geography and Universal Description of the Indies; 1574).[29] No sooner had López de Velasco finished the extensive and no doubt laborious task, did the Council criticize the *Geografía* for being out of date.

López de Velasco would never author another comprehensive cosmography. Instead, he chose to rely on the eyewitness's testimonies to provide the Council of Indies with this information. He prepared a questionnaire that covered the same topics as those he had compiled in the book of descriptions. (On the development of the questionnaire, see Barrera-Osorio in Chapter 11.)[30] Leveraging the empire's bureaucratic machinery to urge compliance, the printed questionnaire was sent to the ranking officials of all jurisdictions in the Indies instructing them to distribute the questionnaires to all the towns in their district inhabited by Spaniards, as well as those "of Indians" (see Figure 3.3). Respondents were to follow the list of questions carefully, answering only the relevant ones. Once completed, the document was dated, the name of the person responsible for the answers recorded, and the response returned to the Council of Indies.

The first ten questions inquired about a region's geographical characteristics. These were followed by forty questions concerning "particular" matters specific to each location. These inquired about the region's native population, local geographical and coastal features, and natural history—or, perhaps more accurately, its natural resources. It continued with questions about the town's buildings, both domestic and defensive, as well as a general inquiry about commerce. Questions 34 to 37 concerned the relationship between the town and the church and asked for a listing of churches, monasteries, and convents. Velasco concluded the questionnaire by asking for any other relevant information about unpopulated areas or other "notable natural features."

The responses to the questionnaires form what is perhaps the most valuable single body of knowledge about colonial Spanish America. The corpus is known as the *Relaciones de Indias*; over two hundred still exist.[31] They contain a treasure trove of descriptions of the peoples living

✷Inſtruction, y memoria, de las relaciones que

ſe han de hazer, para la deſcripcion de las Indias, que ſu Mageſtad man-
da hazer, para el buen gouierno y ennobleſcimiento dellas.

RIMERAMENTE, los Gouernadores, Corregidores, ò Alcaldes mayores, aquien los Vireyes, o Audiencias, y otras perſonas del gouierno, em-
biaren eſtas inſtructiones, las diſtribuyran por los pueblos de Eſpañoles, y de
Indios de ſu juriſdiction, embiandolas a los conceios, ò a los curas ſi los vuiere, y ſino a los religioſos, a cuyo cargo fuere la doctrina, mandandoles de parte de ſu Mageſtad, que dentro de vn breue termino, las reſpondan, como en
ellas ſe declara, y les embien las relaciones que hizieren, juntamente con eſtas
memorias, para que ellos como fueren recibiendo las relaciones, vayã, e embiandolas a las perſonas de gouierno y las inſtructiones y memoias las bueluan a diſtribuyr ſi
fueren meneſter por los otros pueblos adonde no las vuieren embiado.

Las perſonas aquien ſe diere cargo en los pueblos de hazer la relacion, reſponderan a los capitulos de
la memoria, que ſe ſigue por la orden, y forma ſiguiente.

Primeramente, en vn papel a parte, pondran por cabeça de la relacion que hizierẽ, el dia, mes y año
de la fecha de ella: con el nombre de la perſona, operſonas, que ſe hallaren a hazerla, y del Gouernador, y otra perſona que les vuiere embiado la dicha inſtruction.

Y leyendo attentamente cada capitulo de la memoria, eſcreuirã lo que huuiere q̃ dezir a el, en otro
capitulo por ſi, reſpodiendo a cada vno por ſus numeros, como van en la memoria vno tras otro
y en los que no huuiere que dezir, dexarlos han ſin hazer mencion de ellos, y paſſaran a los ſiguientes, haſta acabarlos de leer todos, y reſponder los que tunieren que dezir: como queda dicho, breue y claramente, en todo: ſ irmando por cierto lo que lo fuere, y lo que no, poniendolo
por dudoſo: de manera que las relaciones vengan ciertas, conforme a lo contenido en los capitulos ſiguientes.

¶ Memoria de las coſas, a que ſe ha de reſponder: y de que ſe han de hazer las relaciones.

1. PRIMERAMENTE, en los pueblos de Eſpañoles ſe diga, el nombre de la comarca, o prouincia en que eſtan, y que quiere dezir el dicho nombre en lengua de Indios, y
   po que ſe llama aſſi.

2. Quien fue el deſcubridor y conquiſtador de la dicha prouincia, y por cuya orden y mandado,
   ſe deſcubrio, y el año de ſu deſcubrimiento y conquiſta, lo que de todo buenamente ſe pu-
   diere ſaber.

3. Y generalmente, el temperamento y calidad de la dicha prouincia, o comarca, ſi es muy fria,
   o caliente, o humeda, o ſeca, de muchas aguas o pocas, y quando ſon mas o menos, y los vien-
   tos que corren en ella, que tan violentos, y de que parte ſon, y en que tiempos del año.

4. Si es tierra llana, o aſpera, raſa o montoſa, de muchos o pocos rios o fuentes, y abundoſa o falta de aguas, fertil, o falta de paſtos, abundoſa o eſteril de frutos, y de mantenimientos.

5. De muchos o pocos Indios, y ſi a tenido mas o menos en otro tiempo que ahora, y las cauſas
   que dello ſe ſupierẽ, y ſi los que ay eſtan poblados en pueblos formados y permanentes, y el
   talle y fuerte de ſus entendimientos, inclinaciones y manera de biuir, y ſi ay differentes len-
   guas en toda la prouincia, o tienen alguna general en que habien todos.

6. El altura o eleuaciõ del polo, en que eſtan los dichos pueblos de Eſpañoles, ſi eſtuuiere tomada,
   y ſi ſe ſupiere, o vuiere quien la ſepa tomar, o en que dias del año el ſol no hecha ſombra a nin-
   guna a l punto del medio dia.

7. Las leguas que cada ciudad o pueblo de Eſpañoles eſtuuiere de la ciudad donde reſidiere la audiencia, en cuyo diſtricto cayere, o del pueblo donde reſidiere el gouernador aquien eſtuuie
   reſugeta: y aque parte de las dichas ciudades o pueblos eſtuuiere.

8. Aſſi miſmo las leguas que diſtare cada ciudad o pueblo de Eſpañoles de los otros cõ quien par-
   tiere terminos, declarando aque parte cae dellos, y ſi las leguas ſon grãdes o pequeñas, y por
   tierra llana o doblada. y ſi por caminos derechos ò torcidos, buenos o malos de caminar.

9. El nombre y ſobrenombre que tiene. o vuiere tenido cada ciudad ó pueblo, y perque le vuiere llamado aſſi. (ſi ſe ſupiere) ; quien le puſo el nombre, y fue el fundador della, y por cuya
   orden y mandado la pobló. y el año de ſu fundacion, y con quantos vezinos ſe començo a
   poblar, y los que al preſente tiene.

10. El

in the New World some fifty years after colonization began. They provide valuable information on the demographic situation of native populations as well as the Spanish populations, their towns, villages, and natural resources. The respondents in many cases included maps—most drawn with strong Amerindian influence—and accounts of the natural world and medicinal plants.[32] Carefully filed in the Council, they amounted to cosmographical depositions describing the New World.

López de Velasco directed another equally ambitious project, this time aimed at determining the exact location of Spain's overseas domain.[33] The project, also mandated by Ovando's ordinances of 1575, set in motion the first coordinated worldwide project of lunar eclipse observations for determining terrestrial longitude.[34] López de Velasco once again sent royal orders, this time in the form of detailed observational instructions to all seats of government in America and the Philippines. These instructed observers to build a simple instrument to record the shadow cast by the moon at the beginning and end of the eclipse and thus fixing the moon's altitude at a particular sidereal moment. The resulting observations were then sent to the royal cosmographer in Madrid. The instruction did not explain how to use the measurements derived from the moon's shadow to determine longitude. Clearly, the intention was for cosmographers back in Spain to do the mathematical computations—a perhaps unsatisfying outcome for a diligent observer in the Indies—but also a way to keep the potentially strategic information secret. Over the years, López de Velasco received observations from modern-day Puebla, Veracruz, Mexico City, Puerto Rico, Panama, Peru, and the Philippines, but the results of the eclipse project in the form of new cartography were never made public during the reign of Philip II. It would be one of López de Velasco's successors as cosmographer of the Council of Indies, Andrés García de Céspedes, who made public use of the coordinates yielded by the eclipse project in his *Regimiento de navegación e hydrografía* (Regiment of Navigation and Hydrography, 1606).

García de Céspedes served as cosmographer major of the Council of Indies from 1596 until 1611. A university-trained astronomer and mathematical practitioner, during his tenure he practiced almost exclusively the aspects of the cosmography associated with mathematics: cartography, geodesy, astronomy, and astronomical navigation. The other aspects of the discipline that had been integral to Renaissance cosmography—in particular, those that used a textual form of representation (descriptions, ethnography, and history)—ceased to be associated with the Council's cosmographer and became the responsibility of the chronicler major. At the heart of this disassociation was a shift in cosmographical epistemology characterized by an increasing emphasis on using mathematical arguments to ascertain matters of cosmographical fact. This approach was fundamentally at odds with the methods López de Velasco had employed that relied principally in establishing the credibility of an eyewitness testimony and ensuring comprehensiveness.

THE HABSBURG ROYAL COURT

After the death of Alonso de Santa Cruz in 1567, and for the duration of the reign of Philip II, cosmographical activity at court was under the direction of the architect of the monastery and palace of San Lorenzo de El Escorial, Juan de Herrera (1530–1597).[35] (On the role of Herrera's disciples in the development of early-seventeenth-century cosmology, see Sheehan in Chapter 12.) By 1563, Herrera was working as an assistant and draftsman to Philip's architect, Juan Bautista de Toledo.[36] After the architect's death, Herrera became the de facto architect of the many royal projects then under way. He collaborated with other humanists at court, among them the celebrated Benito Arias Montano, to assemble an excellent library at El Escorial. He also accumulated an impressive scientific and technical library, as well as a large collection of astronomical and navigational instruments.[37]

Herrera preferred the mathematical rather than the descriptive or historical branches of cosmography, an interest nurtured by a lifelong fascination with the problem of measuring terrestrial longitude, both at sea and on land. When in 1580, Philip II successfully laid claim to the Portuguese crown, the unification of the two kingdoms eased some of the political tensions that arose periodically about the line of demarcation. It also revealed to the Spanish crown the cartographic material used by the Portuguese to support claims that the Philippine islands fell on their side of the demarcation. While in Portugal, Philip II, with Juan de Herrera by his side, compared Portuguese and Spanish maps and found fundamental and alarming discrepancies between them. Herrera's stay in Lisbon alerted him to the political and administrative relevance of cosmography. In response, he formulated a series of ambitious cosmographical projects designed to place Spanish cosmography on a sound empirical foundation, as well as provide means to educate future generations of Spanish cosmographers. In contrast to López de Velasco's bureaucratic approach, Herrera preferred placing the task of conducting cosmographical experiments in the hands of experienced cosmographers. He was perhaps following the model used in the late 1560s, when the task of compiling a natural history (materia medica) of medicinal plants of Mexico and Peru was placed in the hands of Dr. Francisco Hernández (ca. 1515–1587).[38]

Historians have come to frame the project as an early example of modern natural history, principally because Francisco Hernández's work begins the slow erosion of the reliance on Pliny's *Natural History* for stylistic guidance on taxonomy, replaced by a systematic and experimentally based classification of the natural history of the New World.

In late 1582, while still in Portugal, Herrera planned an ambitious project to send astronomer and mathematician Jaime Juan on a voyage to Mexico and the Philippines to carry out astronomical observations for the purpose of determining the latitude and longitude of Spain's principal overseas territories, or at least to definitively locate key places and landmasses. Juan was to carry out these investigations using instruments designed by Herrera and following procedures detailed in a comprehensive set of instructions.[39] After leaving Seville, Juan was to make positional astronomical observations and measurements of magnetic variation while at sea and in all ports of debarkation en route to the Indies. Once on the American mainland, the instructions specify he should note carefully his position on the east coast of Mexico, in Mexico City, and the in western Mexican port of Navidad. Observations would continue during the Pacific crossing until Juan arrived in the Philippine city of Manila, where he would conclude his expedition. Juan carried out his assigned duties but died shortly after arriving in the Philippines. Unfortunately, except for lunar eclipse observations taken in Mexico City in 1584 (see Figure 3.4), his work has been lost.[40]

Herrera recognized that few persons in the kingdom possessed the skills of a Jaime Juan and resolved to remedy the situation. He became the driving force behind the Royal Mathematics Academy in Madrid, an institution whose mission was to prepare young men to practice mathematics-based professions.[41] Rather than organizing the academy's curriculum around teaching fundamental sciences, as was typical in universities, Herrera instead tailored each course of study to prepare students to practice specific professions that would be "useful" to the empire: architecture, cosmography, surveying, mechanics, and others. In 1584, to publicize the academy's plan of study, Herrera published the academy's charter, *Institución de la Academia Real Mathemática.*[42]

The *Institución* records Herrera's interest in setting up a pedagogical institution for teaching the empire's future problem solvers the fundamentals of the technical arts. In formulating this program, Herrera followed the tradition of the "practical mathematicians" oriented toward solving problems rather than philosophical speculation. By locating the academy in the courtly context, he leveraged on his behalf the king's authority and enthusiasm for the endeavor. Moreover, by displacing learning from the universities, Herrera had the flexibility to construct

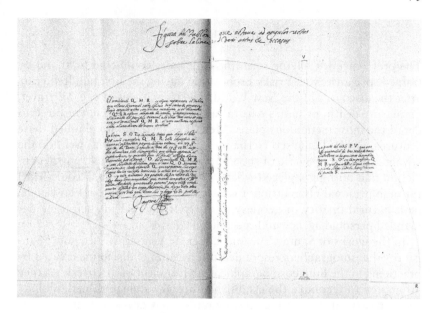

FIGURE 3.4. *Lunar eclipse observation by Jaime Juan, Mexico, 1584*
SOURCE: Archivo General de Indias, MP, México 34, P5-2. Ministerio de Cultura, Spain.

an educational project free from the humanistic overreliance on classical authority and the universities' rigid statutes. Herrera's action also
alerts us to an increasing divergence between the descriptive and the
mathematical aspects of Renaissance cosmography. His description of
the cosmography curriculum at the academy equates descriptions with
maps and makes no mention of studying Mela or Strabo.[43]

Plans for the academy began during Herrera's two-year stay in Portugal. All the courses were to be taught in Castilian, and in 1581 Pedro
Ambrosio de Onderiz, a student of classical languages, was sent to Lisbon to study mathematics and begin work on a series of translations of
classic mathematical texts into Castilian.[44] In late 1582, the king hired
Portuguese cosmographer Juan Bautista Labaña (or Lavanha; 1555–
1624) to teach in the academy and to serve as court cosmographer.[45] The
academy was in continuous operation from 1583 to 1604 and, after a
three-year hiatus, until 1625—the year its pedagogical mission was del-
egated to the Jesuit's Colegio Imperial de San Isidro in Madrid. In 1628,
the Council of Indies began to rely on the Jesuit College for its cosmographical advice, naming the professor of the mathematics academy as
cosmographer major of the Council of Indies.[46]

CONCLUSION

The preference for having scientific endeavors conducted firsthand by trained specialists was a risky proposition during the first hundred years after the discovery of the New World. The vicissitudes of oceanic travel, premature death, or man's natural unpredictability often conspired to derail the most meticulous plans. Men like Herrera who planned these ventures, or like Francisco Hernández and Jaime Juan who went on the voyages, were well aware of these difficulties, but they held a firm conviction that nothing replaced what a well-trained individual could learn through personal observation. Constructing a new body of knowledge, be it natural history or geodesy, was a pursuit that in their view demanded personal agency and specialized education.

In the works of Santa Cruz and López de Velasco we see how Spanish royal cosmographers responded to crisis in the discipline created by practicing in the fluid cosmographical environment of discovery. As they attempted to reconcile the established literary style and methodology characteristic of Renaissance cosmography, they developed an epistemic criteria that demanded facts based on empiricism and corroborated eyewitness testimony, ways of ascertaining matters of fact that became the hallmark of seventeenth-century scientific practices. The questionnaire of the *Relaciones de Indias* and the instructions for observing lunar eclipses employed novel methodologies to collect a vast amount of cosmographical information in the context of discovery and colonization. López de Velasco leveraged the empire's administrative machinery to put the programs in place and encourage compliance. At the Casa de la Contratación, cosmographers designed a set of scientific practices in the form of astronomical navigation that they hoped would serve as a substitute for praxis. They articulated these in the very popular genre of navigation manuals, one of the first European works to wed science and technology, that earned theoreticians the intellectual authority to define what constituted best practice in the craft that was navigation.

The sites of knowledge production—the *Casa, Consejo,* and *Corte*—and the imperatives the imperial program required of each institution inevitably shaped the cosmographical practitioner's response. Managing the largest empire the world had ever known demanded pragmatic cosmographical practitioners. The only knowledge products these institutions of empire needed were those useful for expanding, establishing, and maintaining Spanish primacy abroad. Therefore, the impetus each institution placed behind securing increasingly specialized cosmographical products operated to dissolve Renaissance cosmography into the specialized areas we recognize today as geography, cartography, and natural history, with only the historical narrative becoming the purview of humanist. By the seventeenth century, the cosmographer was ideally a mathematician, cartographer, and astronomer. The New World burst the seams of a discipline conceived and nurtured during the fifteenth century when the known world was smaller and humanists could aspire to encompass the world between the covers of a book.

CHAPTER FOUR

Science During the Portuguese Maritime Discoveries

A Telling Case of Interaction Between Experimenters and Theoreticians

ONÉSIMO T. ALMEIDA

Portuguese navigators of the fifteenth and sixteenth centuries produced a significant body of literature—mostly logbooks and manuals—that is not well known outside Portugal. Some of these documents have never been translated. Others have received little attention because they appeared in editions with limited circulation among historiographers of science in the English-speaking world.[1] Taken as a whole, this literature constitutes an important contribution to our understanding of the transitional years between the Middle Ages and modernity. A close analysis of these writings, of the trials and errors of the agents directly involved in the process, confirms that certain important developments in Portugal helped pave the way for modernity and the development of an experimentalist attitude toward the world. If so, the grand narratives that usually jump from the Islamic developments of science to the English contribution leave out a missing link, a remarkable web of activity and thinking that helps us realize that the seventeenth-century scientific revolution actually took a few centuries to develop, with the Portuguese navigators at the forefront of the process. Taken as a whole, these works support the idea that the transformation of the medieval worldview into the modern one was slow and geographically more widespread than is usually considered.[2] I dare say that if Francis Bacon's *Novum Organum*,[3] a classic considered to be the great herald of the modern mind, is read in light of the developments that took place in Portugal prior to its publication in 1620, it will appear more as a synthesis of the transformations that had been occurring in Europe rather than the revolutionary work it is assumed to be.[4] (On Bacon and the reception of early-seventeenth-century Spanish geographical discoveries, see Sheehan in Chapter 12.) Bacon himself said of the sea travelers of the times before him:

We must also take into our consideration that many objects in nature fit to throw light upon philosophy have been exposed to our view, and discovered by means of long voyages and travels, in which our time have abounded. It would indeed, be dishonorable to mankind, if the regions of the material globe, the earth, the sea, and stars, should be so prodigiously developed and illustrated in our age, and yet the boundaries of the intellectual globe should be confined to the narrow discoveries of the ancients.[5]

In another passage, Bacon includes navigation in his enumeration of advancements unknown to the ancients, a list that counts printing, gunpowder, and the compass among strides that, "have changed the appearance and state of the whole world." From these novelties, "innumerable changes have been thence derived, so that no empire, sect or star, appears to have exercised a greater power and influence on human affairs than these mechanical discoveries."[6]

During the majority of the fifteenth century, the Portuguese improved, to a remarkable degree, inherited navigational technologies. Necessity was the impulse for the finding of new means to carry out navigation on the high seas, for the observation and registration of prevailing patterns of hitherto unknown winds, for the discovery of Atlantic currents, and for the mapping of the stars of the Southern Hemisphere to guide their way in the waters thereof (see Figure 4.1). They were obliged to construct better *naus* (ships) to adapt efficiently to new circumstances. Maps of the African coasts were elaborated, registering innumerable and important details for future voyages. All this was possible only thanks to an accentuated development, a progression of a spirit of observation, great care in research, and recording of a wide, diverse range of useful and important information. Over time, the idea was enshrined that data collected by seamen and through experience possessed greater validity than that of the "mathematicians" on land, the intelligentsia with ties to the royal court, or those who obtained most of their knowledge of maritime exploration from ancient tomes with little in the way of fact. This new mentality was firmly instilled in a small elite that elaborated notable theories about the epistemological superiority of empirical knowledge.

Key among the Portuguese entrepreneurs of the maritime discoveries were Duarte Pacheco Pereira (ca. 1455–1530?),[7] D. João de Castro (1500–1548?), Pedro Nunes (1502–1578),[8] Garcia de Orta (1490–ca. 1570), and Fernando Oliveira (ca. 1507–ca. 1585).[9] The work of other figures such as Francisco Sanches (1551–1623) should also be read in the context of

FIGURE 4.1. *Astrolabe, 1555*

SOURCE: António Estácio dos Reis, *Medir estrelas* [Measuring Stars] (Lisbon: CTT, 1996).
By permission of the author. From the collection of McManus Galleries, Dundee, Scotland.

the maritime discoveries.[10] Of this elite set, the names Duarte Pacheco
Pereira, Fernando de Oliveira, and Garcia de Orta stand out for having
written largely about the importance of experience. Francis Bacon himself
was not aware of their work. He was informed about the great impact of
the discoveries, but he lacked closer information that would allow him
to understand that such a complex process did not happen by accident,
nor simply thanks to the courage of intrepid navigators, but because in
their enterprise they developed and followed ground rules elaborated and
improved upon by their own experience.

The leitmotif in authors such as Duarte Pacheco Pereira is, "experience is the mother of things," with frequent statements being made that
point to the superiority of knowledge acquired from firsthand experience—"experience has shown us that this is not so,"[11] or, "experience
has disabused us of the errors and fictions which some of the ancient
cosmographers are guilty of."[12]

Aristotle had of course theorized clairvoyantly on experience, bringing the Greek conception of science to an impressive level. It was not,
however, the Aristotelian empirical outlook that prevailed in the late
Middle Ages and that became the established philosophical dogma.
His philosophy had been somewhat turned on its head, and experience
disappeared from its foundation. There were various brilliant spirits
throughout medieval times who, on their own, recovered experience as
a key foundation of the edifice of knowledge (Roger Bacon and Robert
Grosseteste are just two examples), but when the Portuguese entrepre-

neurs looked in the books of the Ancients for information about the oceans and lands overseas, they mostly gathered myths, or rather facts that they demonstrated through their own experience to be mere fables and myths. Aristotle, for instance, was a great biologist because he based his writings about animals on empirical evidence. However, he did not venture much out of Athens, and as a result most of his cosmological descriptions were the result of armchair philosophy or hearsay.

Having written elsewhere about the overall contributions of these authors,[13] I wish to concentrate here on the particular working relationship that existed between two of the above-mentioned figures, Pedro Nunes and D. João de Castro. Their example will challenge the received view that so-called scientific activities in the 1400s and 1500s were conducted in isolation and with no bearing on real-life problems.

Indeed, the figure that stands out the most in formulating a "new" theory of knowledge is that of D. João de Castro (1500–1548) (see Figure 4.2).[14] Born in Lisbon, he was a naval commander who wrote a cos-

FIGURE 4.2. *Portrait of Dom João de Castro, Paris, 1833*
SOURCE: Dom João de Castro, *Roteiro em que se contem a viagem que fizeram os Portuguezes no anno de 1541* (Paris, 1833). John Hay Library, Brown University.

mographical treatise, *Tratado da sphaera por perguntas e respostas a modo de dialogo* (Treatise on the Sphere Through a Dialogue of Questions and Answers), probably written before 1538 (see Figures 4.3 and 4.4). Of greater importance for our purposes are his three *Roteiros* (logbooks), in which important events occurring in his voyages are recorded. The voyages are from Lisbon to Goa (1538), from Goa to Diu (1538), and from Goa to Suez (1540–41). In 1545 he returned to India as a viceroy.

Castro not only theorizes about the role of experience in the process of knowledge acquisition, but he also reaches further, pointing to the limits of the senses whose errors must be corrected by judgment. Castro is indeed the most modern of the maritime travel writers of this period. He reveals a strong empirical mind, takes meticulous notes, undertakes experiments alone and with his crew, takes measurements and double-checks them, and attempts to make sense of the data collected. These are impressive signs of a rather modern mind at work. He goes further, however, reflecting a concern with theory as well as with experience, and advocates for a dialogue between both.[15] (For the Spanish aspects of this dialogue between experience and theory, see Portuondo's essay in Chapter 3.)

The other figure to be highlighted here is Pedro Nunes. Born in Alcácer do Sal in 1502 to a New Christian family, he became a mathematician and was considered one of the greatest of his time. Best known

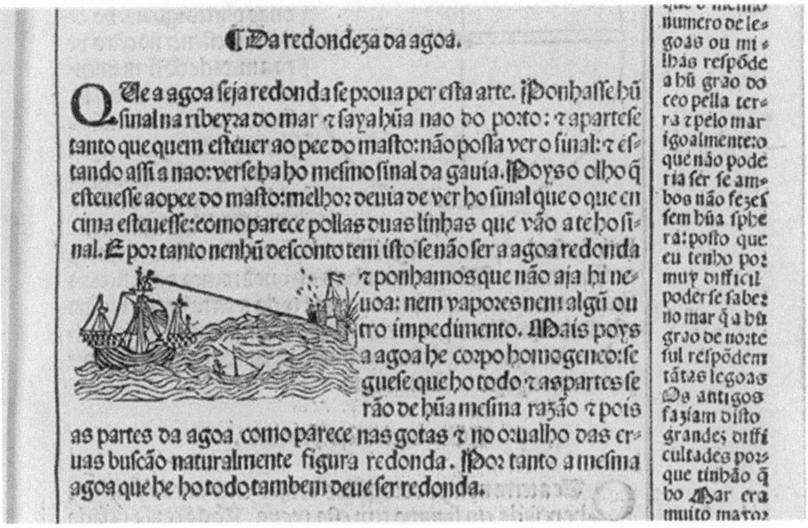

¶ Da redondeza da agoa.

Que a agoa seja redonda se proua per esta arte. Ponhasse hũ sinal na ribeyra do mar ⁊ saya hũa nao do porto: ⁊ apartese tanto que quem esteuer ao pee do masto: não possa ver o sinal: ⁊ estando assi a nao: verseha ho mesmo sinal da gauia. Poys o olho q̃ esteuesse ao pee do masto: melhor deuia de ver ho sinal que o que en cima esteuesse: como parece pollas duas linhas que vão ate ho sinal. E por tanto nenhũ desconto tem isto senão ser a agoa redonda ⁊ ponhamos que não aja hi neuoa: nem vapores nem algũ outro impedimento. Mais poys a agoa he corpo homogeneo: seguese que ho todo ⁊ as partes serão de hũa mesma razão ⁊ pois as partes da agoa como parece nas gotas ⁊ no orualho das eruas buscão naturalmente figura redonda. Por tanto a mesma agoa que he ho todo tambem deue ser redonda.

numero de legoas ou milhas respõde a hũ grao do ceo pella terra ⁊ pelo mar igoalmente: o que não poderia ser se ambos não fezessem hũa sphera: posto que eu tenho por muy difficil poderse saber no mar q̃ a hũ grao de norte sul respõdem tātas legoas. Os antigos faziam disto grandes difficultades porque tinhão q̃ ho Mar era muito mayor

FIGURE 4.3. *Page from D. João de Castro,* Tratado da sphera, *1537*
SOURCE: John Carter Brown Library, Brown University.

grao:po2quehealargura do nacimeto os mefmos. rriij: graos
⁊ meo.E no mefmo dia aos que viuem em.rrrv.graos defta ban
da do no2telhes nace o fol ao no2defte ⁊ quafi quarta ⁊ mea õ lefte
po2quefam.rrir.graos: ⁊aosde noffa altura q̃ fam.rrrir.graos
nacera em no2defte quarta õ lefte:com dous graos ⁊ tres quartos
mays pera lefte:po2que fam.rrrj.graos de largura do nacimeto
do fol: q̃ he ho arco do o2izonte:que nacendo fe aparta da linha pa
a parte do no2te.E daqui fe tirara ligeiramete onde fe po2a no mef
mo dia.⁊ onde nacera ⁊ fe po2a eftando no outro tropico:como pa
rece na figura.E tudo ifto fe demoftra fer affi po2que a p2opo2ção

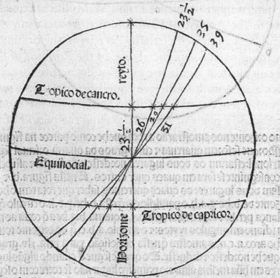

que tẽ o fino do comp2imẽto da altura em qualquer região: cõ o fi
no vniuerfal do circulo:efta mefma ha do fino da declinação q̃ tem
o fol em qualquer dia:ao fino do rumo em q̃ nace: o que c2aramẽte
fe p2oua per Tolomeo:no fegũ do do almagefto. Do qual fe fegue
quam facil coufa feja:refguardando pella menhã o fol no feu naci
mẽto:com a agulha bem verificada: ou cõ linha meridiana:fe
fo2 na terra:faber per conta fem mais inftrumento a altura do polo
em q̃ nos achamos:o que eu em todo tempo fem faber a ho2a q̃ he
nem ter linha meridiana:cõ inftrumẽtos faço:mas os pilotos fabẽ

FIGURE 4.4. *Page from D. João de Castro,* Tratado da sphera, *1537*

for the invention of the *nonius* (an instrument named after his Latin name), he made important contributions in the area of technical navigation, but he was also the most theoretical of all the authors and thinkers involved in the process of the discoveries. The interactions and exchanges he had with D. João de Castro provide for an impressive example of coordination between theoreticians and practitioners, with Nunes staying on land, collecting data and information, theorizing, and attempting to make sense of things, while Castro was out at sea, facing problems of all sorts. These interactions are emblematic of the relationships maintained by the leaders and masterminds, thinkers and entrepreneurs involved in the overseas voyages of discovery.

The philosopher of science Ernan McMullin affirmed in a presidential address to the American Philosophical Association that, "the goal of technical control played virtually no part in the origins of science."[16]

A reading, even a superficial one, of the *Roteiros* of D. João de Castro would constitute a legitimate point of departure for us to question McMullin's assertion. It will serve equally as a call to attention for the historians of science to the scientific activity (in the modern sense of the term) that took place in Portugal during the period of the discoveries.[17] The focus of this essay will be limited to the interaction between theorists and navigators and will evaluate their success or failure in achieving greater certainty as to the movement of ships, as well as making further advances in the dominion of the seas. Let us begin with the *Roteiro de Lisboa a Goa* (Rutter from Lisbon to Goa). Pedro Nunes had conceived of two "shadow instruments"—one to measure the altitude of the sun and the other to calculate the magnetic declension (see Figure 4.5). The mathematician was not, however, a sailor; he sought only to respond to the needs expressed by the navigators, who, upon returning from their voyages, brought him questions, qualms, and new data that called into question previous explanations and, in some cases, revealed that the instruments created by Nunes himself were difficult to use, if not totally useless. The description of the voyage in the vicinity of the island of Porto Santo reads:

Saturday, April thirteenth, at dawn we saw the palm, that is one of the Canary Islands, and at once I made ready the lamina and shadow instrument, which I owe to the graciousness of his highness prince D. Luís,[18] with a great desire to verify two things: the first, if on these islands the needles varied or not, as the practice of many pilots at this place and meridian would make the north of their needles on the true pole of the world; and the second, if it was true and accurate the rule given to us by the doctor Pero nunez that, at all hours of the day in which shadows appear, we know the elevation of the pole; with which instrument I made the following observations, the wind being calm all day, such that the ship did not sway.[19]

Was hilfft der wechter in der statt/
Dem geweltigen schiff im meer sein fart/
So sie Gott beyde nicht bewart.

FIGURE 4.5. *Measuring the height of the sun, 1557*

SOURCE: Hans Staden, *Warhaftig historia* (Marburg, Germany, 1557).
John Carter Brown Library, Brown University.

Later on, Castro records two measurements and adds that he then, "ordered the pilot to calculate the altitude of the sun at midday, and [moved] to the armillary sphere, in order to verify the elevation of the pole of this day."[20] Following this is a meticulous description that is worth transcribing in detail, as it reveals a truly rigorous methodological approach:

Firstly, on the graduated horizon of the armillary sphere, I recorded the variation made by the shadow of the pointer from the first altitude to the second, the variation of which was seven degrees, and the very first of these seven degrees were set on the horizon, I recorded the first altitude; and it was 57 degrees for one graduated meridian above, and in the place where the number of these 57 degrees ended, I put a point; and going over the same meridian on the other extremity of the variation of the shadow that I recorded on the horizon, I counted by the meridian upwards to the second altitude, which was 61 degrees and 1/2, and in the place where they ended I put another point.

He continues to describe his method, and concludes by conferring with his pilot in order to confirm his findings:

I then sent this altitude to the pilot in a sealed note, so that after he calculated the altitude of the sun at midday we could see both of them together, without suspicion (and thus) we would be able to determine how much mine, taken in the morning, differed from his, taken at midday. Now the pilot, having finished calculating his altitude, came to me and said that we were at an altitude of 29 degrees and 1/3, and immediately opened the note and saw mine, at which he was quite startled.[21]

Particularly noteworthy is Castro's preoccupation with temporarily concealing the data he had obtained, in order to compare it later with that of the pilot without danger of influencing the measurements of the latter. Meanwhile, he proceeds with his own measurements, making clear that he utilizes the aforementioned shadow instrument in order to verify the declension of the needle, exactly as Pedro Nunes had advised him.[22]

Further on, on April 21, D. João writes:

Because I make mention many times of the altitude taken by many people, and expect to do so later on, already it may be that those who read this *Roteiro*, finding considerable the difference between some altitudes and others, could think that this diversity would come from having tables of different declensions, or from erring in the calculations; therefore I will always make mention of the altitudes they take of the sun at the horizon, and thus declare that all of the altitudes written here, will be done using the book and tables of declension of Doctor Pero Nunez, so that the difference will come only from each person's own judgment, or from the defects of the astrolabes; and so that this art discovered for the purpose of helping the navigators does not become discredited due to this diversity of altitudes, we should take into consideration how much we owe to it, for teaching us how we should not ignore proximity to truth, how human the things are that men are able to understand. Because, as Monte Regio says in his triangles, of all good fortune it is better to know the approximate truth rather than to ignore completely that same truth, which one should not judge only by virtue of hitting the mark, but also by coming close to it.[23]

Here, again, should be noted the care Castro takes in always utilizing the same solar tables, published by Pedro Nunes, so that the data could be better controlled, and the eventual differences attributed exclusively to defects in the instruments of measurement or to the error of the measurer. No less noteworthy is the consciousness he displays as regards the importance of describing in detail the process utilized in obtaining the data—even assuming that this deviates from the "truth." In addition, we must above all consider, yet again, his collaboration with the mathematician Pedro Nunes, the author of said tables.

At times, Castro gives up on finding explanations for certain phenomena or incidences, deeming that he does not possess any hypothesis for obtaining sufficient data to interpret them. With a light touch of humor he accepts this fact with resignation:

They may argue that we miscalculate the altitude of the sun; to this I respond, that it would be as strange for this to be a miscalculation, as for us to turn back around, having wind and prow to go forward, since five people calculated the altitude on this day, and all of us found ourselves in precise conformity with the sun. What this appears to me, is that we had a torrent of water that made us turn around, and if it was not this, may Apollo determine the cause.[24]

In still other cases, it seems to him rather that his question deserves to be studied by someone like Pedro Nunes, to whom he attributes a broader vision. For instance, after describing the steps he took in making his magnetic observations, which generated significant anomalies, he writes that, "the defect appears to be with the armillary sphere or demonstration, and because at present I cannot determine the cause, may the question be left to the doctor Pero Nunez."[25]

Near the end of his *Roteiro*, Castro again observes an anomaly for which he has no explanation. Although accustomed to unanswered questions, he reveals his intention to consult his teacher Pedro Nunes. Thus he concludes, "some aspect would mislead us that day, which we cannot ascertain, as befalls us in the majority of things and secrets of nature; but, however it may be, may the resolution of this question be left to the doctor pero nunez."[26]

At times, certain discrepancies are attributed to the still-imperfect instruments: "the armillary sphere is not suitably round, and the meridians prove poorly graduated, and the horizon does not move closely together

with the sphere, but all of these things are made poorly and imperfectly."
However, he adds that, "with all of these defects, much of the time [his
findings] proved correct, as has been shown."[27]

When practical reasons contrast with theoretical instructions received
on land prior to departure, if these appear ill-advised, D. João de Castro
does not hesitate to follow his own instincts. Still, he does not neglect to
record the data from his measurements, as well as provide explanatory
examples for those who may not understand him otherwise. He writes:

> We had shortened the sailing course from that which the designation of the point
> of the compass was showing us, although the rules by which we steer presume
> that we must travel by true courses. And so that this may be better understood,
> I will put here some examples, without deviating from the route that we take
> departing from Lisbon toward India, as this document is not made for the doctor
> Pero nunez, but for those who know nothing of mathematics, nor have experience
> with the sea.[28]

In Castro's correspondences we find equal reference to his close collaboration with Nunes. In a message informing D. João III of the location
of the Moluccas, he explains that "the report of this should be given by
me to your highness in the presence of Pero Nunez, its cosmographer."[29]
In the same letter he also makes specific reference to the instrument that
he himself had made, "by [his] own hand, and that Doctor Pedro Nunez
had invented." This instrument allows "one . . . traveling eastward for
each degree of altitude how many leagues correspond to the degree."[30]

Luís de Albuquerque, upon evaluating the "innovations" of Pedro
Nunes, pointed out the failure of one of his inventions, tested by D. João
de Castro, and counterbalanced this with the success of another. As
Albuquerque relates, the determination of the latitudes according to the
methods recommended by Nunes seemed at first to produce satisfactory
results. However, Castro later went on to note various types of deficien-
cies, coming to the conclusion that the instrument was not of great nautical interest, so much so that the pilots scarcely took any interest in it.[31]

Doubt exists as to Pedro Nunes's role as the original creator of the
shadow instruments.[32] For the sake of the argument developed in this
essay, the question of the paternity of the instrument is secondary to
the fact of the cooperation between theorists and navigators. But more
important still, and what should be underscored, is the trial and error
approach, understood by some as indispensable for obtaining, by various
stages, the instrument that best corresponded to the needs of the sailors.

In response to questions presented by Martim Afonso de Sousa, commander of the fleet that sailed to Brazil in 1530, Pedro Nunes points out
the defects of the quadrate marine maps and attempts to get around these
in different ways. This and other practical problems are of central con-

cern in documents such as *O tratado sobre certas dúvidas da navegação* (Treatise on Certain Questions of Navigation) and *Tratado em defensam da carta de marear* (Treatise in Defense of the Sea Chart). Consider the following example from his writing, in which the difficulties and questions presented by the sailors are a continual presence. Contrasting the sailors' supposed path and what he considers to be their true course, he writes: "he who sees well this mental image of mine will understand that from the errors of the navigators one can extract the truth, and that without going anywhere, nor ever navigating, by the false information that they give of it, one can know the truth. But best would be through clear knowledge of the registration of the places: to navigate by art."[33]

The two treatises (the first is a relatively short fifteen pages) are clearly a written response to the navigators who seem to object to the theorist, pointing to his inexperience in the art of navigation. The sailors, as Pedro Nunes himself reveals in his response to the questions of Martim Afonso de Sousa, laugh at his theories: "I am so scrupulous in mixing with the common rules of this art terms and points of science: that the pilots even laugh."[34] This, however, does not prevent him from writing, "For the satisfaction of which it behoves me to bring not only practical things of the art of navigating: but points of geometry and the theoretical part as well."[35]

Pedro Nunes seeks to demonstrate the importance and the reliability of his assertions, but always pays a great deal of attention to the issues raised by the navigators. Furthermore, at the opening of this treatise, he quotes Martim Afonso de Sousa, whose questions are precisely the reason for the text, stating that, after responding to his "questions orally," he decided to write down his own responses, "for those who wish to know how one must navigate by art and by reason."[36]

Pedro Nunes seeks to resolve practical problems in another treatise, titled *De arte atque ratione navigandi* (The Art and Reason of Navigation).[37] The approach remains the same. Pedro Nunes has theoretical concerns, but seeks to base them in the experience gathered by the navigators themselves. They, in turn, listen to him because he appears to possess a broader vision, which allows him to find new hypotheses for solutions. It is up to the sailors to test them and, possibly, to acknowledge that they work well, or at least with a smaller margin of error.

Francisco Teixeira Gomes, a historian of mathematics in Portugal, states that some of Pedro Nunes's ideas are addressed in this text in a more scientific manner than in previous treatises. He concludes by synthesizing the mathematician's contribution in the following terms:

What we can say with certainty is that Pedro Nunes brought the said curve from the field of the nautical empirical, in which it was for the pilots only a route traced by the ship directed by the mariner's compass, to the field of Geometry

where it is a curve traced by a point that cuts the meridians of the sphere under a
constant angle, which showed that it is not generally circular and which cleared
the way for his theory.

The Portuguese cosmographers that preceded him knew of a process for determining the difference of the longitudes of two points of the curve with an approximation however greater or smaller its distance; through applications of this
process to partial arcs in which they discomposed an arc given by the curve, they
obtained the difference of longitudes of the points that limit it with as large an
approximation as they wanted.

This process is equivalent to the use of the equation of the curve. This equation, obtained later by Leibniz, depends on logarithms, an algorithm unknown
in Nunes' time, and the advantage that the use of this equation would have
over the method used in our old navigation to resolve the problem considered,
would be that of reducing the numerical calculation that demands calculation
by logarithms (. . . .)

In order to apply these doctrines to navigation, they continued the regulations
. . . [of the] two numerical tables that gave one the difference of longitudes of
two places of the curve traversed by the ship corresponding to the difference of
one degree of latitude, and that gave the other the length of this arc. However,
Pedro Nunes redid these tables, in the second part of the aforementioned works,
improving them and invented an instrument, which was called a *compasso* . . .
in order to measure the ratio of the arc of the terrestrial parallel corresponding
to a latitude given for the radius of the earth, a ratio that is necessary to know in
order to apply the first table.[38]

Pedro Nunes displays, in fact, an extended vision of events beyond the
comings and goings of the caravels entering and leaving the Tejo River.
At the very beginning of his *Tratado em defensam da carta de marear*,
this consciousness emerges clearly in a passage that, being quite well
known, will be only partially transcribed:

There is no doubt that the navigations of this kingdom from one hundred years
to this part: are the greatest: most marvelous: of higher and more discreet conjectures: than those of any other people in the world. The Portuguese dared to take
on the great sea and ocean. They entered it without any fear. They discovered
new islands / new lands / new seas / new peoples: and what is more: new heavens:
and new stars.[39]

I will not be forced to bring Karl Popper and his classic *Conjectures
and refutations* to bear on these observations, as it is precisely with Popper's mental approach that Pedro Nunes is concerned.[40] Pedro Nunes
himself used the term *conjectures*. Modern science, incipient to be sure,
had already learned and assimilated the rule dictated by experience:
knowledge advances through conjectures and refutations; experience
should be compared with theory, and vice-versa. The process of questioning thus became part of the daily venture of trying to understand

how the machine of the world functioned, so as to be able to duly utilize it. The control instruments (Nunes limited these almost entirely to measurement, but always with the practical objective of improving or facilitating the control of navigation) permitted more exact knowledge of the globe and the heavens that surround it, so that the navigators could better orient themselves within it.

Hence, another one of Ernan McMullin's limiting assertions becomes problematic: "The goal of technical control played virtually no part in the origins of science. The Greek D-science did not depend for its warrant upon the test of consequences. And the gap between natural science, as Aristotle conceived it, and actual technological control was so great that the idea that one might actually inform the other simply would not have occurred to anyone."[41]

McMullin considers untenable the antiquated idea that in classical Greece science was only of the "armchair" variety, since Aristotle was an experimental scientist who was fully conscious of the importance of experience.[42] Nevertheless, Aristotelian science is obviously aimed exclusively at knowledge—not at the transformation, nor the domination—of nature.

Further on, in a discussion on Francis Bacon, McMullin concludes that Bacon continues to emphasize the understanding, not the control, as even he believes that the process of inquiry can be distorted if subordinated very early on to practical application. "Technical control is to be a consequence, and in that sense a touchstone, of the new science rather than a goal in its own right."[43] Later, McMullin insists on the idea that the understanding of nature "in no sense was pursued because of a hope of technological advance."[44]

Such a resolute series of assertions is odd on the part of a philosopher of science who, pages earlier, had taken a much more moderate position regarding the use of the word "science,"[45] advocating its more generalized usage. In addition, relative to other dimensions of science, McMullin had affirmed that evolution always occurred in a slow and gradual manner instead of in gestalt leaps.[46]

Although we can not yet speak of a generalized scientific approach concerning the period in question, no doubt remains—as the previously cited passages seem to demonstrate—that, within a small nucleus of theorists and navigators, there already existed a "communal practice" to which McMullin alludes.[47] It is indisputable that such a nucleus of people was dedicated to developing precise instruments for accurate measurement. These would permit safer and more efficient voyages, and would avoid and control the difficulties encountered by their users. This dedication remained even when they failed to meet their goals, as in the case of the detection of the magnetic force and the demarcation of longitude.

(On the continued importance of the development of navigational instruments in the early seventeenth century, see Sheehan in Chapter 12.)

All of these were incipient phases of the modernity that was gradually establishing itself. Pedro Nunes and D. João de Castro, the first with a more theoretical and abstract vocation, the second with more practical inclinations, though profoundly dedicated to careful observation, are both pioneers of a scientific approach that is characterized by attention to the real, by a preoccupation with rigor and exactness in measurements, by the collection of data (even that which contradicts theory), and by experimentation, questioning, and a clearly formed notion of the unknown world being immense and immersed in the secrets of Apollo for a long time to come.

These two minds were decidedly modern. Science, in the modern sense, had in both of them extraordinary cultivators. They functioned within a medieval vision of the cosmos (Newton and Galileo had not yet effected the great revolution in the conception of the universe) but proceeded in a manner methodologically recommended by modern science. Thus, they clearly contributed to the generalized advent of the still-emergent new paradigm.

One of the very few documents existent in English regarding Portuguese figures connected with science in the period of the discoveries is titled *Pedro Nunes (1502–1578): His Lost Algebra and Other Discoveries.* In the introduction, John R. C. Martyn summarizes the contributions of Pedro Nunes to science, enumerating his contributions and adding afterward: "Thanks to these many inventions, mainly due to his early understanding of algebra, Nunes turned nautical science upside down and gave his country's mariners the necessary equipment for them to cross such vast oceans and discover and map so many distant countries, long before any other European power could do so."[48]

Ernan McMullin will thus also have to apply to the area of "science with the objective of control" his vision, historically correct and easily demonstrable, according to which the history of science did not take gestalt leaps but underwent successive phases of growth, some of which have unfortunately been ignored as much by him as by the great majority of historians.[49]

CHAPTER FIVE

Baroque Natures

Juan E. Nieremberg, American Wonders, and Preterimperial Natural History

JUAN PIMENTEL

Difference causes novelty and awakens desire.
Lope de Vega, *La Dorotea* (1632)

The fate of the Hispanic monarchy in the development of modern science seems trapped within the same paradox as is baroque culture: Many historians have seen both seventeenth-century Spanish imperial science and the baroque arts as foreign bodies in the making of modernity, as deformed entities in the course of events leading to the scientific revolution and the expansion of the overseas empire. A preformationist viewpoint has tended to pervade the history of science, and perhaps even the larger field of early modern history.[1] Because the history of science is so invested in the theme of progress, and expresses itself mostly in English, this discipline perhaps more than any other might reveal that the interests of the Hispanic monarchy, even baroque culture itself, lie in what might be awakened in the rare and the strange: curiosity, novelty, desire.

Is it possible to talk of "baroque science"? Can one talk of "baroque natures" in discussions of anything other than still life painting (see Figure 5.1)?[2] In the absence of the necessary debate to answer such questions, all that can be done here is to point out that the adjective *baroque*, considered rather exotic for the history of science,[3] seems apt to portray the three main subjects of this essay: an ascetic Jesuit writer, seventeenth-century American nature, and the discipline of natural history, which in

FIGURE 5.1. Still Life, *Juan Van der Hamer, 1627*
SOURCE: Museo del Prado, Madrid.

teratological terms looks to us today like a hybrid, just as our modern-day biology might have looked one-legged or one-eyed to an early modern philosopher.

THE JESUIT AND THE EMPRESS

Juan Eusebio Nieremberg (1595–1658) was both a scholar and a polymath, far from uncommon for the European scholarly community of that age (see Figure 5.2). His life reveals something of the Germanic roots of the Hispanic monarchy, attesting that the Habsburgs themselves were not the only ones to pursue a career in Spain. Nieremberg was born in Madrid to German parents, who served at the court of Maria of Austria—daughter, wife, and mother of emperors—and Maximilian II. Upon the death of the emperor, the widowed empress returned to Spain to be close to her brother, King Philip II, and took her court followers with her. Among them were Nieremberg's parents.

A pious and highly cultivated woman, Maria of Austria played an active role at the Spanish court. She entered the Convent of the Descalzas

Reales, to this day one of the main museums of Golden Age Madrid, and later became its patroness. When she died in 1603, Maria left a donation to the Company of Jesus to support the Colegio Imperial, a teaching space intended to assist in two of the great Jesuit undertakings of the day: combating the Protestant Reformation and educating the nobility. In this way, the Habsburg empress had left a legacy that allowed for the foundation of an institution in which Nieremberg, our ascetic Jesuit, was eventually to teach and write.[4]

The Jesuits had had a school in Madrid since 1572, but lacked a center for higher education. When they began to offer advanced courses after

FIGURE 5.2. *Portrait of Juan Eusebio Nieremberg, woodcut, 1659*
SOURCE: Probably by V. Guigon, in Juan Eusebio Nieremberg, *Succus prudentiae sacropoliticae* (Lyon, 1659). Biblioteca Nacional, Madrid.

1625, the order of the Dominicans and the universities of Alcalá and Salamanca quite correctly interpreted the move as a challenge to their own teaching hegemony. These opponents of the Jesuits succeeded in preventing them from imparting the subjects of Suma and Logic; in turn, the Colegio Imperial took over the moribund mathematics academy of Philip II, and thereby assumed a leading role in cosmography, mixed mathematics, and other scientific areas relating to the administration of the empire. Thus a pedagogical center was founded in Madrid where, as in other Jesuit establishments, the oldest nobility lived together with the newest sciences.[5] (On the role of Jesuit scientific opinion and cosmology in the early seventeenth century, see Sheehan in Chapter 12.)

The young Nieremberg had entered the Society of Jesus and studied under the tutorship of Father Aguado, confessor to the Count Duke of Olivares, and the favorite of Philip IV. Nieremberg later studied at the universities of Salamanca and Alcalá, where he acquired the grade of "metafísico"; that is to say, he was an expert in arts and theology. His linguistic proficiency in languages was beyond doubt: he read Hebrew, Greek, and Latin from an early age. In 1623 he took holy orders, and in 1625 he started to dictate classes in Natural History at the Colegio Imperial, where he also taught Grammar, History, Erudition, and Holy Scripture.[6]

Certainly, Nieremberg's life would never serve as the basis of an adventure novel. Surviving accounts describe him as a sickly man of ascetic tendencies. "He broke his bones writing books and martyring his body," wrote Andrade, his first biographer.[7] Totally devoted to teaching and his books, Nieremberg became a prolific and remarkable writer. Most of his work was in Latin and consisted of long treatises on patristics, Marian theology, or ascetic doctrine. He also translated important devotional books, such as Thomas Kempis's *Imitatio Christi* (Imitation of Christ).

Nieremberg's best-known Spanish-language texts were *Del aprecio y estima de la gracia divina* (Appreciation and Esteem of Divine Grac; 1641) and, above all, *De la diferencia entre lo temporal y lo eterno* (On the Difference between the Temporal and the Eternal; 1640), a text worth commenting on.[8] As its title suggests, this work was a comparison—very much to the tastes of its time—between two opposites, the earthly and the eternal, the worldly and the divine. Nieremberg explored the theme of the *contemptus mundi* (rejection of or aversion toward the world) in order to understand the notion of death and defend the primacy of eternity over the temporal. Following Plotinus and the Neoplatonists, and developing a widespread idea in seventeenth-century Spain, Nieremberg argued that the sensible world was no more than "the dream of a shadow,"[9] a term he would also use to define time: an image, a shadow of eternity. This was a familiar theme then, whose greatest expression was formulated by another student of the Jesuits, the great playwright, Calderón de la Barca.[10]

To Nieremberg, man was an imperfect being who lived surrounded by temporal things that were vain, ephemeral, and deceitful; thus the subtitle, *Crucible of Disillusionments*, the last word constituting one of the commonplaces of Spanish Golden Age culture. Man's only chance, therefore, was grace, the very path of spiritual perfection and the only opportunity to achieve knowledge of God and His work. *De la diferencia* procured for Nieremberg a certain celebrity status. Translated into French, English, Latin, and even Arabic, the book was republished in Spanish more than thirty times up to 1800.

Nieremberg also wrote two treatises on philosophy and natural history, which I will consider shortly. For many decades, these texts were not considered relevant enough to remove his name from the sphere of religiosity and ascetics, as it is reflected on the two main biographies written about him, both focused on the world of religion.[11]

This is understandable, given that most of Nieremberg's work dealt with spirituality, divine attributes, grace, sin, and man's chances of knowing God, but Newton also wrote many pages of Biblical exegesis without this ever leading to him being excluded from the history of science. It is true that Nieremberg did not deduce any fundamental laws of nature. His works on natural philosophy and natural history, seen retrospectively, do not contain a single idea that has turned out to be decisive in the development of our knowledge. This may render his intellectual contribution useless in the eyes of today's science, but surely it makes a reading of his work more fruitful to the task of understanding the sciences of his day.

Just as the baroque culture has proved fertile ground for studying the art, literature, and political culture of early modern Spain, so Nieremberg's work seems to belong to a spiritual context, not a scientific one. Only recently have a few historians of science and Americanists shown any interest in it.[12] His name, though, is still associated with the peripheral and the merely colorful.

IN THE REALM OF ANALOGY
AND CORRESPONDENCE

Were we to preserve the externalist/internalist distinction between context and content, we would claim that the natural knowledge of Nieremberg's age was embedded in religious thought. Paraphrasing Bruno

Latour, however, it might be said that there is as much science in the
digressions of Nieremberg on the eternal as there is religion in his treatises on natural knowledge.[13] Let us now consider Nieremberg's two scientific books. On the one hand, *Curiosa y oculta filosofía* (Curious and
Obscure Philosophy, 1649), a treatise containing two different studies
that had its origin in lessons given by Nieremberg at the Colegio Imperial, deals with wonders and curiosities of nature, a widespread literary
theme in this period, and uses material from several genres: compilations of prodigies, books of secrets, and encyclopedic manuals of natural
philosophy.[14]

Historia naturae, maxime peregrinae (History of Nature, the Greatest Pilgrim; 1635), on the other hand, is a Latin work that has never
been translated into any modern language. Despite being a more specialized text, it is a miscellaneous work, somewhere between American
natural histories and the standard repertory of curiosities, rarities, and
surprising facts of nature.[15] From our modern perspective, both books
are difficult to classify, since, as Lorraine Daston and Katharine Park
have pointed out, the preternatural is made up of a "stratigraphy of
heterogeneous phenomena . . . with no internal coherence except their
awkward relationship to *scientia* in the Aristotelian sense."[16] However,
it is not hard to detect similarities with other books of the same period.
Thus, *Curiosa y oculta filosofía* is related not only to the work of Gaspar Schott, Johannes Jonstonius, and other taumatographers, but also to
that of Pedro Mexía, Antonio de Torquemada, and the sixteenth-century
Spanish humanists.[17] The *Historia naturae* lies somewhere in between
travel literature, with its tales of the marvelous, and treatises of exotic
therapeutics, both of them genres enriched by the encounter with the
New World and other geographic discoveries.

The early chapters of both *Curiosa y oculta filosofía* and the *Historia naturae* are given over to justifying these forms of knowledge and
to clarifying certain epistemological presuppositions. But how could an
author be so interested in earthly affairs when he seemed to disdain
them? How could he combine his devotion to the profoundest asceticism,
the perfecting of spiritual life, with such a passionate description of the
natural world? For one who sought eternal truths, what was the point
of studying that which was no more than apparent, vain, deceitful, and
ephemeral? The Jesuits, as is well known, were great masters in the arts
of eclecticism, an ability taken to virtuoso extremes by the followers of
sacred *conceptismo* (conceptism) in Spain, among whom Nieremberg
should be counted.[18] A taste for paradox and contrast, appreciable in his
work, is probably the key to explaining Nieremberg's twin vocations, so
pronounced and so apparently opposed to one another.

A certain approach to the study of nature is recovered in these introductory pages. Shared by many authors working within a physicotheological framework, it is based on the idea that the observation of natural phenomena was a possible means of coming closer to God. Reading the Book of Nature, as Saint Augustine had shown, was an *oratio* (oration) that confirmed Holy Scripture, and for Nieremberg this explained why literature dealing with animals had in the past been the occupation of "very spiritual men and scholars of holy letters."[19] Like many other Christian authors, Nieremberg began by legitimizing that strange passion, which, since Aristotle, had been fueled by the will to know. Both favored and condemned by the scholastics, the desire of knowledge, in Nieremberg's view, was a form of greed, but without doubt the most glorious, the noblest, and the purest. To study facts of nature, or more correctly to decipher them, was to unveil the mysteries of their Creator. The task of the philosopher, therefore, was to read and interpret the universe, which, by definition, was apparent and deceitful. Indeed, the natural world was made of hidden, arcane, but also holy truths. (On reading the Book of Nature, see also Ferreira Furtado in Chapter 9.)

Here the metaphor of the *theatrum mundi* (great theater of the world)[20] came into play—an old Stoic view, central to the Counter-Reformation, which understood human life as a representation, where nothing was what it seemed and everything acquired another meaning or another sense going *beyond* the apparent. This idea runs through the literature, poetry, and painting of the Spanish Golden Age and pervades the work of such illustrious names as Quevedo, Calderón, and Gracián. It is also a feature of the philosophy of the age and of what today would be termed scientific thought. The natural world, like human existence, was also representation, the result of a divine plan, a code to be deciphered, a labyrinth to negotiate and resolve. Nature's manifestations, in Nieremberg's words, were a copy of God: "by them we make Him out and by them we venerate Him, and thus they had to have much that is admirable, much that is incredible, much of which we are ignorant."[21]

Nieremberg's work is also immersed in a cultural universe ruled by similarity, as defined by Foucault,[22] a cosmos governed by analogy, sympathies, and correspondences among things. Nieremberg explores the characteristic relations that Marsilio Ficino and the Neoplatonists had popularized in educated circles since the early Renaissance, a frequent feature of work by Jesuits, who produced the best syntheses of scholastic tradition and Renaissance hermetism. Thus, for Nieremberg, celestial movements are to be found in animals, and animals are found inscribed upon the stars. Stones take organic forms, and the vegetable pharmacopoeia announces through its physical features the illness or organ it is

designed to cure. Dead bodies shed blood in the presence of their mur-
derers; a mother's imagination is the origin of deformations; fear turns
the hair white. The harmonies of the world, the mute or secret virtues
of things, the connections between the spiritual and the physical are, in
short, the kinds of themes developed in great detail and of course with
great erudition. This is the case with music. Taking up an idea proposed
by, among others, Robert Fludd, Nieremberg traces the link between
music and living beings to the age of Creation. He detects and describes
its effects on animals, plants, and men, a very characteristic ploy in his
natural philosophy: replete with neo-Platonic evocations, music embod-
ied the mysterious link between the visible and the invisible, between the
senses and the spirit. In fact, when Nieremberg arrives at a conclusion
concerning the nature of the sympathy and antipathy of things, he refers
to music as the best example: "the sympathy or antipathy of things is the
music of the world."[23]

Close to Nieremberg in so many ways, Athanasius Kircher also
claimed that the world was founded upon postulates that were "much
more intricate and indiscernible than might be supposed by some ingenuous minds of past centuries or the vulgar philosophers of the present."[24]
Giving a free rein to the illusion of universal order, the so-called last man
to know everything also stated that "all things are connected by arcane
knots."[25] Thus the author of treatises on spirituality and the paths to virtue, a man devoted to underlining the contrast and opposition between
the natural and supernatural worlds, also sought out its similarities and
multiplied its correspondences, establishing new syntheses and finding
new proofs. In this cosmovision ruled by analogy and similarity, we
should not forget the most important of them all: nature was—to adopt
Plotinus—God's poetical labyrinth, a reflection of the supernatural, its
correlative.[26] And whether this labyrinth was a pale reflection—a simu-
lacrum, shadow, or dream—or whether it was a demonstration of its
Creator's infinite wisdom, that is, whether Nieremberg chose to illustrate
this link in dark or light tones, there is no doubt that such a mixture of
elements—as dramatic as any baroque chiaroscuro—seems as strange to
us today as it would have seemed natural then.

NATURAL HISTORY BEYOND THE EMPIRE

Natural knowledge, understood in this way, fulfilled an obvious metaphysical objective. After theology, speculation about animate matter was
the foremost of all sciences. None other could be "more certain, more
useful, nor more divine."[27] God Himself had been the first master of this

doctrine, followed by Adam, "the first man, the first disciple," and later by Solomon, the wise king. Nature was Holy Scripture written by the fingers of Omnipotence. It was the first theology and the first writing, where the ray of divinity was reflected "with particular ardor."[28]

Having said this, the notion of the divine inscription of nature may need some nuance. The metaphor of the "Book of Nature," running from Pythagoras to Boyle through Abelardo, Paracelsus, and Galileo, involved so many authors and so many different ways of carrying out science that we need to identify what sort of text nature was to Nieremberg, and what sort of reading he made of it.

First of all, it was a highly written text and a highly read reading. The relation between books and Nature, between words and natural phenomena or living beings, was so close that to talk or write about them was, in effect, to talk and write about what others had talked and written about. Nature in Nieremberg's writing is a set of emblems, signs, and symbols to be deciphered. Their meaning is located in a place beyond the visible. In this sense, we can qualify his natural history as a preterimperial discipline, since his method and objectives, the issues he seeks to resolve, all go well beyond the empire of man over Nature, the old Augustine idea brought up to date by modern science.

A truly preterimperial natural history, the kingdom of Nieremberg's science is not of this world. Or at least not completely. It is located in an intermediate zone, halfway between the emblematic and allegorical natural history of Renaissance humanism, and the morphological and taxonomical discipline that was to impose itself in the age of Linnaeus. Let us think of the great developments in Western natural history. We have Pliny, whose liking for narrating the life, customs, uses, and legends of living beings still remained in the works of Nieremberg and others. It was a discipline concerned with words and tales, singular and curious facts, the extraordinary and the meaningful. It was also a moral history in the sense that it highlighted values and teachings that men might take from animate life.

Then there is of course Aristotle, the great precursor of the taxonomical shift experienced by natural history in the hands of Cesalpino, Tournefort, and finally Linnaeus. The latter's work could be described as a truly imperial discipline, since its aim was to name and classify living beings, represent them, and organize them into a system. Linnaeus sought to legislate over the living beings of the whole world, just as the Mercator projection had brought order to space or as Newton had gathered terrestrial bodies and planets under the same laws of motion. Nieremberg's approach to the Book of Nature combined the old reading between the lines, focused not on brute facts but on their meanings, and the new, silent, succinct reading of forms and figures, the "algorhythmic reading," according to David R. Olson, typical of modern science and focused on the observable, the measurable, and the quantifiable.[29] Including the morphological and new scientific approach together with the humanistic, hermetic, and analogical view of nature, Nieremberg's science articulated a kind of preterimperial natural history. For it went *beyond* the imperial (modern) one, representing a preterimperial discipline more focused on the figural reading of nature than on the literal, more concerned with God than with what we call today natural facts. It is not quite a premodern science. Properly speaking, we are facing a preterimperial science.

ʊ

Does this mean that Nieremberg's science remained alien to the imperial interests of his day? Clearly not, but in his work these interests come across as subordinated to others. One example would be his treatment of the theme of time, or to put it in his terms, the *mudanza* (mutability) of nature. This was no trivial matter for someone who was concerned not only with the difference but also with the connections between the temporal and the eternal. In general terms, reflection on the expiration, change, and alteration of matter constitutes one of the great commonplaces of baroque culture. Here Nieremberg faced a serious problem. On the one hand, he found evidence in support of the hypothesis of change: in antiquity there had existed facts and beings that had later disappeared (the purple, the phoenix, flax cleansed by fire). There were also examples of the opposite phenomenon: new animals, plants, stones, and stars.[30] Nieremberg defended the incorruptibility of the heavens, the immutability of species: the permanent was superior to the mutable. God had created an unalterable and perfect world, not a monster. The emergence of novelties could only be due to an unfurling of the principle of plenitude, to man's influence on the course of events, or to the effect of the coming of Christ and the instauration of the sacraments. The causes of certain losses were seen as the effects of empires, "the changes in kingdoms, the movement of Monarchies."[31] These in turn were due to divine grace, the fall into sin, and the plans of providence. Nieremberg looked at Holy Scripture for that which "proved somewhat the changes of nature."[32] The point is clear: sacred history contains and directs moral history (human history, chronology), and the latter provides a model for natural history.

Let us examine other aspects of *Historia naturae, maxime peregrinae.* As I have said, the book is a combination of bookish learning, encyclopedic knowledge, and erudition, made up of compilations of diverse and heterogeneous reports, intended to assist in the task of understanding the divine. Depending on the subject, Nieremberg quotes the peripatetics or the Neoplatonists; the philosophers, doctors, and naturalists of his day; and even classical and recondite figures. This sort of tactic, common to much other contemporary scholarship, is particularly striking in the work of an author who lacked firsthand knowledge of the phenomena he discussed and who was, moreover, a Jesuit. The scholastic and Jesuitical method had generated a text that circled rhetorically through arguments and counterarguments, eventually providing a number of answers to certain questions. Derived from the Aristotelian *Problemata*, the Jesuits turned this methodology into a style of their own.[33]

Most Jesuit work is marked by an attitude of syncretism, a desire to record alternative theories, different opinions, and hypotheses about the same facts or events.[34] This was not just a question of authorizing legitimate versions or interpretations, but also of discrediting illegitimate ones, in a manner reminiscent of the work of those like Thomas Browne, Nieremberg's English contemporary, who compiled a lengthy treatise on vulgar errors.[35]

True knowledge required a meticulous work devoted to clearing away erroneous readings and authorized falsehoods. This is the reason why, for instance, Nieremberg wrote so many pages to demonstrate that immortal animals could not exist, or that plants or the heavens were not animate, faithfully recording what Aristotle, Empedocles, or the church fathers had written on these subjects.[36] He worked hard, moreover, at avoiding the sorts of pantheistic conclusions to which his neo-Platonic postulates might have led him.

Having to deal with a text written by God, but also read by other men, exegesis lay at the heart of this science of nature, in a twofold sense: exegesis of what God had expressed through His work, and exegesis of what others had read or believed to have understood in it. Related to the issue of interpreting the Holy Scripture, this was a delicate problem in the context of the Counter-Reformation movement.

The *Historia naturae* is thus conceived as a formidable review of many themes that had been studied since the times of ancient Greece up to the early decades of the seventeenth century. Of its eighteen books, the first eight are markedly Aristotelian. In them, first principles are laid down. According to the scholastic tenets, animal life occupied an intermediate place in the hierarchical scheme of Creation. The higher was better than the inferior, animate life superior to inanimate. Although the light of divinity was shed over all of God's works, its splendor increased as one moved up the scale from inert matter (meteors, metals, stones) toward human beings.

Halfway along this scale were weeds, herbs, bushes, and trees, then beasts, birds, and fish, which were provided with "awakened vitality," that is, they were sentient beings. For anyone seeking to show divine harmony and symmetry, the place occupied by animals, therefore, was quite meaningful and unique. They constituted the *diaphragm* of Nature.[37]

The interstices of that long chain of being, where connections could be made, were strategic points for the study and the diatribe. This was the case of sponges, madrepores, and zoophytes, an intermediate grade between animals and vegetables where Nieremberg wielded the names of Aristotle, Pliny, and Rondelet before entering fully upon one of the star themes of the prolonged season of curiosity: generation and hybridization.[38]

As if brought before a cabinet of curiosities, Nieremberg stops to consider everything that is surprising and odd. He records every detail of the case of the ivy that sprang from a deer's horn or the flower that grew from a woman's nostril. He gives credit to tales of fish that ate gold, and records accounts by Acosta on mice and worms born from reeds in New Spain. He also quotes Della Porta's confirmation of Aristotle's claim that eels were born from seaweed under the effect of water. The Jesuit wonders whether air can be fecund or if fire can admit living beings. He discusses the furs and skins worn by Adam and Eve, and considers the question of what might have been eaten by the carnivores on Noah's Ark.

The list is a long one. Each case lends itself to considerable digression and speculation. Nieremberg establishes a match between the natural world and Holy Scripture, the temporal and the eternal. He maintains a proverbial inclination toward those intermediate territories or shadowy areas that contain an enigma but also promise some scientific and theological answers.

DECIPHERING AMERICAN SIGNS AND WONDERS

Few cases were as promising as the New World, a continent whose mere existence had been posing serious intellectual challenges to European scholars for over a century. America, a land associated with the marvelous ever since Columbus, the subject of every possible kind of speculation, still remained a question to be answered.[39] For many reasons, America constituted something like a black hole for a man like Nieremberg, who, as we have seen, interpreted nature as a labyrinth, a code to be deciphered. On the one hand, American themes had all the necessary attributes for evoking and catalyzing the most genuine interests of his natural philosophy and history. On the other hand, we should not forget where Nieremberg worked, and for whom. He was a member of a court headed by a transatlantic monarchy and a member of an institution that had taken over the responsibilities of the main cosmographer of the Council of the Indies. It is understandable that such a man should devote so much of his treatise on natural history to American subjects, when that treatise contained, as its frontispiece announced, so much material on the exotic and far off. In fact, the New World takes up almost half of his text, from books IX to XVI, as well as parts of other books.

It is important to emphasize that Nieremberg never actually went to America. In fact, he never even left Spain. The bookish nature of his work becomes even more apparent when dealing with the subject of America. He relies entirely on the testimony of others, drawing information from the works of José de Acosta, Peter Martyr, Fernández de Oviedo, Cieza de León, and other great chroniclers and historians of the New World. Above all, he quotes from Francisco Hernández, a doctor of the age of Philip II, considered the sharpest observer of novo-Hispanic nature and the true precursor of zoological and botanical studies in the New World. (For a further exposition of the work of Hernández, see Barrera-Osorio in Chapter 11.)

Nieremberg is known to have copied quite a lot of chapters and illustrations from texts by Hernández, whose original drafts and plates he was able to consult at the Colegio Imperial and also in the library of El Escorial, far from the Lincei edition of what would come to be known as Hernández's *Tesoro messicano* (Mexican Treasure). Nieremberg's reliance on Hernández is as widespread as it is explicit. His chapters on botany are a literal transcription from Hernández's *Historia de las plantas de Nueva España* (History of the Plants of New Spain). More interestingly, perhaps, we also know that John Ray learned about Hernández's descriptions through Nieremberg's treatise.[40] But it is not our aim here to argue for the importance of Hispanic scientists in the development of modern science. The object is not to trace what has survived in the history of knowledge, but what is missing; such is the case of most of Nieremberg's preterimperial natural works and ideas.

Nieremberg skillfully deploys the rhetoric of novelty and indulges in the taste for the strange and wonder: "What is certain is that we see so many things which before being seen would have seemed impossible to us that Nature has now gained credit for all kinds of wonders," as expressed in the words engraved in the façade of the convent of San Marcos de León in 1542, *Omnia nova placet* (All new things give pleasure).[41] Humanist interest in novelties had spread in many directions, stimulating the desire to learn about faraway things, distant in both time and space.[42] This interest in the distant past had triggered the recovery of classical culture, making it possible to identify the ancient with the new, as something that had been newly recovered.

The recently discovered lands, their inhabitants, and their fauna and flora had been playing an analogous role in the domain of the material and the spatial. José de Acosta, the Jesuit creator of the genre of moral and natural histories, had noted that "novelty pleases," arguing that it dignified speculation about natural things, "mostly if they are notable and rare."[43] Baltasar Gracián, another Jesuit author, had written: "uniformity limits, variety dilates."[44]

By the time Nieremberg wrote on natural history, baroque culture had exacerbated this tendency even further. What was sought out and appreciated was in fact the strange rather than the new, and the stranger the better.[45] This can be seen in the arts and in literary precepts. Painters prided themselves on representing the difficult and the deformed. Poets were prompted to produce stranger and more admirable inventions. Moreover, this recommended subordination of the baroque writer to the strange was also extended to those who worked in the realm of true facts. Thus a contemporary Spanish historian related truth with the outstanding, which enabled one to "teach and delight through its singularity and strangeness."[46] Nieremberg's science follows this agenda. It seeks to blend "agreeable truth" with "authentic novelty."[47] By the early to mid-seventeenth century, America had been placed definitively within the realm of the strange, the curious, the extraordinary. It was even identified with the "monstrous," in the original sense of being something portentous, a fact showing and meaning something, a sign.[48]

Several chapters of *Curiosa y oculta filosofía* deal with the same themes as any other teratological treatise of the period: maternal imagination, the phenomena of hybridization, the classification of monsters, the verisimilitude of fabulous monsters, and so on. In the *Historia naturae*, Nieremberg alternates between the two most common ways of approaching these topics, reflecting the double orientation of his sources.

On the one hand, there was teratoscopy, the discipline related to the arts of divination, which undertook the study of monsters and prodigies for their presumed relation with the divine, the future, and the course of history. It was a case of interpreting those *téras* (signs), of knowing what they foretold, and of understanding what warning, punishment, or prophecy was boded by those extraordinary facts or events. This was a prominent genre in the preternatural philosophy between 1550 and 1600 in the work of Peucerus, Lycosthenes, Paré, Sorbin, or Gemma.[49]

On the other hand, there was taumatography, an approach toward the marvelous undertaken from the standpoint of natural history and medicine, a form of knowledge that sought to observe, describe, and classify unusual phenomena.[50] This second genre, more characteristic of Nieremberg's period, was represented by the treatises of Jonstonius, Schott, and Licetus.[51] In it, metaphysical hermeneutics was replaced by a physiological and anatomical approach. An interest in working out the meaning of events to come was replaced by a questioning about the past. Its objective was to decipher the origins and causes of the unprecedented.

Nieremberg combines these two approaches in his study of that gigantic wonderland called America. As a result of his heterogeneous and disparate interests, an extraordinary amount of motley information was gathered by the Jesuit. The atrophy of his material derives from its excess. He deals with many of the quadrupeds, birds, fishes, reptiles, insects, zoophytes, plants, and minerals of the New World. Sometimes his remarks are restricted to brief, merely descriptive notes, often transcribed from other authors. On other occasions he digresses more widely, stopping to consider everything that is extraordinary or to look for possible connections with the ancient world.

Nieremberg comments on both real and imaginary beings, the distinction is ours rather than his. For him, all these beings come from the same source, as uncertain as it was authorized: the printed word. Neither excessively credulous nor the opposite, Nieremberg has a sure grasp of the subjects that are likely to awaken curiosity among his readers. And he has a fondness for recalling that "no lie can triumph unless it contains an element of truth."[52] Following the procedure with which Ctesias of Cnido had inaugurated the transference of the imaginary onto the Far East, Nieremberg displaces onto the New World the complete contemporary repertory of fantastic zoology. Perhaps, he speculates, the condor of the Andes is the griffin, or the unicorn will turn out to be the horned horse of which some barbarians are known to have spoken.

A description of the manatee (one of the great sea mammals) or of the axolotl (a novo-Hispanic salamander) follows that of tritons spotted in Cuba and digresses on the ivy and behemoth of the Book of Job.[53] This is the result, so strange to us today, as bizarre as that formed by stones and stars to his contemporary Pascal.

The *manucodiata* (bird of paradise) deserves special attention, since animals inhabiting the skies were "more certain interpreters of divine indulgence and wrath" (see Figure 5.3).[54] Following the commonplace, Nieremberg stated that this miraculous bird lived in the air without ever coming down to touch the surface of the earth. Rather than confirming the sacred place held by America in the plans of Providence, the bird seemed to augur the place which that land might come to hold one day. Instead of constituting evidence that Paradise was once located in America, as it was claimed by Nieremberg's near-contemporary León

MANVCODIATA

CAPVT XIII.

De manucodiatá.

Admirati fumus aues manucodiatas siue paradifiacas. Indigenæ putant é cælo defcendiffe, ideô appellarunt manu-codiatam, id eft, auem paradifi, fiue Dei. Hugo Lincoftanus & nofter Iarricus fcri-bunt, à Lufitanis vocari *paffaros de fol*, & non nifi in folem volare. In meo Sigalione inge-

FIGURE 5.3. Manucodiata *(Bird of Paradise), woodcut, 1635*

SOURCE: Juan Eusebio Nieremberg, *Historia naturae, maxime peregrinae* (Antwerp, 1635), vol. 10, chap. 13, p. 210. Museo Nacional de Ciencias Naturales, Madrid.

Pinelo,[55] the prodigy seemed to hold tidings of the future. America would be the new paradise, a promised land of the future, more than the true and biblical paradise of the past.

This, indeed, was Nieremberg's main concern: to provide the New World with meaning, to readjust the correspondence between the American temporal and the divine, to make a reading of American nature tally with Holy Scripture. This was a question that continued to occupy many of the finest minds, not only among students of nature but also in the fields of history, law, and politics, as was the case of Juan de Solórzano Pereira, whose *Política Indiana* (1647), an essential text for any understanding of the structure of the baroque monarchy, opened with an explicit statement of the same thorny issue.[56]

Nieremberg offered genuinely preternatural explanations in his attempts to resolve some of the most controversial issues concerning the New World. The question about the origin of its fauna was one of them, which, naturally enough, he explained by reference to Noah's Ark. In his opinion, American species had traveled from the New World and back again, transported by guardian angels. In this providentialist vision, God had disposed that each species should be protected in this way.[57]

Even more delicate was the issue of the origin of American people. This was an ongoing debate since the times of Acosta, and in Nieremberg's lifetime had even been the subject of monographic treatises such as that by Gregorio García.[58] Nieremberg reviews several contemporary hypotheses. He mentions Plato's Atlantis, Solomon's Ophir, and voyages to America, presumably by Carthaginian or Phoenician sailors. He also introduces the idea of a possible lost tribe of Israel, thus opening the way to the well-known identification of American natives with the Hebrew people.

The human nature of the American man, halfway between demonization and mystification, demanded his inclusion in the Bible story. It thus became necessary that Saint Thomas or other apostles had traveled to the New World. The alleged relationship with the Hebrews also permitted the king of Spain to be portrayed as a second Moses, a savior.

In Nieremberg's view, the king of Spain was invested with great taumaturgical powers. Demons were particularly averse to him, because the Catholic king was their greatest enemy on earth.[59] As usual, Nieremberg records both sides: the humanity, virtues, and knowledge of the Indians, but also their satanic and superstitious rites. It is not difficult to guess the propagandistic intentions of this second reading. To Spain had fallen the task of completing an evangelical mission on a planetary scale. (For the implications of this in terms of expansion in the Pacific, see Sheehan in Chapter 12.) It occupied a strategic mediating position between the temporal and the eternal. For example, before true faith was introduced to the island of La Española (Santo Domingo), hurricanes there frequently ripped the trees from the ground. But once Holy Communion had been instituted, the violence of those hurricanes ceased.[60] Nature, which had once been mutated by the coming of Christ, now experienced in America the beneficial consequences of Christ's second coming through the people of Spain.

America, the discovered land, was also a land of revelation, as has been pointed out by Domingo Ledezma.[61] Nieremberg deploys all his talent to reconcile holy texts, nature, and history. Deciphering American signs and prodigies contributed to the knowledge of God, His work, and His plans. It is a preterimperial natural history lesson that justifies and proclaims the empire from a holy standpoint.

The dedication heading Nieremberg's text, addressed to the Count Duke of Olivares, royal favorite and defender of the Miles Christi during the Counter-Reformation period, seems paradigmatic. It is a classic example in court literature and a veritable display of symbolic combination. Its theme is the olive tree, with all its profound imperial, Christian, and specifically Jesuit resonances. Its fertile and growing presence in Bética (Andalucía) was interpreted by Nieremberg as a sign of the role that Spain had been called upon to play in the course of history.

The Company of Jesus, whose branches extended over the earth like that holy tree,[62] tried to decipher American nature through a reading that hybridized a symbolically displayed and heterogeneous set of beliefs. The search for remnants of true religion among pre-Columbine cultures leads Nieremberg to validate news brought back by Peter Martyr about the adoration of the cross among certain tribes of the Yucatan.[63]

This was no mere detail. Nor was the attention Nieremberg paid to the *granadilla,* or *maracuyá,* and the image accompanying its description (see Figure 5.4). This plant was the passionflower, in which Nieremberg is able to pick out the chalice, nails, and other signs of the death of Jesus Christ.[64] It is a natural reflection of a culminating moment, the crucifixion, that point where sacred history had redirected human history and altered the course of nature itself. It was the Eucharist, the sacramental motif brandished by the Counter-Reformation as the main emblem of the true faith and the right exegesis of holy texts. The transubstantiation of bread and wine into the body of Christ was not only the most controversial episode in theological disputes between Catholics and Protestants but also an entirely preternatural motif, the maximum expression of how the finite contained the infinite.[65] As far as the baroque Hispanic monarchy was concerned, the promise of the resurrection implied by the Eucharist carried within it the reading of a regeneration both necessary and possible.

GRANADILLÆ RAMVS.

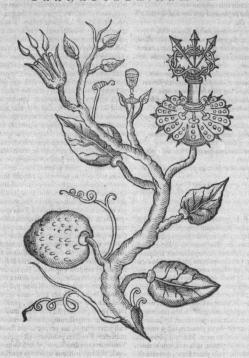

CAPVT X.

De granadilla.

IN Peruanis montibus sponte nascitur granadilla. Nomen hoc imposuerūt Hispani, ob similitudinem quam cum nostris malogranatis habet: nam fructus eiusdem ferè est amplitudinis, atque etiā coloris, quando maturus est, nisi quod coronā careat: nunc in siccato si agitetur, sonitū edit semen intus occlusum, quòd pyri semini simile est aliquantò maius, tuberculis quibusdam eleganter elaboratum, & aspectu iucundum. Pulpa candida est, & saporis expers. Planta

hunc fructum ferens hederæ similis est, eodemque modo repit & scandit quocumque loco ponatur. Elegans est, cùm fructu onusta conspicitur, propter eius amplitudinem. Florem habet albę rosæ persimilem, in cuius folijs passionis Christi figuræ delineatę conspiciuntur, quas magnā diligentiā istic pictas existimes, eam ob causam elegantissimus est flos. Fructus ipse sunt granatula iam dicta, quæ matura, acidulo liquore abundant, & semine plena sunt. Aperiuntur vt oua, & liquor ille cū magnā voluptate sorbetur cùm ab Indis tum ab Hispanis, hęc, etiam si multos sorbeas, vllam vētriculi grauedinem senties, sed potius aluū emolliri. Herba rara est, quæ

FIGURE 5.4. Granadilla *(Passion Flower), woodcut, 1635*

SOURCE: Juan Eusebio Nieremberg, *Historia naturae maxime peregrinae* (Antwerp, 1635), vol. 14, chap. 10, p. 299. Museo Nacional de Ciencias Naturales, Madrid.

PART III

Knowledge Production: Local Contexts, Global Empires

CHAPTER SIX

Cosmopolitanism and Scientific Reason in New Spain

Carlos de Sigüenza y Góngora and the Dispute over the 1680 Comet

ANNA MORE

Recently, scholarship on early modern European science has revised the narrative of a seventeenth-century scientific revolution enacted by lone visionaries, emphasizing instead the collaborative and institutional aspects of scientific debate and change during this period. From this new perspective, it has been increasingly easy to relate changes in scientific theory and practice to the discursive and political context in which European scientific activity took place. Studies of patronage, for instance, have shown how scientific endeavors often reflected the whims of aristocratic patrons and the slippery politics of court culture.[1] Likewise, scholarship on early scientific academies has argued that the rules of etiquette that arose to arbitrate internal scientific disputes provided models for an emergent civil society.[2] Not only have these new studies emphasized the contingent nature of scientific activity and its dependence on institutional and discursive contexts, they also have shown the ways in which scientific discourse and practice restructured these same contexts and thus contributed to broader political and cultural changes during the period.

Pierre Bourdieu has systematized this new sociological history of science as the study of the "field of scientific reason," in which "a most specific social logic is at work, affirming itself more and more to the degree that symbolic relations of power impose themselves that are irreducible to those that are current in the political field as well as to those instituted in the legal or theological field."[3] The value of Bourdieu's notion of the "scientific field"—as opposed to analyses that contextualize science either through the history of institutions or through supra-institutional discursive regimes—is that it allows us to perceive the characteristics of an early modern public sphere modeled specifically on scientific reason.[4] The public sphere of science is based on a specific form of authority, structurally distinct and more or less autonomous from traditional forms of knowledge. Above all, scientific reason disrupts previous structures of judgment and authority by establishing facts that are "universally known and recognized."[5] According to Bourdieu, this structure of recognition in scientific knowledge creates an inherently contradictory form of public knowledge. As opposed to the closed and solitary universe of the alchemist, which was "secret, unchecked, and uncontrollable," the early modern scientific field emerges "only from the moment when a scholarly city has been instituted that is simultaneously open and public (as opposed to hermetic and private), as well as closed and selective."[6]

Initially, therefore, scientific authority "rests upon the recognition granted by the peer competitors" that only afterwards becomes parlayed into various forms of institutional and statutory authority, such as through the educational system.[7] In his emphasis on the struggles for dominance among the very scientists that form its selective ranks, Bourdieu's analysis consciously breaks with the stereotype of the "irenic scientific community."[8] Recent studies have noted this social volatility in the propensity for debates among members of early modern scientific communities to devolve into resentment or violent, ad hominem attacks, but equally, when conflict is avoided, to serve as models for civil society.[9] Bourdieu's analysis also suggests a structural reason behind the affective regime of modern science. By breaking with older forms of authority, scientific reason loosens social bonds and exposes the practitioner to a thicket of friends and foes. In place of moral communities based on theological and juridical authority, science offers only the compensatory illusion of a disinterested universal truth. The agent of scientific reason is at once public and individual, subject to a certain lonely anomie that Bourdieu opposes to the hermetic comfort of the alchemist's laboratory.[10]

Not only does Bourdieu's structural paradigm elucidate the conditions of emergent communities of science in early modern Europe, it also provides a basis for understanding the conditions of science on Europe's colonial periphery. In the seventeenth century, the Spanish viceroyalty of New Spain was just such a periphery. Heavily indebted to Jesuit educational institutions and scholarship of the Habsburg empire, it was also marked by its geographical distance from Europe and the particular political conditions of a viceroyalty with a heterogeneous colonial populace.[11] One of the most notable scholars to emerge from this setting was

Carlos de Sigüenza y Góngora (1645–1700), a polymath who was an accomplished astronomer and university mathematics professor. He is also responsible for one of the most explicit defenses of scientific reason in the seventeenth-century Americas, often cited as evidence of New Spain's scientific "modernity."[12] In 1681, Sigüenza entered into a public dispute with the Austrian Jesuit Eusebio Kino, a missionary who had stopped in Mexico City on his way to the northern frontier of the viceroyalty. The dispute concerned the significance of the comet seen worldwide for several months during the previous year. Shortly after it disappeared from view, Sigüenza published a pamphlet titled *Manifiesto filosófico contra los cometas despojados del imperio que tenían sobre los tímidos* (Philosophical Manifesto against Comets, Stripped of the Empire They Have Held over the Timid) in which he sought to reassure the New Spanish public that the comet did not portend ill omens for the colony.[13] Either ignoring this publication or, as Sigüenza claimed, deliberately contesting it, Kino published his own treatise, *Exposición astronómica* (Astronomical Exposition), upon arriving in New Spain in June. In sharp contrast to Sigüenza's position, he argued that comets were divinely authored signs of imminent threat, either political or natural.[14] Whether or not Kino intended his own treatise to supersede Sigüenza's publication, the latter understood the Jesuit's text to be a refutation of his argument. He responded in kind with a scathing analysis of Kino's logic in a publication titled *Libra astronómica* (Astronomical Balance) in imitation of the title of the Jesuit Orazio Grassi's published attack on Galileo Galilei earlier in the century.[15] Sigüenza did not immediately publish this second treatise, instead giving it to his friend Sebastián Guzmán y Cordova, a Spanish mathematician in the viceregal court, who finally published it in 1690.[16]

Although Sigüenza has rightly been understood to pose an early defense of scientific reason, over authority and against popular knowledge, why he so ardently took up this cause and what this meant in his particular context has not been sufficiently explained. For the most part, Sigüenza has been seen as a contradiction and anomaly in a landscape of baroque language and Neoscholastic learning, a contradiction that in the past has been analyzed somewhat crudely as a "proto-Enlightenment" stance.[17] Sigüenza's treatise, and the events of the dispute, indicate more than simply a competition between two versions of scientific modernity, one already antiquated and the other resonant with new forms of European science. Instead, the debate with Kino over the comet of 1680 shows the extent to which scientific reason in colonial Mexico was related to the geopolitics of core and periphery in early modern science. Sigüenza's public debate with Kino, in its intention at least, was the first time that scientific reason had been structured and defended as a cosmopolitan knowledge, that is, as the basis for equality between Mexico and Europe in an imagined scientific "republic." This defense reflected not only Sigüenza's desire for recognition by a European colleague but also the importance that he attributed to scientific reason as a method and mode of colonial governance. By turning to the metascientific commentary in Sigüenza's treatise, it is possible to assess the relationship between these two seemingly distinct goals and to understand the particular emotional force with which he treated Kino's rejection.

Rather than locked in a battle for scientific modernity, as they have been seen to be, Sigüenza and Kino represented two extremes of what might be understood as a transatlantic split in the value and meaning of scientific reason. Whereas Kino's stance reflected the culture of Habsburg learning, with its reliance on the symbolic logic of the political and religious authority of the court, both the content of Sigüenza's argument and the metascientific commentary in his treatise show the extent to which in the Iberian colonies empiricism and mathematical models had created a "field" able to disrupt institutional forms of social and political authority. Sigüenza's text also indicates why a scientist on the European periphery would wish to disrupt these forms of authority. In fact, if the emergence of scientific reason as a field of symbolic power depended on a community of peers, conditions on the colonial periphery were even more precarious than those created by patronage rivalries in European courts. Rather than a sign of scientific modernity for its own sake, Sigüenza's treatise allows for a glimpse of how scientific reason, particularly Jesuit mathematics and astronomy, was instrumental in forming an idealized community of intellectual peers, at once cosmopolitan and outside of established political institutions such as the viceregal court, to which Creole scholars could aspire.

This ideal cosmopolitan community of scientists presupposed a particular notion of intellectual or ethical "friendship" by which Sigüenza seems to have measured Kino's actions. In fact, the affective force with which Sigüenza debated Kino, the sarcastic and stinging rebukes of what he perceived to be a rebuttal to his argument, indicate more than the resentment of a scientist seeking recognition from a European colleague. It is evident that science was an arbiter of inclusion in an idealized community of equals that would overcome the peripheral geography of New Spain. Equally evident, however, is that Sigüenza sought to participate in this community of scientists not as a universal subject of reason, abstracted from his geographical location, but as someone privy to a perspective that those in the center could not attain. If not an ideal polity or civil society, this cosmopolitan scientific community provided a compensation for the limitations of theological and political institutions in seventeenth-century New Spain, as well as personally, in Sigüenza's case, for a life led outside of the Jesuit order, from which he had been expelled at an early age.[18] Furthermore, the ideals of a cosmopolitan community based on scientific reason were consonant with Sigüenza's advocacy of reason as a means to manage what he increasingly represented as the irrationality of popular culture in New Spain.[19] In this sense, rather than a defense of scientific method for its own sake, Sigüenza's interventions around the comet reflected another of his particular obsessions, namely, the delineation and defense of a local homeland, or *patria*, in the context of Spanish imperial decline in which the central administration of the viceroyalty was notably weakened.[20]

ANATOMY OF A "LITERARY DUEL"

One of the most remarkable aspects of the dispute between Sigüenza and Kino is the fact that the Austrian Jesuit makes no mention of the polemic. Its details are owed entirely to Sigüenza y Góngora, who describes them in his counterattack to Kino's imputed provocation.[21] According to Sigüenza, the comet had been visible in Mexico from mid-November of 1680 through the first week of January of 1681. Shortly after this, he had published his popular broadside, the *Manifiesto*, intended to assuage fears that the comet portended natural and social calamities for the viceroyalty. In the *Manifiesto* he excuses himself from citing authorities in Latin because "I do not want pretension [*latines*] in what is meant to be popular [*vulgar*]" but promises to publish a second treatise in which he will enter into more technical detail on the comet. This second publication may have been the *Belerofonte matemático contra la quimera astrológica* (Mathematical Bellerophon against the Astrological Chimera), a section of which Sigüenza also includes in the *Libra*.[22] In the *Belerofonte*, Sigüenza attacks Martín de la Torre, a Flemish astronomer living in Campeche and the author of the first of four public critiques of the *Manifiesto*.[23] Apparently, Sigüenza did not feel that the other critiques warranted such elaborate replies. To the second, written by Josef de Escobar Salmerón y Castro, Sigüenza declares "I never intend to respond, as his extraordinary treatise and shocking proposition that the comet was formed from the exhalations of dead bodies and human sweat do not merit [a reply]."[24] In his prologue, Sigüenza mentions an anonymous third manuscript written by "a certain concealed [*oculto*] mathematician, with whom I have never had any relationship whatsoever." Titled *Examen comético* (Cometary Examination), the text would be made public, he is told, only when Kino published his own critique of Sigüenza's broadside.[25] This rumor appears to be the first Sigüenza hears of Kino's attack on his published argument.

In fact, of the four responses to his *Manifiesto*, only Kino's *Exposición astronómica* appears to have piqued Sigüenza's ire, and he dedicates the majority of the *Libra* to countering the Jesuit's logic and argument. Overall, the treatises differ substantially in both tone and content. Whereas Kino published a short and relatively superficial tract, documenting the comet's course in a celestial map and arguing strenuously for its metaphysical meaning, Sigüenza's *Libra* is at least four texts in one: (1) the *Manifiesto*, in which he establishes his argument, (2) the *Libra*, in which he defends the same in a point-by-point refutation of Kino's stance, (3) his own observations of the comet, in which he disputes the accuracy of Kino's, and (4) a section of the *Belerofonte*, in which he responds to Martin de la Torre's defense of astrology. In both its scope and tone, the *Libra* goes well beyond a statement of Sigüenza's position. Employing a double-edged attack on Kino, the treatise simultaneously argues against the missionary's overall logic and points out contradictions in his stated positions and method. In addition, Sigüenza utilizes his bibliographic resources to bury Kino's paltry authorizations in a heap of citations from authorities both ancient and modern. The excessive and decisive nature of Sigüenza's attack on the *Exposición astronómica* parallels a strong tone of opprobrium and at times outright sarcasm as he points out the fallacies and inconsistencies in Kino's argument.

In the prologue to the *Libra astronómica*, Sigüenza explains the particular circumstances that warranted such an elaborate and biting defense of his position. The narrative evokes his sentiments of betrayal and resentment when his expectations for an intellectual friendship with the missionary, who was reputed to be a "very eminent mathematician," did not come to fruition. Sigüenza recounts that his own prejudice toward European learning led him to believe that Kino would prove to be a suitable match for his scientific interests. Based on this assumption, he welcomed the mathematician to Mexico City in an act of intellectual hospitality:

Stimulated by the insatiable desire that I have to communicate with such men and blinded by my assumption that only those sciences learned in remote provinces are perfect, I entered his chambers by the front door, I befriended him, I took him into my house and regaled him there, I introduced him to my friends and defended him among his own, being able to write a long list here of those who asked me what the reverend father knew and to whom, even against the dictates of my own conscience, I responded that very much and all with perfection.[26]

Among his other acts of generosity, Sigüenza lent Kino his collection of maps of the northern territories, to which the missionary was headed, but which "were returned to me incomplete and in shreds" and then thanks only to the diligence of Francisco de Florencia, the rector of the Colegio de San Pedro y San Pablo where Kino had been staying. Sigüenza qualifies the latter as the "glory of our Creole nation and a very singular friend of mine" in clear contradistinction to the Austrian Jesuit's crudely ungrateful behavior, thus emphasizing the contours of an ideal intellectual friendship based on both equality and reciprocity.[27]

Whether or not Sigüenza's version of the events is accurate, it clearly shows the lens through which he viewed Kino's actions and which framed his own response.[28] Of all the affronts that Sigüenza claimed to have suffered following his initial act of hospitality toward the Jesuit, the worst of all was Kino's pointed provocation on the question of the comet. In his preface, Sigüenza recounts that Kino had stopped by his house to present him personally with a copy of the recently published *Exposición astronómica*: "and taking leave of me to go that very afternoon to the provinces of Sinaloa, he asked me what it was that was occupying [my time] at that moment. And [since] I responded to him that I did not have anything in particular claiming my studies he insisted that upon reading his book I would not lack anything to write about or fill my time."[29] Not only did Sigüenza find in this farewell gesture a confirmation of "the truth of which they had warned me"—that is, Kino's intentions to publish a tract against him—but even further, a formal invitation to enter into what he calls a *duelo literario* (literary duel).[30] The sentiments with which Sigüenza came to the dispute may be seen in this strangely contradictory term: while the term "duel" evokes a context of violence after the breakdown of social bonds built on a code of honor, the epithet "literary" returns the dispute to a verbal, rather than physical, contest. Implicitly, it suggests the use of reason to combat an aggression made on social standing or honor. As Sigüenza says, "in the literary field it has always been licit to correct one another when necessary."[31] In fact, as Sigüenza notes, the Jesuits were particularly fond of this type of dispute.[32]

Rather than describing a formal scholastic debate, however, it is clear that Sigüenza envisions the literary duel as an antidote to the underhanded and private nature of Kino's attack. For this reason, he distinguishes the "I don't know whether to call it mockery or derision with which [Kino] speaks of me when referring to my opinions" from dissent made in the context of a friendship. Although Kino's attack has not been made against Sigüenza's social standing or virility, but rather against his reputation as a mathematician and astronomer, Sigüenza defends his equation of social and physical violence by stating that "I value my fame as [if it were] my life"[33] and claims that his own counterargument has been made merely in self-defense. As the epigraph to the *Libra* reads, "If I am forced to write in my defense, the blame lies with you who provoked me rather than with me who was obliged to respond." Finally, establishing the dispute as a "literary duel" brings Kino's provocative comments, made in private, into a public arena. This issue of publicity is for Sigüenza one of the most important aspects of the dispute with Kino. In fact, part of Sigüenza's anger has to do with the indirect form in which Kino attacked him, by association likened to that of the "concealed mathematician" who waited until Kino published the *Exposición astronómica* to publish his own argument against the *Manifiesto*. Publicity is the only way to establish truth and reason and return the duel to its proper "literary" field. As Sebastián Guzmán y Córdova writes in his own prologue to the 1690 publication of the *Libra*, "if what is said in it appears wrong and inaccurate, do not respond with anonymous sonnets nor place objections where they cannot be answered but rather publish these in print so that we can all hear them."[34]

A "literary duel" conducted in print, and therefore in public, would be the only means to counteract the double-edged sword of an attack declared in the privacy of a personal conversation. It would also redress an implicit advantage afforded to Kino's treatise. In 1680, Sigüenza had just been awarded the title of royal cosmographer, a title that Kino's argument implicitly questioned.[35] As Sigüenza writes, "moreover, finding myself in my homeland with such credits as my studies have gained for me with a salary from the king our lord, as his professor of Mathematics in the Mexican University, I do not wish that at some future time it appear that the reverend father arrived here from his Bavarian province to set me straight."[36] Even though it did not mention Sigüenza by name, the mere fact that the *Exposición astronómica* had been written by a European mathematician, had directly contradicted Sigüenza's own, and had been dedicated, as Sigüenza ironically notes, to the Condesa de Paredes when Sigüenza had dedicated his *Manifiesto* to her husband the viceroy Conde de Paredes, automatically tarnished Sigüenza's local reputation. In fact, of the two hidden attacks against his *Manifiesto*, Sigüenza had been told by friends not to worry excessively about that of the anonymous mathematician. Of Kino's, however, they "urged me to be on alert . . . if only because of the prejudice that all shared, thinking that only because he had recently arrived from Germany to this New Spain that the reverend father must be a most accomplished mathematician."[37] Citing Salviano, Sigüenza comments on this prejudicial value attached to Kino's origin, which he himself had already admitted to having shared: "So superficial and almost without value are the judgments of many in this time that those who read do not consider so much what they read as who it is that they read; nor do they think about the force and value of what is said as much as the stature of he who professes it."[38]

The prejudicial advantage awarded to Kino's publication also raised the dispute above a question of a personal quarrel between scientists to one with implications for all of Mexican scholarship, and Sigüenza reserves some of his most scathing remarks for what he sees as Kino's suggestion that Mexico could not produce correct science. Throughout his treatise, Sigüenza defends Mexican learning, comparing his university favorably to Ingolstadt where Kino had studied ("I am not referring [to my understanding] as German and cultivated in the very celebrated University of Ingolstadt but rather as American and roughly hewn in the still little-celebrated [University] of my Mexican homeland"[39]) and equating his own work with proof that there are "mathematicians outside of Germany, even stuck among the rushes and reeds of the Mexican lake."[40] Referring to Kino's dedication of the *Exposición astronómica* to the Condesa de Paredes, he argues that it appears that the Jesuit "had come from Germany to this septentrional America to free her excellence from the deceit and error in which I had placed her [when I said that] comets should not be feared, as it was false that they foretell calamities and destruction."[41] Finally, in one of his most impassioned outbursts, he accuses Kino of associating Creoles with the presumed irrationality of the indigenous inhabitants of the Americas:

May the very reverend and religious father live a thousand years for the high esteem in which he held us Americans when he wrote these words! In some parts of Europe, and most especially in the North which is more remote, they think that not only the Indians, original inhabitants of these countries, but also those of us who happen to have been born here of Spanish parents, either walk on two legs by divine dispensation or that even by making use of English microscopes, one barely finds reason in us. Accordingly, it is clear that the reverend father was educated in the one of the furthest of all and that neither his stay for months on end at our court nor his conversations with those who were born and reside in it have led him to set aside his ideas on these topics and, judging from what he writes, he feels that we cannot read and that as a consequence we are incapable of understanding what words plainly say.[42]

Sigüenza thus declares that his silence would discredit "not only myself but also my homeland and nation," leading others to believe that "the Spanish had as a public professor of Mathematics in the University of Mexico a madman who held an opinion that no one else shared."[43] One of the most important aspects of a literary duel, then, will be its ability to establish "reason" as the arbiter of the dispute, rather than prejudice according to geographical origin. As Sigüenza says, "there is no other way to free ourselves from that doubt than to place it on the scales of reason."[44]

Although, as Víctor Navarro has written, Sigüenza and Kino shared many sources and coincided for the most part in their assessment of the physical properties of comets, their arguments differ greatly in the value they assign to reason.[45] Peter Dear has shown that, over the course of the seventeenth century, the practice of applied mathematics among Jesuits had created a subterranean version of the mechanistic natural philosophies of Descartes, Gassendi, and Pascal, along with a method and language to establish certainty and truth through "reason" rather than authority.[46] Despite its suspect categorization as an Aristotelian discipline, by the end of the century the hybrid term "physicomathematics" was being used to describe the field by such eminent scientists as Kircher and Riccioli.[47] Sigüenza was well versed in this literature and peppered his writing with terms culled from Habsburg and Jesuit writings on mathematics.[48] Even in his popular treatise, Sigüenza advocates mathematical reason as the arbiter of physical truth: "until now no one has been able to know with physical or mathematical certainty from what and where comets are engendered, much less been able to predict them."[49] In Sigüenza's fideistic argument, reason is the limited province of human beings and, even while it may establish certain truths, cannot hope to "inquire into God's motives."[50] Although he cites many of the same authors, Kino's argument is quite distinct. Overall, he emphasizes the providential and unexpected nature of comets in order to support his interpretation that they are divine signs or causes. There is perhaps no greater measure of the distance between the two arguments than their respective positions on "monsters." Whereas Kino echoes classical and modern literature in defining comets as monsters outside the natural order, Sigüenza sarcastically declares that monsters signify nothing beyond themselves: "because if the idea is laughable that a monster, even one born in the public square, foretells the end of reigns and the death of princes and changes in religion, how would it not also be that a comet signifies these things when the origin of the former and the latter can be attributed to one reason?"[51]

Rather than a defense of the moral or metaphysical grounds of scientific method, Sigüenza's stance responds to an immediate context of social prejudice and authority in which reason promises to level an uneven geopolitical playing field. This defense of reason does not preclude his own practice of citing authorities, one of the strongest methodological foundations of Neoscholastic learning. Although he bases his argument on reason and logic, Sigüenza continues to refer to authorities' opinions, complaining when Kino prefers Seneca over Aristotle in an argument on the nature of comets and questioning Kino's interpretation of the Hebrew term *mazaroth* as an indication that the Bible had mentioned comets as divine signs.[52] He reserves the possibility that authorities may be overturned, however, by reasoned argument and knowledge gained through the use of modern scientific instruments, such as the telescope. In an oft-cited passage, he attacks Kino's reliance on classical authorities: "Given this, I ask: would it be prudent (surely it would be imprudent) to affirm in these times that the heavens are solid and incorruptible only because the most ancient authors affirm this? That the Moon is eclipsed by the Earth's shadow, that all comets are sublunary because they also teach this?"[53] One of the more difficult sources for Sigüenza was Athanasius Kircher, a respect for whom he shared with Kino. While Kino cites a passage from Kircher's *Iter extaticum* (Ecstatic Journey) to support his argument that comets were signs of imminent catastrophe, Sigüenza counters with another passage in which Kircher states the opposite.[54] The challenge that Kircher posed to Sigüenza's argument is an example of the bind that he faced in promoting reason over authority. Even while he did not share the more mystical aspects of Kircher's Neoplatonic conclusions, Sigüenza himself was heavily indebted to the Jesuit scholar's post-Aristotelian version of natural philosophy. Furthermore, Kircher was a central figure in the Jesuit and Habsburg scientific community to which Sigüenza aspired to belong.[55]

Sigüenza's contortions to defend Kircher are an indication of one of the most delicate aspects of his position in the dispute. Above all, Sigüenza was very concerned with the insinuation that his defense of reason could be taken as an attack on the Jesuit order itself, from which he had been expelled after completing his initial studies. By 1680, Sigüenza's relationship to the Jesuits had moved from one of obligation to affect, a bond that ironically he perhaps felt more strongly than that of obedience, upon which the order was founded.[56] Throughout the *Libra*, Sigüenza emphasizes that the dispute takes place outside of any obligations that he has to the Jesuit order and that he was arguing against Kino "not as part of such a venerable whole but as a particular subject and mathematician."[57] In an even stronger attack, he aligns reasoned argument with the Jesuits, thus further dividing Kino from the order:

With all certainty I was convinced that the most reverend and learned fathers of the Company of Jesus, who are such patrons of the truth, would not take this controversy badly, since it takes place between two individuals and two mathematicians without going any farther than that; and all the more so because the duels that occasionally occur between those who pore over books are not only common but indeed licit and even necessary since one is assisted in them only by reason and since they almost always contribute many truths to the literary republic.[58]

The crux of Sigüenza's self-defense has to do with the nature of Kino's attack. Unlike reasoned dissent, which preserved the social bonds of friendship and an affective relationship to the Jesuit order, Kino's attack had been underhanded, breaking with the implicit rules of honor structuring formal disputes. The violence of an attack on the social bonds symbolized by honor justified Sigüenza's "self-defense" and trumped any obligations owed to the order.[59] Even more interesting, however, is Sigüenza's argument that while the Jesuit order should establish certainty, this could be done only by reason and public argument. Sigüenza's argument trod a thin line: implicitly breaking with the obligations of an authority internal to the Jesuit order, to which he still professed a debt, while at the same time underscoring the Jesuit approval of reasoned argument in order to show that he, and not Kino, was more in line with the order's practice.[60]

Overall, the differences between Sigüenza's and Kino's positions might best be understood according to their views on the relationship between natural events and social upheaval. Whereas Sigüenza admits that various calamitous events had coincided with the 1680 comet, from the religious wars in Europe to the recent uprising against Spanish governance in New Mexico, he denies that the comet may be linked to these either as cause or sign.[61] Implicitly, he thus negates both a metaphysical meaning and a salvational solution to social events.[62] Kino's position that comets provoke calamities and even predict an imminent final judgment is consistent with the zeal and salvational ethos of an Austrian who dreamed from an early age of becoming a missionary in distant lands.[63] The respective frontispieces of their treatises indicate the distinct discursive and practical contexts in which each was writing. While Kino adorns both his frontispiece and the map of the celestial sky with the Virgin of Guadalupe, to whom his patroness the Duchess of Aveiro was particularly devoted, Sigüenza decorates his with his personal device, the Pegasus, whose banner reads *sic itur ad astra*.[64] Although it is certainly clear that Sigüenza was irritated by Kino's dedication of his treatise to the Condesa de Parades, the difference between their positions should not be reduced to a question of patronage politics.[65] Nor should Sigüenza be understood as defending a position of scientific reason for the sake of modernity and progress. Rather than an absolute and programmatic stance, the defense of scientific reason in the *Libra* is pragmatic and circumstantial, reflecting the difficulties faced by a colonial scientist wishing to penetrate the cosmopolitan community of Habsburg science, difficulties which, given the particular circumstances of his life, Sigüenza perhaps felt more acutely than any other scholar in the Iberian colonies.

SCIENTIFIC REASON AND THE
POLITICS OF FRIENDSHIP

Although he defends reason as an arbiter of his dispute with Kino, some of the most insistent terms in Sigüenza's *Libra* are those related to "friendship." Surely, this has to do with the particularly fraught relationship between friendship and reason.[66] As Sigüenza notes several times, reasoned dissent should be permitted in relations of obligation and affect, as for instance among members of the Jesuit order or between friends. The relationship with Kino, however, is another circumstance. Instead of a friendship outside and apart from reason, Sigüenza desires a relationship with Kino *based* on reason. In this "intellectual friendship," forged through reciprocity and hospitality, affect follows from reason rather than being suspended for reason's sake. The expectation of reciprocity outside of calculation, the ethics of hospitality, is what Kino betrays when he fails to recognize Sigüenza's reason.[67] The paradox underlying the debate between Sigüenza and Kino, which explains the tone and extent of the former's defense, has to do not only with the contradiction between the pretensions of reason as a power independent from other forms of authority and the dependence of the scientific field on peer recognition, but also with the inherent tensions between calculation and virtue in early modern ideas of friendship.[68] This appears to have been the geopolitical problem created for Sigüenza by his own advocacy of reason: whereas reason could counter the prejudicial authority of geographical origin, it still depended upon recognition by other mathematicians such as Kino for its social authority.

Sigüenza's expectation that an intellectual friendship would overcome the geopolitics of early modern science indicates the logic behind what he appeared to have considered most insulting in Kino's argument. In his prologue, Sigüenza presents the *Exposición astronómica* as a litany of insults against the position he had taken in the *Manifiesto*:

That I come armed with the authority and the foresight of the prophet Jeremiah; that I anchor myself to a foreign opinion; that I am endeared of comets as if enamored of their starry bleariness; that the opinion contrary to mine is universally followed by mortals both high and low, noble and plebeian, learned and unlearned, from which it might be inferred that in the judgment of the most religious father that I must be nothing at all since I would not be mortal, whether high, low, noble, plebeian, learned, or unlearned, but rather the entity of reason disputed by the metaphysicians. But none of this is so worthy of censure as the fact that after having referred in his *Exposición astronómica* to the imagined fatalities caused by some comets, he ends his opinion with the following exact words: "I close this proof, truly an idle one (if not for some belabored minds), of this opinion which is not so much mine as it is that of all." Those who understand the Castilian language know well that to say to someone that he has a belabored mind is to judge him mad; and if this is true, which it surely is, then may the reverend father live many years for the very singular praise with which he honors me![69]

Of all these insults, two appear to have been the ones that Sigüenza most deeply resented. First was the argument that "no one" held his opinion ("that the opinion contrary to mine is universally followed"), which, as he says, left him without a social identity ("from which it might be inferred that in the judgment of the most religious father I must be nothing at all") or at most reduced him to an alienated or even disembodied thought ("a foreign thought" or "the entity of reason disputed by the metaphysicians"). Second was the complementary claim that those who thought as he did betrayed a "belabored mind," since that, as Sigüenza argues, is tantamount to calling someone irrational. According to Sigüenza's gloss, then, Kino had reduced him to two complementary and untenable positions: to reason without a social persona or to be a person without reason, irrational and mad.

Together these insults exclude the qualification that Sigüenza apparently expected to receive from his European peer: that of a subject possessing a correct opinion based on scientific reason. As Irving Leonard comments, it was perhaps natural that a scientist practicing in the loneliness of the colonial periphery should have hoped to find in Kino a kindred spirit.[70] But Leonard's explanation does not account for the sense of betrayal with which Sigüenza responds to Kino's insults. As Sigüenza's citation goes on to indicate, this was closely related to expectations established by his acts of hospitality toward the Austrian missionary:

But I ask, in what did he experience my madness? In the words that I spoke to him? In those I communicated respect and submission. In writings of mine which he read? All have been published with the approval of very learned men. In my actions? He never saw me, nor will he ever see me (God willing), throwing stones. In the way that I praised him? In the festivities with which I received him? That just might be; that just might be.[71]

To understand the affect with which Sigüenza treats Kino's lack of correspondence, his interest in a friendship based on scientific reason must be analyzed in relation to the larger context of his project. Expelled from the Jesuit order, Sigüenza was in a particularly precarious social position. Like his contemporary, Sor Juana Inés de la Cruz, Sigüenza depended on his work and reputation to establish himself within New Spanish society, in which corporate ties defined social relations.[72] In the absence of corporate ties based on obligation, Sigüenza appears to have emphasized ones of sentiment and affect, as he continually associated his projects with the "love that one owes his *patria*"[73] and emphasized his ties of friendship to other scholars. He was, literally, a "particular mathematical subject" who relied on recognition not only for the calculation of patronage but also for affective ends, as a substitute for his tenuous corporate ties. The two insults that he underscores as most biting indicate what lack of recognition from a scientist such as Kino could mean for Sigüenza. Not only was he cut off from a social identity, but worse, he was in danger of association with an irrationality attributed to his origin: the imputed lack of reason of indigenous subjects amid the mud and reeds of the lake upon which the university was built.

Accordingly, the cosmopolitanism that Sigüenza envisions in his debate with Kino is an attempt not only to compensate for geographical limitations of a scientific periphery, but indeed to overcome a colonial context deemed "irrational" and in need of governance by reason. Unlike the humanist recognition that founds the early modern Republic of Letters, scientific cosmopolitanism was not based on a preestablished code of learning and etiquette.[74] Rather than a set content, as cultural capital, reason promised to be self-regulating and productive of a community of peers, as Bourdieu argues. The need for early modern scientists to work within institutions such as the court or to create their communities based on the ideals of reasoned debate has been studied for the case of Europe.[75] In the context of colonial Mexico, however, these models do not suffice. Although Sigüenza dreams of an intellectual community of peers, it is one that extends beyond the regional context of the viceregal court, which by 1680 he had sufficiently penetrated. His list of "friends" includes both local and metropolitan figures, and it is precisely this supraregional aspect that he appears to have relished. His claims to have corresponded with European mathematicians such as Juan Caramuel and José de Zaragoza are an attempt to link his work to the greater context of Habsburg science and should be distinguished from those of his contemporary Alejandro Favián, whose correspondence with Kircher was made in the guise of a disciple.[76] The community of which Sigüenza dreamed was based not on "gifts" made in exchange for recognition by a scientific elite but rather on reciprocity established through the mutual exchange of scientific information.[77]

Throughout his life, Sigüenza was able to establish these types of cosmopolitan friendships. In the *Libra*, he describes an encounter with Peter Van Hamme, a Jesuit missionary who arrived some years after Kino and whose willingness to exchange astronomical observations on the 1680 comet presents a sharp contrast to Kino's reluctance:

The 1687 fleet brought to these parts the Reverend Father Pedro Van Hamme, of the Company of Jesus and of the Flemish nation, a subject truly worthy of esteem for his affable manner, his courtly discretion and solid religion; as someone with a fondness for mathematics, he visited me and I found that he knew the subject exactly as should one who teaches it, which is to say perfectly and without affectation. In the very few days that he stayed in Mexico he shared with me some of the observations of this comet that had been made in Europe and, among those compiled by Juan Domingo Cassini, mathematician of the King of France, I found one by a Monsieur Picard made on the same day of January 18th.[78]

Sigüenza would later make the acquaintance of the Neapolitan traveler Giovanni Gemelli Careri, with whom he shared his collection of codices and other antiquities and to whom he presented a copy of the then-printed *Libra astronómica*.[79] These exchanges appear to have reflected what Sigüenza envisioned by the term "friendship." This should be distinguished from the early modern ideal of friendship espoused by writers from Aristotle to Montaigne, in which the friend would be a virtuous mirror or "twin soul" for the self. While the classical ideal of virtue in friendship emphatically rejected the possibility of calculation or reason, and therefore of reciprocity or exchange, Sigüenza's version of friendship is closer to what Jacques Derrida refers to as the "politics of friendship," a fraternization or naturalization of social bonds bordering on notions of citizenship. Rather than twin souls, Sigüenza envisions a friendship in cosmopolitan guise, combining calculation, or the preservation of particular interests, and virtue in an inherently unstable symbiosis.[80]

Scientific reason, and in particular mathematics and astronomy, served as an ideal foundation for this type of cosmopolitan friendship. As Steven Harris notes, astronomy depended on precision and multiple observations, demands that made it naturally suited to the global network of the Jesuit order.[81] Sigüenza's library included some of the most complete compendia of astronomical observations from Europe, including those of Oldenberg and Hevelius.[82] Sigüenza could hope to be included in this effort as a representative not of an abstractly universal mathematics but rather from the particular perspective gained from his geographical position in Mexico. This appears to have been the ethos behind his offer of observations: contributions to a universal and therefore global science that depended not only on the exotic particulars of natural history but also on unique geographical perspectives.[83] As Sebastián Guzmán y Córdova notes in his introduction to *Libra*, there were observations possible from the Americas that could not be made from Europe.[84] The reciprocity that Sigüenza imagines, therefore, is based on a scientific need for empirical data rather than an abstracted scientific reason. Nonetheless, it is the precision and accuracy of his observations and his general correspondence with European science that gives value to these observations. Several times in the *Libra,* Sigüenza offers to exchange his astronomical observations with those of his scientific peers, ending his treatise on this note: "If, in order to certify them or for other purposes, any mathematician would like to send me either his or another's observations of eclipses, especially lunar from [the year] 1670 on, I will reciprocate with complete liberality by sending him mine from the same period."[85]

This combination of a desire to participate in a universal order and a defense of particularity is what takes scientific reason in the debate between Sigüenza and Kino into the realm of an ideal cosmopolitanism. It also indicates a historical shift in New Spain in which governance was increasingly aligned with scientific reason. With the decline of early missionary humanism and an indigenous elite, urban indigenous and popular culture was increasingly defined as irrational and in need of governance by reason.[86] Yet this scientific "governmentality" is not the only value reason had for Sigüenza.[87] His vision of an ideal scientific community of equals also defined the contours of both a regional and cosmopolitan citizenship. It is clear that Sigüenza did not distinguish between these two projects and that the affect with which he treats his *patria* is no different than the respect he shows for mathematicians from abroad. Indeed, it is precisely the cosmopolitan aspect of science that excuses Sigüenza's vision from competition with the viceregal court. Both outside of viceregal governance and superior to it, the "republic of letters" he has in mind will form the guiding body to address natural and social crises in colonial governance.[88]

In Sigüenza's understanding, the community formed through shared scientific reason will be beyond that of local or regional politics and will be defined by a specific virtue of honor and reciprocity, not for personal gain but for the good of a shared and universal enterprise. It is this specific blend of science and politics that defines Sigüenza's expectations for his interaction with Kino. It is perhaps ironic that he should have looked to the Austrian Jesuit for such a fraternity, given the very distinct framework with which Kino had arrived in the Americas to missionize on the northern frontier, at that time the epitome of a colonialist notion of barbarous irrationality.[89] Or perhaps it is an intuition of this very element in the Austrian Jesuit that led Sigüenza to defend so fiercely a Creole cosmopolitan reason. A sense of futility also seems to have afflicted the Creole scientist as he waited for years to publish the *Libra astronómica* in which he vindicated his argument from Kino's attack. Although the delay in publication has been interpreted as a reflection of his fear of the Jesuit order, a speculation not without merit, it is perhaps also possible that Sigüenza sensed the inability of an invective to restore a friendship idealized in terms of hospitality and respect for a shared scientific project.

Medical *Mestizaje* and the Politics of Pregnancy in Colonial Guatemala, 1660–1730

MARTHA FEW

Pregnancy and childbirth were significant events in the lives of families and communities in colonial Latin America. To have a notably difficult pregnancy at a critical moment—one that resulted in a deformed birth for example, or that led to a postpartum illness in the mother—would be seen as especially significant.[1] This essay considers a series of community conflicts over difficult pregnancies between 1660 and 1730 in Santiago de Guatemala, capital city of colonial Central America, and the primarily indigenous surrounding towns.[2] In each instance, a difficult pregnancy resulted in a miscarriage, stillbirth, or deformed birth or led to long-term illness, insanity, or the death of the mother. Family members, husbands, neighbors, employers, community elders, priests, and the pregnant women themselves contested the causes and meanings of problem pregnancies and their outcomes in gendered and racial terms. Participants made claims that a female midwife/ritual specialist, usually an indigenous or *casta* (mixed-race) woman described in the sources as "witch" and "sorcerer," cast an illness in order to intervene in a pregnancy and subvert it.

Debates surrounding difficult pregnancies and their interpretations form a central but often overlooked part of the larger history of reproduction and women's health in colonial Latin America.[3] This is in spite of the fact that from the mid-seventeenth century on, key institutions of

The History Department at the University of Arizona provided funding to help support this research.

the colonial state paid increasingly explicit attention to pregnancy, child-birth, and early infant health from a political, religious, and medical per-spective.[4] Difficult pregnancies, especially in the extreme versions found in Inquisition proceedings, criminal records, and contemporary histories considered here, reveal the politics of pregnancy: cultural conflicts and community tensions surrounding problem pregnancies and competing interpretations for their occurrences.

Through evidence containing descriptions of problem pregnancies and their interpretations, I develop the idea of medical *mestizaje* that took place in the critical area of illness and healing, conceptualized as a con-flictive process that unfolded as new equilibriums of changing definitions of health and healing were continually reworked over three hundred–plus years of Spanish colonial rule in Guatemala.[5] This is in contrast to previous research on medical practice in colonial New Spain that has tended to see this process as one of indigenous acculturation to European norms, or as a top-down process of colonialism via the professional-ization of medicine.[6] Viewed through the lens of problem pregnancies, Mesoamerican, European, and New World interpretive medical frame-works at times appeared to borrow from each other and other times challenged each other for legitimacy, reflecting the same *mestizaje* pro-cess that occurred at other levels of culture.[7] These frameworks informed and interacted with official medical, scientific, religious, and political conceptions of health and illness and shaped the practical interpretation of colonial policies.

Examples from the often ambiguous and culturally charged borders of human reproduction reveal the outlines of the process of medical *mestizaje* under Spanish colonialism. This is not to say, however, that it created a single, unified "colonial" conception of health, though some common themes of the process of medical *mestizaje* are explored below. Instead, the various cultural traditions interacted with developing local ideologies and practices, producing new cultural and medical meanings for health and illness that varied over time and place, generated in part by community-level conflicts such as those considered here.

The chapter's first section explores descriptions of problem pregnan-cies and competing diagnoses for their causes, revealing the continued resonance of Mesoamerican conceptions of illness as they intertwined with European healing cultures, though always in power-laden ways that reflected colonial social hierarchies. The sick, community members, and healers utilized diagnostic illness categories described as "from nature" or "not the word of God," each with their own sets of physical symptoms and material evidence that necessitated specific healing strategies.[8] The second section probes the religious overtones of community conflicts over difficult pregnancies, and the role of men in mediating these conflicts. In colonial Latin American historiography, care for women during pregnancy and childbirth practices is depicted as female labor, and the birthing room as a gendered female space. Men are primarily depicted as not involved in pregnancy and childbirth until the gradual shift to the professionalization of medicine at the end of the eighteenth century.[9] These examples, however, reveal that in addition to women, men played gendered roles in diagnosing and mediating problem pregnancies and took advantage of or promoted the stereotyping and scapegoating of indigenous and *casta*, female healer-midwives as having the power to subvert pregnancies. As participants sought explanations and resolutions for especially graphic problem pregnancies by appealing to church officials, employers, indigenous community leaders, or Spanish colonial authorities, men often played central roles in ideologically policing the boundaries of reproduction. Competing representations of difficult pregnancies thus reveal the intersection of religious, medical, and community conflicts over health and illness under Spanish colonialism, elaborated in gendered and racial terms through a process of medical *mestizaje* in Guatemala.

"NOT THE WORD OF GOD":
DIAGNOSING DIFFICULT PREGNANCIES

While pregnancy and other women's health issues were not a specific focus of early colonial medical efforts, the Spanish colonial state saw health and healing in general as a key pillar in the establishment of colonial rule and its institutionalization in the New World.[10] Even though the colonial state did not formally address issues related to the health of women during pregnancy and childbirth in this early period, colonial-era chroniclers did make claims that women in the New World in general, and indigenous women in particular, had special knowledge about controlling reproduction. This included the use of plant-based abortifacients. Spanish historian Gonzalo Fernández de Oviedo described indigenous women using abortifacients in his *Historia general y natural de las Indias* (General and Natural History of the Indies), suggesting that from the point of contact reproduction and pregnancy were socially charged issues for the Spanish: "There are others so friendly with lust that if they become pregnant they take a certain herb, that later stirs up and casts out the pregnancy [from a woman's body]."[11] In this quote, Oviedo tied the knowledge of abortifacients with representations of some indigenous women as particularly full of "lust," sexually overactive and sexually available. Adding to this, Oviedo depicted indigenous women as having special knowledge of herbs and plants that allowed them to take control of their pregnancies.

Bartolomé de Las Casas similarly commented on the practice of abortion and the use of abortifacients in his *Historia de las Indias* (The History of the Indies), noting that "They hung [any] doctor or female sorcerer who gave potions to expel infants from the womb, and they did the same to pregnant women who took something to achieve the same end."[12] Here, the account of Las Casas shows the development of stereotypes of women who administered abortifacients as *hechiceras*, or female sorcerers, as part of European expansion into the New World. In contrast, the "doctors" here (implied male and presumably Spanish) receive no such association.[13] The writings of Las Casas also show the development during the conquest period of stereotypes of Indian women as especially fertile from an early age, with the ability to give birth easily: "Because of the nature [of] some secret remedy, or the work that all the women there commonly perform together, in as much as it is moderate, or because their foods are not delicate, that even though they are very young they give birth without danger [of dying], and almost without pain."[14] Interestingly, Las Casas provides cultural explanations for making this assertion—that indigenous communal work patterns and diet led to the ease of birth and early fertility of indigenous women in the New World.

In colonial Guatemala, *curanderas* (female healers) and *parteras* (midwives) played important social roles in community life, and for much of the colonial period acted as ritual and medical specialists who ministered to pregnant and postpartum women and their infants. Men and women consulted healers and midwives as one health care option within a range of strategies for determining the causes and cures of various illnesses in daily life, and in caring for pregnant women through pregnancy and childbirth.[15] Sick pregnant women and their families initially turned to midwives for assistance during a difficult pregnancy. Before the appropriate cure could be enacted however, the illness had to be diagnosed properly. This could take time, as the pregnant woman, her family members and neighbors, her midwife, and others observed the illness symptoms, discussed its details, and weighed in on the causes—which in essence reveals a process of diagnosis via community consensus.[16]

Conflicts over problem pregnancies considered here provide evidence for only one category of supernatural illness: diagnoses that the participants depicted as caused by a malevolent being, someone drawing on malevolent powers, or through an association with someone who did.[17] Men and women who asserted supernatural origins often used the phrase "it is not the word of God" to distinguish this type of illness from others. It was diagnostically important for the sick and their family members to make this distinction as early as possible during the illness event, as supernatural illnesses had different causal frameworks and curing options. This distinction had roots in Mesoamerican healing cultures.[18]

Illnesses diagnosed as supernatural in origin or "not the word of God" fit within both Mesoamerican and Catholic evangelical frameworks that saw the operation of good and evil in the world in part through illness, and that illness could be cast in others by knowledgeable individuals and ritual specialists. In 1660 in the rural indigenous town of Santa Catarina Pinula, located to the east of the capital city of Santiago de Guatemala, two Indian husbands came forward to accuse the married Indian midwife, Marta de la Figueroa, of casting illnesses to interfere in the pregnancies of their wives.[19] Andrés García complained "in the name of his wife," twenty-four-year-old Catalina Gómez, that Figueroa caused her to become ill during her pregnancy.[20] García pointed to his wife's refusal to engage the midwife as marking the beginning of the illness. From that point on, the pregnant Gómez "fell sick." Her abdomen and belly swelled. She expelled *gusanos* (worms) from her "lower parts." And every night, Gómez felt or dreamed (both meanings are implied here) as if she were suffocating. Juan Maldonado, an Indian scribe who witnessed García's accusation of Figueroa to Pinula's political officials, confirmed these details, including the swollen belly and expulsion of worms.[21] He added that while the pregnant Goméz slept, in her dreams she saw that Figueroa wanted to strangle her. The illness lasted for all nine months of the pregnancy. No mention was made in the documents of what happened to the mother or infant.

Andrés Yos, described as Indian, also came forward at the same time and told Pinula officials that Figueroa had also cast an illness on his pregnant wife, Juana Candelera. The illness led to her death and presumably to that of her unborn child.[22] Yos charged that Figueroa used an *hechizo* or "spell" to cast the illness.[23] He also depicted his wife's illness as incurable except by Figueroa herself, the illness caster. Yos brought Figueroa to his wife's bedside as she lay dying, in the hopes that the midwife would cure her. As the midwife approached her sickbed, Candelera reportedly said to Figueroa "for the love of God, take this evil that is in my body, because this is not the word of God, this illness that is killing me."[24] Illnesses in colonial Guatemala, even those involving pregnant women, were public events, as family members, neighbors, and other people important to the family gathered in the sick room. Here, Yos backed up his assertion that his wife diagnosed her illness as "not the word of God" by noting that others were present at the time, including members of the *cofradías* (religious sodalities) of Santísimo Sacramento, Santa Cruz, and San Nicolás, presumably all respected men in Pinula society.

Juan Maldonado, the Indian scribe, concurred that "it is said" that Figueroa cast an illness with a "spell" that killed Juana Candelera. Maldonado pointed to a conflict between the two women as the point where the illness began. Figueroa cast the illness, according to Maldonado, when she formed dirt balls mixed with animal fat and hurled the dirt balls at the pregnant woman, saying certain ritual words. Andrés Maeda, an Indian *regidor* (alderman), also described hearing Figueroa confess to using the bewitched dirt balls to cast the illness in the pregnant woman, empowering them to cast illness not by ritual words, but instead by passing them "between her legs by way of her lower parts."[25]

In these representations of problem pregnancies, the two husbands García and Yos carefully related the details of their wives' dramatic illnesses that led them to diagnose a supernatural illness that interfered with their pregnancies and, in Candelera's case, caused her death. Both García and Yos attributed the start of the illnesses to a conflict between the pregnant women and the midwife, implying that before the illnesses their wives had been healthy. In García's case, he described an illness that caused, among other physical symptoms, a dramatic expulsion of worms from his wife's body, especially from her genital area. Yos represented the illness as a supernatural one as well, recounting his wife's self-diagnosis that her illness was "not the word of God." Yos legitimized his wife's self-diagnosis through the use of witnesses, the *cofradía* members present at the bedside. To further support their arguments that Figueroa possessed the knowledge and ability to cast illnesses, both men noted that their wives could not be cured by other healers or midwives in the area.[26]

The men's depictions of supernatural illness in the context of their wives' pregnancies persuaded the indigenous male political officials and town elites that Figueroa had indeed caused the illnesses in the pregnant women. They arrested her, extracted a confession by hanging her from the ceiling by her ankles, rubbing spicy chili peppers on her body, and burning the spicy peppers to cause her pain and difficulty breathing. Authorities kept her imprisoned in Pinula's jail and did not allow her to communicate with anyone.[27] Clearly, Figueroa's reputed ability to cast illness to cause problem pregnancies was no minor matter, given the violent response of Pinula's officials. And, as those involved represented and diagnosed the problem pregnancy as supernatural in origin with the requisite dramatic symptoms, the male elites of Pinula intervened in the matter, supporting the diagnosis of supernatural illness made by the husbands Andrés García and Andrés Yos. Other community members, all men, came forward around the same time to confirm that Figueroa had cast a supernatural illness on them or one of their family members, giving further weight to the husbands' claims and further legitimizing the actions of Pinula's political officials.

Cristobal Silva, an Indian *principal* (elder; member of the indigenous hereditary nobility) asserted that Figueroa cast illnesses in both of his sons in the context of a personal conflict between them. The illnesses proved incurable, killing one son and crippling the other. A traveling Indian male healer from the town of San Juan Sacatepequéz unsuccessfully attempted to cure the crippled son, Matías, with an extractive procedure common in Maya medical practice. The healer cut into Matías's body using a *lanceta* (lancet with a small blade) to remove the offending material from his knees, feet, and throat, all medically key physical points in indigenous medical theory.[28] The father, Cristobal, witnessed the healer extract the items from his son's body: maize, wooden sticks, and wooden chips. The Indian healer, however, was unable to cure him. Cristobal's other son, Andrés, self-diagnosed his illness as supernatural in origin. According to his father, he said while dying "this is not the word of God, Marta has bewitched me."[29]

Matías Vicente claimed that five years earlier, Figueroa had cast an illness on him in the context of a *temescal* treatment for *frío* (a cold illness). A *temescal* is a ritual steam bath used in indigenous curing methods and in fortifying women during pregnancy, and is still in use today in Guatemala and Mexico. Vicente asserted that his illness only became worse after the *temescal* treatment, transmitted, he said, when Figueroa put her finger in his anus and he began to feel extreme burning sensations "that felt like chili pepper[s], that felt like ant[s]." Later, Figueroa returned to cure him by administering a healing drink made from a variety of plant materials including nettles and peach tree leaves. Vicente said that this treatment did not cure him, although it did cause him to expel "a multitude of worms" from his body.

The accused midwife Marta de la Figueroa, however, was not without connections of her own to assist her in countering these assertions. Her husband, Pedro Luis, himself a former alcalde (community official, member of the town coucil) in Pinula, together with the help of a scribe from Cazaguastlán, brought a petition before the *alcalde ordinario* (member of the cabildo exercising official judicial authority) of Santiago de Guatemala, Captain Zeledón de Santiago.[30] Luis complained that Pinula's officials "publicly tormented" his wife with burning chili peppers "until she was on the verge of death." He further asserted that the charges against her were "frivolous and without merit." Zeledón, however, eventually found Marta de la Figueroa guilty, and punished her with one hundred lashes given to her in the street in the pueblo of Pinula.

This example shows the interactions of indigenous medical diagnoses of supernatural illnesses with Spanish conceptions of sorcery. All of the above representations reveal common characteristics of supernatural illness diagnosis in men and women: an unexplained, often graphic or extreme illness in the context of a community conflict; the appearance of insects or worms in sores or expelled from the body; and the incurability of the illness by other medical specialists. The healing strategies undertaken by the participants to cure their illness reveal the continued operation of Mesoamerica medical cultures, along with a sophisticated array of indigenous healing therapies: those deployed via the Indian male healer lancing key parts of the body to extract offending materials from the body; the use of the *temescal*; and the role of indigenous female midwives in prenatal care in addition to the childbirth process and postpartum care of mother and infant. In this instance, indigenous medical cultures appear to operate fairly comfortably within the changing frameworks of illness diagnosis under Spanish colonialism, integrating into it the language of witchcraft and the association of women as witches or sorcerers.

Supernatural illnesses did not exist just in rural areas and indigenous towns but also in multiethnic urban centers. In the capital Santiago de Guatemala in 1693, Cecilia de Arriola, a married, twenty-nine-year-old *mulata blanca* (light-skinned mulata) described two problem pregnancies as she denounced a mulata woman named Gerónima de Varaona to the Inquisition.[31] Arriola described how a female acquaintance of hers suffered a miscarriage every six months.[32] She attributed the miscarriages to Varaona, who reportedly used certain rituals to cast illness that caused the woman to repeatedly "give birth to clumpy matter," coagulated material that resulted from miscarriage. The only places I have found extensive discussions of the biological material that resulted from miscarriages are in cesarean operation manuals that were written in Guatemala and circulated there starting in the 1780s. The *congelos* referred to in the testimony are similar to "a fleshy lump from the uterus," or a "portion of blood or other liquid that has congealed," which are described in these manuals in the context of diagnosing a miscarriage.[33]

To place the illness in the supernatural category, Arriola described the casting of the illness as *le [h]echo cosa*, that Varaona "did something to her [friend]." Arriola constructed this argument about the miscarriages and their cause within the context of an Inquisition proceeding, mediated by Guatemalan priests. To back up her assertion of Varaona's role in the miscarriages, Arriola connected their origin to an affair her friend had with Varaona's lover, described as "an old *mulato*." Here, in the context of a rumored love triangle, Varaona took revenge against her rival by supposedly preventing her pregnancies from coming to term.

Varaona also cast an illness on her mulato partner, "covering him with sores." Arriola then contextualized the miscarriages within ongoing community social relationships. Colonial-era illness descriptions were often contextualized within local social relations in which the intimate behavior of men and women and their link to illness were outlined in detail. Those thought to have the tendency and the knowledge to cast illness tended to be indigenous, black, or mixed-race healers, and also often women.

Arriola brought up a further description of Varaona casting illness to interfere in another woman's pregnancy. She made a vague reference to a "fight" that Varaona had with Josefa María (no surname given), a married mulata woman, after which Varaona reportedly "put something on the door so that she would die in childbirth." Arriola asserted that Josefa María did in fact die in childbirth, and her husband "lost his mind to this day." It is interesting to note that, in both illness descriptions, Varaona is accused of casting illness to both the pregnant women and to their sexual partners.

Not all difficult pregnancies with dramatic outcomes were judged by the participants to be the result of the intervention of a "sorcerer" or malevolent being. In the following account of the birth of conjoined twins, those involved attributed the deformity to natural causes, that is, from nature. In his monumental history of colonial Guatemala, Francisco Antonio Fuentes y Guzmán noted a difficult pregnancy that resulted in the stillbirth of conjoined twins and the death of the mother in childbirth.[34] He characterized the twins as a "natural monster," writing: "On the 12th of August 1675, a natural monster was born from an Indian woman from the pueblo of Santo Domingo Sinacoa . . . monstrous and admirable in the formation of its body, of a beautiful shape and perfectly human in its perfection and physionomical symmetry."[35] Fuentes y Guzmán noted that the mother in this case was also an indigenous woman. He provided no accusations of sorcery, instead labeling the twins' deformity as "natural," an object of keen interest and wonder rather than one of repugnance and/or fear.

Fuentes y Guzmán's attribution of the twins' deformity to the outcome of nature and natural causes, rather than to sorcery, was a key diagnosis that early modern Europeans used to explain the occurrence of deformed births and the ascription of natural origins that acted as a competing explanation to divine origins and sorcery.[36] Fuentes y Guzmán's characterization of the conjoined twins as a natural monster suggests that he was aware of multiple categories circulating between Europe and the colonial Americas for explaining the occurrence of deformed births considered "monstrous," and that he applied one or the other category based on the specific circumstances.

Fuentes y Guzmán's account focused his descriptive gaze on the twins, taking care to fully describe its exceptional physical nature:

From one single womb were born two distinctly perfect bodies, separate and apart for most of the trunks of their bodies, each one of them with two perfectly formed arms and hands, with pleasing and beautiful faces, with the same similarity and appearance overall, two well-proportioned legs, and at the waist another very short *pernezuela* (little leg), also accompanied like the other two with its corresponding foot and toes. They [the twins] do not show members[37] that indicated their true class and natural to their sex, because from that part of the body there has sprouted and born one of the three legs, that of the small one.[38]

In the context of this deformed birth, Fuentes y Guzmán insisted on their humanity, labeling the twins "perfectly human," even though they appeared, according to his description, to have three legs and no genitalia. While stillborn children were not technically supposed to be baptized, the parish priest of Santo Domingo baptized the twins after they died in childbirth. The twins' body and their mother's were eventually all buried in consecrated ground in the parish church of Santo Domingo.

While Fuentes y Guzmán's account of the "natural monster" provides information that indicated he was aware of early modern European conceptions of monstrous births, he was also aware of Mesoamerican ideas regarding deformity and monstrous births that continued to circulate in postconquest Guatemala, and he included these in his account. For example, Fuentes y Guzmán ended his detailed description of the twins' physical body with the following comparison of the language associated with conjoined twins and monsters in both Mesoamerican and European cultural traditions: "This has to be what the Indians call *chachaguates* and what we call *gemelos*. And they were born united as I have described, that which they [the Indians] call *nannosos* and what we [the Spanish] call *monstruos*."[39] *Gemelo* is the standard Spanish word for twin. Fuentes y Guzmán also provided the word commonly used by the local indigenous peoples, *chachaguates*, a Hispanicized version of a Nahuatl word, *chachahuatl*, which means binding or stuck together, indicating conjoined twins.[40] The indigenous word for conjoined twins (translated as "stuck together" or "binding") was even more specific than the word *gemelos* in Spanish (twins with no indication of being physically bound together). *Monstruo* in Spanish means "monster," defined in the seventeenth century as "a newborn infant or production against the regular order of nature." Possibly, the word *nannoso* is related to the Spanish word *enano*, which means "dwarf."[41]

COLONIALISM AND GENDERED AUTHORITY
IN THE CONTEXT OF PROBLEM PREGNANCIES

As we have seen, illness experiences enmeshed and interconnected with local social relations, including ongoing personal conflicts and romantic relationships. While pregnancy and childbirth in colonial Latin America have been depicted as a process primarily involving female specialists, especially before the professionalization of medicine, evidence from problem pregnancies indicates that men became intimately involved as well. Their central roles in both can be seen in the above examples, where men claimed authority as political officials, priests, husbands, and village elders to interpret, explain, or diagnose problem pregnancies, as well as lend their support to one diagnosis or another.

In the instance of the conjoined twins, men also attempted to control, mediate, and explain the outcome, not only through diagnosing the origin of the deformity but also through physical control of the "monster" itself. After the birth, Santo Domingo's parish priest removed the twins' corpse from the family and circulated it among elite social circles and the medical community in the capital. Here, the conjoined twins' body functioned as an object of cultural and medical curiosity carefully mediated by the priest. According to Fuentes y Guzmán, "there was not a distinguished or admirable citizen that [the priest] did not visit." The priest also brought the body to Fuentes y Guzmán's own house, where he carefully examined the body and had a portrait painting made of it.[42]

Difficult pregnancy accounts also reveal evidence about the operation of gender and racial ideologies of colonial rule, and about stereotypes used and reinforced in assessing blame. The Catholic church as an institution, and also its priests and parishioners, attempted to monopolize their interpretations within frameworks of Christian evangelization of colonial rule. This did not mean, however, that all agreed on the same interpretation. The following account of a half-human, half-toad baby born to an elite Spanish woman shows that pregnancy and childbirth, especially in the case of a deformed infant, became an important flash point for competing interpretations by various social groups in colonial society, in this case the Guatemalan Inquisition, male members of missionary orders active in the capital of Santiago, and male and female community members themselves.

In 1729, Doña María Cecilia de Paniagua, a twenty-nine-year-old elite Spanish woman married to a public scribe, related to Inquisition authorities an account of her pregnancy and childbirth in which she gave birth to a stillborn, deformed infant that she described as half-human, half-toad.[43] When de Paniagua was five months pregnant, she became involved in a conflict with María Savina, a single *mestiza*. De Paniagua had frequently contracted the ritual services that Savina offered for spells for sexual witchcraft and to bring economic wealth, in return for cash payments.

After the conflict ended, Savina brought a pot of beef stew to de Paniagua as a peace offering. De Paniagua had suspicions about Savina's motives and did not want to taste the stew, but after Savina pressed her she relented and ate a mouthful. From then on through the last four months of her pregnancy, de Paniagua asserted that every night from dusk until dawn she heard a toad croaking, and the sound followed her through the house even when she changed rooms. Other family members and servants reported hearing the croaking sounds as well. De Paniagua later found the stew filled with worms, which underscored her unease.

After a nine-month pregnancy, de Paniagua described giving birth to a "dead monster," whose top half was shaped like a toad's with a toadlike head, with very long arms that reached almost to the infant's feet. The bottom half appeared human, except for its rough skin that appeared toadlike as well. Furthermore, the blood and afterbirth "was not like natural blood [but] like mud from a lake," the kind of natural environment that a toad lived in. When the Dominican priest Tomás Serrano viewed the deformed dead infant, he diagnosed it the result of a "spell." Finally, de Paniagua asserted that as soon as she gave birth, the toad's croaking ceased.

The use of Inquisition records and other sources generated by colonial institutions and elites as evidence for illness diagnosis is difficult, because the illness descriptions found in the documents are framed by those enmeshed in policing arms of various aspects of colonial rule such as the Catholic church. Furthermore, these kinds of records usually allow the historian to view the process only at the end of the story, after the establishment of a loose consensus between the parish priest, the Inquisition, and the afflicted and their family members that sorcery was at the root of the problem pregnancy or deformed birth. Presumably, however, as the birth mother, family members, and neighbors first learned of the deformed birth, sorcery was just one of a number of options that could be considered at the root. Often, natural origins for the deformity and other explanations had already been ruled out by those involved. Here, de Paniagua used the term *infezion de maleficio* (infection spell) to describe what happened to her and to self-diagnose a supernatural origin to her infant's deformity and stillbirth.

At the moment of birth or shortly thereafter, the unnamed birth attendant(s) showed the deformed infant to the Dominican priest, Tomás Serrano. Presumably, Serrano was present to baptize the infant. He passed judgment on the "half-toad" baby, declaring the deformity to be the result of a spell. The presence of toads and croaking through the last months of pregnancy, and the toadlike appearance of the stillborn baby, suggested that the toad acted as a vehicle for casting illness, and shaped the infant's deformity to resemble its physical likeness. In Maya culture, toads, snakes, and frogs were often associated with sorcery and death.[44] Similar beliefs about toads and supernatural interventions existed in Africa as well.[45]

As a priest, Serrano was not exceptional in his role in diagnosing the origin of the infant's deformity. Religious handbooks in Guatemala and elsewhere in colonial Latin America contained guidelines about whether and how to baptize in the case of miscarriage, deformed births, and other difficult births. In a miscarriage handbook for priests published in 1765, Francisco Sunzin de Herrera counseled that deformed births, here called "monstrous births," should in fact be baptized, whatever the diagnosis for the deformity: "It is not against the custom of the Church nor against [its practices] to baptize *fetos* [fetuses] if they live; if it is alive, baptize. Nor is it [against the custom of the church] to baptize monstrous births whose *naturaleza* [nature] is in doubt."[46]

Official church policy on the issue of baptizing miscarriages and deformed births called for baptism if they were born alive, whatever their "nature." Priests did not necessarily follow this in practice, as we saw in the above example of the conjoined twins. In that instance, the priest baptized the twins even though they were born dead. Here, Serrano chose not to baptize the stillborn deformed infant, and he reportedly attributed the cause of the deformity not to God or nature, but to a sorcery spell cast by a mixed-race female sorcerer during a personal conflict with one of her clients, an elite Spanish woman.

Men and women involved in diagnosing and explaining deformed births and other pregnancy-related illnesses, however, did not always agree on the cause, nor did they always publicly present what they judged to be the real cause to those involved. José de Baños y Sotomayor, *comisario* (head) of Guatemala's Inquisition, wrote a letter in July of 1694 to the Inquisition officials in Mexico City, confessing that he had neglected to relate the use of a "curse" during a recent Inquisitorial inspection.[47] Doña María Limón, wife of Dr. Miguel Fernandéz, a medical professor at the Universidad de San Carlos, went into labor during her eighth month of pregnancy and was dying.[48] Baños y Sotomayor hurried to the woman's house and found her experiencing contractions, and in great pain. Noting that Limón seemed "close to losing her mind," he quickly administered the last rights and left to return home.

From Baños y Sotomayor's letter, however, it is clear that up to this point he did not see anything "supernatural" or troubling about Limón's difficult childbirth or her descent into insanity. He appeared to diagnose her symptoms as a natural outcome of a difficult childbirth. Two things, however, changed his mind. As he left the dying woman's house, her servant Doña Ana approached him and asserted that a "spell" caused her employer's difficult childbirth. She vaguely described a lover's triangle between Limón, her neighbor Doña Magdalena de Medrano, herself an elite woman and widowed wife of a *fiscal* (Audiencia official), and an unnamed male, a "person of high status." De Medrano became jealous, according to the servant, and cast the illness to interfere with Limón's pregnancy by showing her a toad in the drawing room. Apparently, the toad's association with casting illness was well known; Doña Ana did not need to explain the association to Baños y Sotomayor when she made her diagnosis, nor did he need to explain it in his letter to Inquisition officials.

Baños y Sotomayor's initial response to the servant's diagnosis of supernatural illness was scorn, and he told her that she should not say such things about a woman of reputation. He returned to his own home but was called back to Limón's bedside soon after. By that point, she had miscarried the infant, described as "dead and completely corrupted," and her *locura* (insanity) continued. It is clear that Limón's physician husband was neither present in the birthing room nor even at home at the time of her difficult childbirth and near death. In her delirium, Limón apparently began to shout about her flirtation. Baños y Sotomayor wrote: "And I knew the danger that Doña María Limón was putting herself in if her husband returned while she shouted this out, publicizing the flirtation. I persuaded him [her husband, the physician] to bring his wife to the Convent of Santa Catalina Mártir in this city [of Santiago de Guatemala], and that all that she suffered was from *melancólias* [melancholies]."

Here, the priest Baños y Sotomayor did two interesting things in the context of Limón's problem pregnancy, both of which challenged the patriarchal and medical authority of the husband. He publicly diagnosed Limón's illness as *melancólias*, defined in a colonial-era dictionary as "signifying a great and permanent sadness, originating from when the melancholy humor dominates, and that the person who suffers from this does not enjoy or want to do anything."[49] This must have been a persuasive diagnosis in the context of the stillbirth and Limón's postpartum symptoms. Limón's husband apparently agreed with the diagnosis and took his wife to recover in the convent. Baños y Sotomayor had in effect protected her within the walls of the convent that her husband could presumably not enter.

Even though the priest publicly attributed Limón's illness to a humoral imbalance caused by the difficult pregnancy and stillbirth leading to *melancólias*, he confessed in his letter that ultimately he diagnosed the deformity and illness as supernatural in origin, the result of a spell: "I confess to [you] that I was ingenuous, that even though I believe in [the] faith, [and] that there exist *maleficios* [curses], never would I have been persuaded that someone would make use of such notorious methods as those [used] by Doña Magdalena de Medrano."

Baños y Sotomayor went on to lump de Medrano as "one of those worthless women who use spells and tricks." This case allows a glimpse into competing illness diagnoses for problem pregnancies. It also shows that illness diagnosis—especially for a problem pregnancy and in the socially charged context of a sorcery accusation made amidst a possible amorous indiscretion—could shape the public diagnosis as well as the private one.

CONCLUSION:
GENDER, RACE, AND MEDICAL MESTIZAJE

Evidence from colonial Guatemala suggests that difficult pregnancies acted as significant signs with varied meanings in the New World. While it is possible to uncover separate Mesoamerican and European conceptions of healing that continued to operate into the seventeenth and eighteenth centuries, it is important to note that these medical traditions interacted and came into conflict with each other, and were reshaped over time by local contexts and social relations under colonialism. (On the development of medicine within the Portuguese colonial context, see Walker's essay in this collection.) Representations of problem pregnancies reveal a language of sorcery and magic that operated in ways that evoked the powers of female healers to cast illness to intervene in pregnancy, the difficulties families faced caring for the sick, and how these difficulties could play out in terms of the gender, race, and class conflicts of community life.

Moreover, illness events could also provide the opportunity for colonial authorities to intervene directly into community life. Gendered authority of men in these extreme circumstances to diagnose the origins themselves, or give weight to one explanation over others for the problem pregnancy often scapegoated female caregivers and the pregnant woman herself. Thus the conflictive dialogue surrounding difficult pregnancies can be used to begin to outline and contextualize local understandings of illness causation and conceptions of health and healing that are neither wholly European or Mesoamerican, nor completely religious or scientific. Shared understandings of health and healing in colonial Guatemala changed over time through the process of medical *mestizaje* in this multi-ethnic colonial society.

CHAPTER EIGHT

"Read All About It"

Science, Translation, Adaptation, and Confrontation in the Gazeta de Literatura de México, 1788–1795

FIONA CLARK

The study of the formation of the periodical press has long played a significant role in our understanding of the dissemination and reception of ideas in the early modern world, as the rich tradition of histories written on European journals demonstrates. The *Gazeta de Literatura de México*, which is the focus of this study, is one of the first and most valuable literary-scientific periodical publications to come to light in the Spanish Americas. While there has been increased interest by historians of science over the past two to three decades in the work of its editor, José Antonio Alzate y Ramírez (1737–1799)—one thinks here of Roberto Moreno, Juan José Saldaña, and Patricia Aceves Pastrana, to name but a few—these studies have focused primarily on the scientific and technological aspects of Alzate's work. As such, his ideas have been examined within the national context and the relevant sociopolitical questions pertaining to late colonial New Spain, but as yet little attention has been paid to the actual practical formation of the periodicals to which Alzate devoted so much of his life or to their function in the wider Republic of Letters, never to mention the use of foreign material within their pages.

I would like to thank the Bakken Library and Museum, Minneapolis, and its staff for the generous funding and unfailing help that first gave me the opportunity to study the French periodicals involved in this article.

Alberto Saladino García, in his in-depth study of the work of Alzate and F. J. Caldas, began to address some of these issues by providing a long list of authors and works that are mentioned within the periodicals.[1] This process, however, does not make a distinction between the articles written by Alzate and those pertaining to other authors, thus leading to a potentially false sense of the breadth of literature involved. Moreover, no attempt has been made, to date, to verify the sources referenced by the editor in an effort to understand the variety of material he had at his command that originated outside of Mexico. This study, therefore, is an attempt to deepen our understanding of the use and presentation of material from across Europe (although centered in the French periodical press) and, in a more limited sense, from Spanish America, in the *Gazeta de Literatura*, demonstrating the way in which Alzate played an active part in the international Republic of Letters and setting the *Gazeta de Literatura* in the context of the developments in the Spanish American periodical press.[2]

JOSÉ ANTONIO ALZATE Y RAMÍREZ IN THE REPUBLIC OF LETTERS

Born in 1737, son of a native Spanish father and Mexican-born mother, the secular priest José Antonio Alzate y Ramírez stands as one of the most illustrious figures of eighteenth-century Mexico (see Figure 8.1). This fact refers not only to his prolific career but also to his insatiable desire for scientific investigation and the propagation of new ideas. A controversial polemicist, even in terms of his relationship with the various viceroys of New Spain, Alzate was a formidable adversary. While expressive in his praise of those people or ideas he felt deserving, he was also inclined to interpret opposing views as personal affronts and react accordingly with mockery, ridicule, and sarcasm.

Shortly after Alzate's death in 1799, Manuel Antonio Valdés published a eulogy in the *Gazeta de México* in which he described Alzate as a *buen filósofo*, *buen patricio*, and *buen sacerdote* (good philosopher, good countryman, and good priest)—thus covering the three defining areas of his life; natural philosophy and literature, his nation, and his vocation.[3] This latter aspect of Alzate's life has been largely overlooked in the belief that his scientific and literary pursuits were of greater importance to him than his ecclesiastical status. Yet his vocation is an intrinsic part in our understanding of his role within eighteenth-century enlightened Catholicism. The religious convictions underlying his reading of people and nature is often clearly expressed in the pages of the *Gazeta de Literatura*.[4]

His arguments for the natural wealth of Mexico are inseparable from manifestations of adoration of the *benéfico creador* (beneficent creator) and clear evidence of divine wisdom as discerned in nature.[5] Within this Spanish American context, the *buen filósofo*, as described by Valdés, was characterized by those who worked between theology and new scientific knowledge and looked to new advances in science following an eclectic approach that did not alter traditional dogma.[6] This approach can be clearly seen in Alzate's work, and although the encyclopedic nature of his beliefs have led some, including Alexander von Humboldt, to comment on weaknesses or superficiality in understanding, his breadth of knowledge was also aimed at opening the study and interest of science to a wide audience.

With this in mind, this study aims to focus primarily on Alzate as the *buen filósofo*, in order to explore his work as part of the eighteenth-century Republic of Letters and to consider how his understanding of

FIGURE 8.1. *Portrait of José Antonio Alzate y Ramírez (1737–1799)*
SOURCE: Nettie Lee Benson Library, The University of Texas at Austin.

his role impacted his use of the French periodical press for his own publication.

In order to judge the extent to which Alzate truly belonged within the Republic of Letters, we must first consider the general characteristics that comprised this "republic of scholars" and the demands and expectations laid upon those who professed their allegiance to, or membership of, the same. Anne Goldgar, in her study of conduct within this learned community, underlines particular features commonly expressed by those who felt they formed part of the Republic of Letters, many of which are also highlighted by Françoise Waquet and Hans Bots.[7] Primary among these was the fact that those who belonged to the Republic did not consider themselves, when relating within it, to be subject to the norms and values of the wider society. As such, it was in essence an international community in which all members were equally permitted to criticize the work and conduct of the others, regardless of nationality or religious affiliation. Yet as Goldgar also points out, while it was not a monarchy, this feature did not necessarily make it a democracy, for a strictly maintained hierarchy did exist within the levels of respect and authority recognized by the members.[8] Waquet and Bots have argued that the Republic was a collective term for those who were interested in literature in its broadest sense, including the sciences; those who cultivated knowledge. A common ethos was held to be the objective of working for the benefit of humanity, seeking out the secrets of nature, and widening man's opportunities within this scope.[9] While it may seem that everyone with an interest in literature and learning could in essence belong to the Republic, even without publishing any work of their own, participants' status seems to have depended largely on their contact with other scholars. Thus, scholarly correspondence, the *commerce de lettres*, was key to maintaining the communal bonds on a national and international level. Scholars felt at liberty to ask for assistance from those in a position to facilitate their pursuit of knowledge, mainly in accessing published works that were unavailable in a given city or country. It was a system that operated on a network of obligation and an ethos of cooperation, which in turn strengthened the communal identity.[10] In this way, as Goldgar states, "a wide *commerce de lettres* brought status in part because it proved a scholar was in the centre of the community."[11] The creation of literary journals facilitated this process of disseminating information and at the same time continued the dependency on scholarly correspondence as the source of much of the printed material. As a result of its central role within the learned community, the journal became at once the ideal member and also the means of regulating the Republic.[12] Let us turn briefly to consider Alzate within this description, first in terms of the *Gazeta de Literatura*, and then with regard to his network of correspondents.

Alzate frequently sets his arguments and responses to his critics within the framework of the ethos of the Republic of Letters. His statements in the *Gazeta de Literatura* indicate that the only acceptable method for criticism was directed by the values of said Republic. Namely, that those who voiced their opinions should do so in an open manner, primarily through publication, not by murmuring against their subject in closed circles or spreading rumors. In this way, the critic or critics would be obliged to state their argument within the public sphere, thus allowing the criticized individuals an appropriate means by which to defend themselves. The two sides of any polemic would be open to the reading public who, according to Alzate, would constitute the best final judge.[13] It must be admitted, however, that while he also argued that any criticism published should be presented in an impersonal manner, aimed at the work and not at the author, Alzate seldom stayed within his own guidelines, publishing scathing attacks on many of those with whom he disagreed. Furthermore, his claim to work within the egalitarian ethos of the Republic was also to lead him into conflict with Juan Vicente de Güemes Pacheco, the second Count of Revilla Gigedo, viceroy of New Spain (1789–1794) over a literary criticism relating to the Duke of Almodovar. While Alzate did eventually publish an apology, he made it equally clear that in stating Almodovar's name without the normal forms of address he was merely acting in accordance with the etiquette of the Republic that granted equality of reference to all scholars within its community.[14] Evidently, therefore, within the confines of the *Gazeta de Literatura* Alzate believed he was a member of the international scholarly community. Yet can we say the same for his relationships outside of the periodical, and if so, what impact did this have on the content of the articles that are under discussion in this study?

On a national level, Alzate was connected through his work to the highest echelons of the colonial ecclesiastical and secular government. He related directly with the various archbishops and viceroys in Mexico City under whom he served, providing them, among many ecclesiastically and scientifically related appointments, with extensive reports on a wide variety of subjects (ranging from meteorological, mineralogical, cartographical, mining, agricultural, and botanical observations to historical treatises).[15] Significantly, several of these local opportunities were to open doors for him to the world of European scientists and governmental recognition. In 1769, when the French and Spanish monarchies had organized an expedition to California under the guidance of astronomer Jean Chappe d'Auteroche along with Joaquín Velázquez de León,

Alzate was given the opportunity to join Juan Ignacio Bartolache (and observers from sixty-two different nations who were taking part in similar observations) in charting the transit of Venus. Unfortunately, the entire expedition fell ill from typhus, and the final list of the dead included the expedition's leader, Chappe d'Auteroche. His reports were to be returned to Paris with Pauly, one of the surviving members of the original group, a circumstance that Alzate used to his advantage by including his own observations of the transit. This report was sent along with a map of the viceroyalty of New Spain, samples of seeds, fruits, plants, minerals, artifacts, and a letter on the natural history of Mexico to the Paris Royal Academy of Sciences. As a result, Henri Louis Duhamel du Monceau (1700–1781) proposed his nomination for membership, and in 1771 he was elected as a corresponding member to Pingré, with the new title Socio Correspondiente de la Real Academia de las Ciencias de Paris (corresponding member of the Paris Royal Academy of Sciences).[16] Despite publicly indicating his status as member of the academy on the frontispiece of every volume of the *Gazeta*, it appears that after 1786 Alzate's name was no longer included on the list of correspondents, although he does not seem to have become aware of this fact until some years later.[17] This introduction to the world of the Academy of Sciences also created avenues for the publication of Alzate's reports in two of the main literary scientific periodicals in France, the *Journal des Sçavans* (1771, 1773) and the *Journal de Physique* (1773).[18] No further references to Alzate's work in European periodicals appear in the years that follow, save for a republication of his criticism of the *Voyageur François* by Joseph de la Porte in the Spanish *Memorial Literario* (1788).[19] This brief period of European success, however, seems to have been sufficient to allow Alzate to consider himself an active and valuable member of the international Republic of Letters, even after he had ceased active correspondence, equal in every right to his European counterparts.

Membership in the Academy of Sciences was also to play a highly significant role in Alzate's life in his relationship to other learned individuals in Mexico and beyond. For the main part, as the only Creole scholar and scientist in New Spain to hold such a title, it granted him a level of authority that he used to maximum advantage when defending his own ideas and opinions to his Mexican contemporaries. Equally important, it opened up avenues of correspondence with various European scientists, among whom we find Antonio Ulloa (with whom he worked in New Spain as part of the Malaspina expedition), Antonio Pineda, Pedro Franco Dávila, and Casimiro Gómez Ortega. Academy membership provided the basis for his initial contacts with the latter two of these correspondents as well as a common identity prior to either offering his knowledge and expertise to the royal cabinet (as in the case of Dávila),[20] or establishing scholarly correspondence (as in the case of Gómez Ortega).[21] Admittedly, these examples are "correspondence" in only the most limited of senses, as there is no record of a response from Dávila and it is not clear how much, if any, correspondence was maintained between Alzate and Gómez Ortega, who proposed his name as correspondent of the Royal Botanical Garden in Madrid. The register of the Botanical Garden demonstrates, nevertheless, that Alzate continued to send seeds to Madrid until 1795.[22]

These few examples of some of Alzate's links with the wider governmental, scholarly, and scientific worlds indicate quite clearly why he considered himself to be so well placed within the international Republic of Letters. Furthermore, his connections with the Academy of Sciences also create an important context within which to understand his use of the articles originating in the French literary-scientific periodical press. Alzate was not merely passively taking information and transferring it to the Mexican reader; rather, he was participating in a scholarly community, judging literary production from the point of view of an equal within this community, and assessing the importance of an international body of work from the basis of the existent body of scientific and literary production in New Spain. With this in mind, we now turn to consider the place of the *Gazeta de Literatura* within the Spanish American periodical press.

THE GAZETA DE LITERATURA AND
THE SPANISH AMERICAN PERIODICAL PRESS

The social structure of New Spain had changed significantly over the two hundred years leading up to the 1700s. The territory had doubled, and production and population had tripled due to an influx of immigrants coming from Asturias, Galicia, and the Basque Country, creating a greater desire for news from Europe, and particularly Spain.[23] In 1722 Juan Ignacio María de Castorena Urúa y Goyeneche established the earliest of the eighteenth-century Mexican periodicals, the *Gazeta de México*. This periodical was to reappear at various periods throughout the century under different editors and slightly amended titles. Yet, after 1743 no regular periodical was published in Mexico until Alzate y Ramírez began his *Diario Literario de México* in 1768, which ran for a year. Another lull ensued until 1772, when both Alzate and Juan Ignacio Bartolache undertook new, short-lived periodicals of a medico-scientific nature—*Asuntos Varios sobre Ciencias y Artes*[24] (Alzate) and *Mercurio Volante*[25] (Bartolache).

The contents of many of the articles used by both Alzate and Bartolache, especially at the beginning of their journalistic careers, were taken from foreign sources and translated into Spanish. This practice does not always seem to have been readily accepted by their readers, who demanded more original productions and exerted a significant amount of public pressure.[26] By the time Alzate came to publish the *Asuntos Varios*, the entire collection featured only three or four translated articles. From Alzate's point of view, the reader was encouraged to interact with the articles in the pursuit of knowledge and the invigoration of intellectual life, even if this were by criticism of his work.[27]

In their scientific writings, Alzate and Bartolache aimed to make the world a better and more humane place in which to live. In this way, science remained a functional element in their advice and not an end in and of itself. In the initial stages leading up to and including the publication of the *Gazeta de Literatura*, scientific periodicals had slowly become the bridge between theory and practice. Although the apparent lack of public support, as shown by constant financial struggles, may raise questions concerning the acceptance of these scientific ideas in Mexican society of the late eighteenth century, there can be no doubt as to the value of these printed archives in our attempts to understand the impact of scientific innovations in pre-Independence Spanish America. The interests of the periodicals reflected a new trend in thinking within the society, an expansion of interests and publication needs, and a greater shift toward cultural, natural, scientific, and historical themes that indicate a nascent Creole mentality. As John Browning has noted:

An important aspect of the Spanish American Enlightenment . . . evident particularly in the last two decades of the eighteenth century and stimulated to a large extent by de Pauw and others, was this journey of local discovery on which many Creoles embarked. [On the development of Creole science, see also More's essay in Chapter 6.] Thanks to the newspapers they became aware of their countries as individual cultural entities in an international context. They came to look with greater knowledge, pride, and appreciation upon their native lands, and their patriotism acquired a more self-confident air.[28]

The lengthy silence in periodical publications between 1773, when these two publications ceased, and 1784, when Manuel Antonio Valdés once more initiated publication of the *Gazeta de México*,[29] suggests that Mexican society had not reached conditions for the establishment of a regular periodical. Yet at the same time, for Creoles who were actively pursuing changing ideas within literature and the sciences, publications such as those by Bartolache and Alzate were a sign that they could compete with printed materials on the other side of the world.[30] Ten years after the *Asuntos*, Alzate was once again to undertake a periodical publication with the *Observaciones sobre la Física, Historia Natural y Artes Útiles* (1787), and the following year, his most important publication, the *Gazeta de Literatura de México* (1788).[31]

By the late 1700s, then, there was an increasing expression of Creole loyalty to their homeland in the face of attacks from the Old World, alongside an increased critical awareness of their own potentiality and a desire to both enlighten and improve their society. Works written on the Americas by a number of European authors voicing theories of corruption, weakness, and degeneracy—such as de Pauw, Buffon, and Raynal—led to an indignant backlash against the erroneous picture painted of the New World and its inhabitants and the image they conveyed to those who might never experience the Spanish Americas for themselves. The various responses of leading Creole literary figures during this time, based both in the Americas and in Europe (especially the exiled Jesuits), have been extensively explored and charted by various historians, such as Antonello Gerbi,[32] David Brading,[33] and Jorge Cañizares-Esguerra,[34] as examples of Creole patriotism and the forging of patriotic epistemologies. (On the development of Creole identity and colonial science, see Goodman's essay in Chapter 1.) Alzate's arguments and concerns expressed through the pages of his periodical publications raise many of the same issues and themes that dominated what has come to be known as the "dispute of the New World,"[35] yet at the same time largely maintaining a very practical quality grounded in the everyday experience of the Mexican reader. In Alzate's hands the periodical press became the tool through which he attempted to undertake a scientifically accurate defense in an effort to overcome the damage incurred by the European theorists and awaken his readers to the privileges and challenges of their own particularly Mexican reality.

ACQUIRE, ACQUAINT, AND ATTACH

Within the first two pages of the prologue to volume 1, Alzate uses three phrases that place his periodical and its contents firmly on the map in terms of geography and nationhood: *la Metrópoli del Nuevo Mundo, la voz México*, and *nuestra Nación Hispano Americana*. As such, he sets his work in the geographical context of the City of Mexico, "the metropolis of the New World"; he determines for whom he is speaking, with "the voice of the Mexican people"; and he clarifies that which he is representing, "the Hispano-American nation."

This discourse of identity, as presented by the *Gazeta* and its editor, is in effect a continuous journey of *acquisition, acquaintance,* and *attachment.* Alzate acquires the ideas that form the basic content of the periodical through his observations, experimentation, reading of literature, and conversation and contacts with other individuals. Once garnered, he acquaints the reading public with the information and ideas through various means, including the publication of translations, adaptations, correspondence, reports, and polemics. The goal of acquainting the public with the information is not merely to provide knowledge but to foment attachment to their particular world, the world of New Spain. Overall, this process encourages the formation of a national common consciousness among the readers and inspires a practical response that would, ideally, lead to the betterment of society and the defense/promotion of New Spain. One might argue, in fact, that Alzate is waging a war on two fronts: internally, that is, within Mexico, he is struggling to change outmoded systems and structures, backwardness, and lack of awareness among the people; externally, looking out toward the rest of the world, he is promoting the great wealth and resources existent in the country and defending Mexico against ill-informed attacks and inapplicable systems. Nowhere is this double-edged approach more visible than in his use of the material he incorporates into the *Gazeta,* particularly articles from periodicals; it is to this subject that we now turn.

THE "UNIVERSAL STOREHOUSE"

Unlike many of the important periodical publications at this time, the *Gazeta de Literatura* was, to a large extent, the work of one individual and not a body of literary figures.[36] Its content can be divided into three general categories: reports written by Alzate; reports contributed by his network of correspondents within Mexico;[37] and material taken from non-Mexican sources, mainly periodicals, that, if not published in Spanish, have been translated and at times adapted by Alzate. It is the latter category that forms the focus for this study.[38] As we saw earlier, Alzate held important links with the European, and especially the French, scientific world that led him to perceive of himself as an active participant with them in the Republic of Letters. Given the important role that the periodical press played within this community, it is essential that we more fully understand the impact of these foreign periodical publications upon Alzate's work as an editor. In order to do this, we must first turn to the immediate predecessor of the *Gazeta,* the third of his earlier publications, the *Observaciones sobre la Física, Historia Natural y Artes Útiles* (1787–1788).[39]

As Roberto Moreno has already indicated, this title was taken directly from the influential French periodical printed in Paris by Jean François

Rozier, *Observations sur la Physique, sur l'Histoire Naturelle et sur les Arts*, which after the first two introductory volumes became more commonly known as the *Journal de Physique*.[40] The Prologue to Alzate's *Observaciones* consisted of a translation of the presentation given by Bernard le Bovier de Fontenelle to the Paris Royal Academy of Sciences, in 1699. In a footnote explaining his choice of material, Alzate argues that "this prologue, praised by many wise critics as being one of the author's principle works, although badly translated, will be of much greater use to the public than any other idea that I, in my weakness, could conceive."[41]

Both Fontenelle's outline and the Prologue to Rozier's *Journal de Physique* reflect the wishes and desires that Alzate was to express throughout the *Gazeta de Literatura*, namely, that it should provide a deposit of knowledge that would also announce discoveries in the different branches of science, promote the progress of new ideas in technology, and serve the good of humanity. Alzate's admiration for Rozier's publication is clear, as is his aspiration that the *Gazeta* should similarly become a "universal storehouse for observations on nature."[42]

As the reader progresses through the *Gazeta,* there are continued references to, and extracts from, articles found in what seem to be a great variety of European publications. These range from reports by academies in cities such as Harlem, Stockholm, Toulouse, Mannheim, Philadelphia, St. Petersburg, and London to periodicals in countries as geographically varied as Martinique, Scotland, Switzerland, Naples, Italy, France, Spain, Portugal, Cuba, and Peru. In fact, as the following tables illustrate, Alzate's sources comprised relatively few journals. On several occasions we find circumspect references in the short commentaries and footnotes to certain journals as the origin of these articles. These references are not always efficient or correct. Some mistake the year of publication, others the exact pages; yet for the most part, when cited, they indicate where approximately within the journals we can find Alzate's source material. Translated extracts are also frequently published without any clear indication of their origin, creating the impression that Alzate has accessed them directly rather than taking them from the pages of a journal.[43] In the absence of evidence to explain such editorial choices, we can merely speculate as to his reasons. It may be that, in tandem with the publication tendencies of the period, it was simply not considered important to state where the information originated. The effect, however, is an increased sense of breadth of material at Alzate's disposal. While such an assertion may be accused of pushing conjecture to its limits, we should bear in mind that throughout his publications Alzate, as one who was largely self-educated in the sciences and who did not belong to the established institutions, was at constant pains to defend the value and merit of his position. To this end, Alzate made great use of the authority of European scientific arguments as a defense for his own observations over and against the established elite in Mexico. Access to the latest scientific information was a tool he wielded whenever possible to his opponents' disadvantage.

The remainder of this study will focus on the further investigation of these sources in two main areas; first, to analyze what we can learn about Alzate's process of gathering information through the practical details that are available to us; second, based on his comments to the articles, to try to understand the main objectives that lay behind Alzate's choice of material. Although the tables of journal articles with this study provide a very concise summary of the amount of each text translated and some of the ensuing changes, there will, regrettably, not be sufficient space within the current work to address more than superficially the important issue of the translation process followed by Alzate and to note the effect of the linguistic changes on the impartation of information.[44]

ACQUIRING THE MATERIAL

As is immediately clear from the table of journal articles, the vast majority of those references, extracts, and reviews published in the *Gazeta de Literatura* can be tracked back to Rozier's *Journal de Physique*. Forty-nine of the sixty-seven articles verified for this study can be strongly linked to this periodical between the years 1771 and 1789 (see Table 8.1 and Figure 8.2). The next most significant title in terms of frequency of reference is, in Alzate's words, the *Biblioteca Económica*, or *Bibliothèque Physico-Économique*,[45] with six articles limited to the years 1785 and 1788 only (Table 8.2). In equal measure, the next two journals— *Diario de los Sabios*, or *Journal des Sçavans*[46] (Table 8.3) and *Diario de Bouillon* or *Journal Encyclopédique*[47] (Table 8.4)—receive three references each: the former for the years 1771 and 1785 (two of which are references to the same article written by Alzate in 1771), while the latter is limited to the years 1785 and 1786.

Two points are worth noting related to the dates of publication as indicated by the tables. First, material taken from the French periodicals terminates in the first years of the Revolutionary period, possibly due to increased pressure from governmental censorship after this point and the resulting difficulty in obtaining the material; second, that although Alzate frequently complains about the length of time it takes for literature from Europe to reach New Spain, we notice on many occasions a delay of only

(text continues on page 171)

TABLE 8.1.

Articles from the Journal de Physique published in the Gazeta de Literatura de México, 1788–1794[1]

GLM	Journal de Physique	Article Title
Volume 1, number 4 (15 February 1788), p. 34	February 1775, pp. 89–120	Éloge de M. Commerson par M. de la Lande, de l'Académie Royale des Sciences[2]
Volume 1, number 9 (28 June 1788), pp. 73–78	Reprinted on various occasions throughout the 1780s	Prix extraordinaire—L'Académie avoit réservé, en 1778, une Médaille de 300 livres, de la fondation de M. Christin, pour un Prix extraordinaire . . .[3]
Volume 1, number 21 (12 May 1789), p. 95	April 1786, pp. 244–52, May 1786, pp. 352–64	Dissertation sur les Montagnes et des terrains a mines en général, qui a été couronnée par l'Académie de Manheim en 1785, par M. Monnet . . .[4]
Volume 1, number 23 (14 August 1789), p. 117	January 1785, pp. 68–70	Tableau méthodique des minéraux, suivant leurs différentes natures, & avec des caractères distinctifs, apparents ou faciles à reconnoître; par M. Daubenton . . .[5]
Volume 1, number 23 (14 August 1789), p. 120	January 1788, pp. 61–63	Lettre de M. le Baron de Marivetz a M. de la Metherie sur la Nomenclature Chimique (Château de Vincennes, 10/11/1787)[6]
Volume 1, number 11b (2 August 1790), p. 86	March 1772, pp. 630–31	Mélange de Physique & de Médecine, par M. le Roi, Professeur en Médecine au Ludovicée de Montpellier, Membre de la Société Royale de Londres, Correspondant de l'Académie Royale des Sciences . . .[7]
Volume 1, number 14b (22 March 1790), pp. 110–11	November 1771, pp. 329–40	Médecine Vétérinaire par M. Vitet, Docteur et Professeur en Médecine, en trois volumes . . .[8]
Volume 1, number 14b (22 March 1790), pp. 111–12	December 1771, pp. 419–21	Traite abrégé des plantes usuelles de Saint-Domingue, par M. Pouppe-Desportes . . .[9]
Volume 1, number 15b (12 April 1790), pp. 115–17	November 1777, pp. 379–81	Extrait d'une lettre de M. Pistoj, Professeur de Mathématiques à Sienne, du 25 Avril dernier[10]
Volume 1, number 19b (8 June 1790), p. 148	July 1776, pp. 56–60	Lettre adressée à l'Auteur de ce Receuil, par M. Maupertit, Prier de Cassan, sur la petit vérole

(continued)

TABLE 8.1.
(*continued*)

GLM	*Journal de Physique*	Article Title
Volume 1, number 19b (8 June 1790), p. 151	January 1776, pp. 64–66	Tableau de mortalité de Londres, depuis 1667 jusqu'à 1772 (Letson & Pringle)
Volume 1, number 19b (8 June 1790), p. 152	August 1788, p. 153	Observations médicales & politiques sur la petite Vérole . . .par M. Mahon
Volume 2, number 2 (21 September 1790), pp. 15–17	September 1779, pp. 225–27	De M. Magellan, Membre de la Société Royale de Londres, à un de ses Amis de Paris. (Re. Mudge & 'taux catharrale')[11]
Volume 2, number 2 (21 September 1790), pp. 17–18	October 1776, pp. 298–304	Mémoire sur une manière de communiquer du mouvement à l'eau d'une bagnoire ordinaire . . .par M. le Comte de Milly[12]
Volume 2, number 5 (2 November 1790), p. 41–42	February 1773, pp. 221–223	Observations tirées d'une lettre écrite de Mexico à l'Académie Royale des Sciences, par Dom de Alzate y Ramyres sur des poissons vivipares, & quelques autres objets d'histoire naturelle[13]
Volume 2, number 8 (13 December 1790), pp. 61–63	November 1773, pp. 381–82	Extrait d'une lettre de M. Franklin à Miss Stevenson; sur les Experiences relatives a la chaleur communiquée par les rayons du Soleil[14]
Volume 2, number 8 (13 December 1790), pp. 63–64	June 1776, pp. 509–14	Premier mémoire d'optique, ou explication d'une expérience de M. Francklin Par de Docteur de Goddart . . .[15]
Volume 2, number 8 (13 December 1790), p. 65	April 1780, pp. 319–33	Observation de Soleil, totale, avec demeure, & annulaire, du 24 Juin 1778, par Don Antonio de Ulloa, . . .[16]
Volume 2, number 9 (30 December 1790), pp. 67–73	January 1787, pp. 47–55	Extrait d'un Traité in- 4 sur l'amalgamation des Métaux nobles; par M. le Chevalier de Born . . .Précis historique de l'Amalgamation usitée en Amérique[17]

GLM	Journal de Physique	Article Title
II, 9, 72–73 (30-12-90) Volume 2, number 9 (30 December 1790), pp. 72–73	April 1787, pp. 289–93	Traduction d'une lettre écrite à M. de Baron de Dietrich, par M. de Trebra, au sujet du nouveau procédé de l'Amalgame de M. de Born; lue à l'Académie des Sciences au mois de Mars 1787[18]
Volume 2, number 9 (30 December 1790), p. 74	November 1785, pp. 362	Mémoire sur la platine, ou or blanc par M. L.[19]
Volume 2, number 10 (11 January 1791), pp. 77–78	November 1774, pp. 360–61	Moyen de calmer les vagues de l'eau avec de l'huile[20]
Volume 2, number 10 (11 January 1791), pp. 78–82	November 1774, pp. 362–69	Extrait d'une lettre du Docteur Franklin A Docteur Brownrigg[21]
Volume 2, number 11 (25 January 1791), pp. 83–85	November 1774, pp. 366–67	Extrait d'une Lettre de M. Tenguagel, à M. le Comte de Bentink, écrite de Batavia, le 15 Janvier 1770[22]
Volume 2, number 13 (12 February 1791), pp. 103–6	November 1778, pp. 410–11	Nouvelles Littératures—Description du monument élevé à la gloire du Czar Pierre 1er, par le comte Marin Carburi . . .[23]
Volume 2, number 14 (8 March 1791), p. 112	February 1778, pp. 104–18	Suite du Mémoire Intitulé: Analyse de Pastel, et examen plus particulier des mouvemens instestins de la Cuve en laine par Quatremer d'Isjonval[24]
(19-4-91) Volume 2, number 17 (19 April 1791), pp. 131–32	February 1789, pp. 108–9	D'une Substance gélatineuse ramassée par M. Dombey, sur une espèce d'Opuntia de la Province de Huanuco au Pérous, près de fleuve des Amazones; par M. Sage[25]
Volume 2, number 19 (17 May 1791), pp. 147–48	April 1784, pp. 321–23	Correspondance entre M. de Michaelis, Professeur du Langues Orientales à Gottingue, et M. Lichtenberg, Professeur en Physique, sur un trait de l'Histoire ancienne, au sujet des Conducteurs[26]
Volume 2, number 19 (17 May 1791), pp. 148–49	October 1784, pp. 297–302	Seconde Lettre de M. Michaelis a M. le Professeur Lichtenberger; Traduite par M. Eysen, Ministre du Saint Evangile à Niederbronn[27]

(continued)

TABLE 8.1.
(continued)

GLM	Journal de Physique	Article Title
Volume 2, number 19 (17 May 1791), p. 149	October 1784, pp. 302–3	Réponse de M. Lichtenberg a la seconde Lettre de M. Michaelis, Traduite par le même
Volume 2, number 19 (17 May 1791), p. 149	February 1785, pp. 101–3	Adition a la Seconde Lettre de M. Michaelis[28]
Volume 2, number 19 (17 May 1791), pp. 149–50	February 1785, pp. 105–6	Réponse de M. Lichtenberger
Volume 2, number 19 (17 May 1791), p. 154	December 1788, pp. 401–12	Notice des Observations faites sur le col de géant; par MM. de Saussure
Volume 2, number 19 (17 May 1791), p. 149	March 1789, pp. 161–80	Suite de la notice des Observations faites sur le col du géant; par MM. de Saussure[29]
Volume 2, number 23 (12 July 1791), pp. 185–86	August 1775, pp. 145–47	Observations de M. Willemet, doyen des Apothicaires de Nancy, sur les Perles qu'on trouve en Lorraine[30]
Volume 2, number 30 (1 November 1791), pp. 235–42	May 1774, pp. 317–24	Lettre sur la comparaison des anciennes & des nouvelles découvertes faites dans la mer du sud, au midi de l'Equateur; par M. Pingré . . .[31]
Volume 2, number 39 (15 May 1791), pp. 307–12	June 1785, p. 482 See also September 1788, p. 237	Académie de Toulouse—sujet du prix—de proposer un vernis simple pour recouvrir la poterie, destinée aux usages domestiques, sans nul danger pour la santé[32]
Volume 2, number 39 (15 May 1791), p. 313	Supplement 1788, pp. 128–30	Description de la mine de fer natif, nouvellement découverte dans la Sibérie; par M. P. Simon Pallas[33]
Volume 2, number 39 (15 May 1791), pp. 313–14	July–August 1776, p. 135	Lettre sur une masse de fer natif par M. Stehlin[34]

GLM	Journal de Physique	Article Title
Volume 2, number 46 (11 September 1792), p. 364	August 1785, pp. 81–83 September 1785, pp. 161–66	Supplement à mon Mémoire sur les volcans et les Tremblements de terre, par M.C.D.L.L.C.A.C.R.D.G.[35]
Volume 3, number 2 (12 November 1792), p. 11	April 1788, p. 304	Géometrie souterraine, élémentaire, théorique & practique de M. Duhamel, de l'Académie Royale des Sciences de Paris, Inspecteur général des Mines & Professeur de l'École Royale desdites Mines . . .[36]
Volume 3, number 6 (22 January 1793), p. 43	March 1783, p. 178	Observations sur les insectes polypiers qui forment le tartre des dents; par M. Magellans, de Londres[37]
Volume 3, number 8 (19 February 1793), p. 59; Volume 3, number 9 (23 March 1793), p. 68	December 1793, pp. 426–33	Recherches sur l'origine des Mattamores, par M. le Baron de Servières[38]
Volume 3, number 13 (28 May 1793), pp. 97–100	March 1778, pp. 262–64	Mémoire sur un cataracte artificielle qu'on peut produire sur les yeux des Cadavres & des Animaux vivans; par M. Troja . . .[39]
Volume 3, number 14 (11 June 1793), p. 105; Volume 3, number 15 (3 July 1793), p. 114	December 1773, pp. 453–57	IIe lettre de M. Franklin au Docteur Lining, sur la Rafraîchissement produit par l'évaporation des liqueurs.[40]
Volume 3, number 15 (3 July 1793), pp. 114–16	January 1776, pp. 82–83	Traité de la petite Vérole, tiré des Commentaires de Van-Swieten . . .[41]
Volume 3, number 24 (15 November 1793), p. 25; Volume 3, number 25 (2 January 1794), p. 191	June 1782, pp. 474–82	Description, usages et avantages de la machine pour réduire les fracteurs de jambes, inventée par Dom Albert Pieropan, de Vicence, par l'Auteur de ce Journal[42]

(continued)

TABLE 8.1.
(continued)

GLM	*Journal de Physique*	Article Title
Volume 3, number 34 (17 October 1794), p. 268; Volume 3, number 35 (17 October 1794), p. 272	January 1776, pp. 83–85	Réflexions sur les dangers des exhumations précipitées, &sur les abus des inhumations dans les églises, suivies d'observations sur les plantations d'arbres dans les cimetières. Par M. Pierre-Toussaint Navier . . . 1775[43]
Volume 3, number 35 (3 December 1794), p. 275; Volume 3, number 37 (22 December 1794), p. 289	September 1779, pp. 229–37	Mémoire sur la méthode singulière de guérir plusieurs maladies par l'Emphysème artificiel; par M. Gallandat . . .[44]

1 For conciseness the *Journal de Physique* is referred to as JP throughout the notes, although Alzate always referred to it as the *Diario de Física* or at times, *Observaciones sobre la Física*.

2 Three short sections of Commerson's text (pp. 106–7) of the original are extracted, translated, and used as part of Alzate's criticism of the Linnean system of nomenclature. No reference to origin.

3 From the original Alzate uses the short introduction to the prize and the fifth section of the announcement as the starting point for an article based on his own observations. It is not possible to say which of the various announcements he might have used as his only reference points to the fact that the Academia de Leon de Francia published the prize and that it was founded by Mr. Christin.

4 Uses only the last of four points that appear in the French but he adds in a footnote that the problem has been published by the Academy of Mannheim at greater length. He wishes the readers to respond with their own solutions but states that in the absence of such he will present his own observations.

5 As Alzate makes only a passing reference to the title of this work we cannot be definitely sure that he is using this precise article. Yet, based on the number of articles taken from numbers of the *JP* around this time there is a strong possibility that he had at least read it in the journal. This extract is used to criticize and ridicule Daubenton's mineral classification system and forms part of the polemic between Alzate and Vicente Cervantes, director of the Royal Botanical Garden in Mexico City.

6 Reference to origin of article is given as *JP* 1788. Translates the original title and the entire letter. This extract also forms part of the Alzate-Cervantes polemic.

7 Referenced as coming from the *JP* Paris, 1772, p. 640. Translates two short paragraphs of the original text. States that he is republishing it as it presents ideas that are useful to public health.

8 Translates one long paragraph of the original text stating that it was printed in Paris in 1771. Alzate uses Vitet's argument to back up his own observations.

9 Introduces the quotations by recalling what he has read in the *Memorias de la Academia de las Ciencias de París* and providing a direct translation of the title of Pouppe-Desportes' work. The quotes in *GLM*, pp. 111–12, are taken directly from pp. 419 and 420 of the *JP*.

10 Alzate translates the entire letter by Pistoi and provides a translation of exactly the same title as the French original. No added comments.

11 Mudge becomes Mugde in the Mexican translation; this error only occurs once in the French version. Sections of the original are extracted, changes are made to the phrasing and to several expressions, and references made in the French; he also adds extra footnotes.

12 States that the extract was presented at the Paris Royal Academy of Sciences. Provides a brief overview of the text stating that Milly writes more extensively on the subject. Also provides an illustrated example of the process in question; changes the labelling of the sections of the diagram of the machine.

13 Mentions that the Paris Royal Academy of Sciences had printed the observations that he is making in this article regarding the Ajolotl and that the present information is part of a work that he has completed on the history of New Spain that he hopes Antonio de Sancha will eventually publish. This extract is included in order to refute the arguments presented by some Naturalists regarding viviparous and oviparous fish.

14 Provides translation of the original title. Text translated almost in its complete format, including translations of the original footnotes alongside comments added in separate footnotes by Alzate.

15 Provides a title that contains some of the information from the original but is not an exact translation: "Experimentos de Francklin acerca de la sensación que los nervios ópticos reciben de los objetos luminosos." The French text also begins by saying: "On lit dans le premier Cahier de l'Esprit des Journaux, de l'an 1774, page 129, l'article suivant: *Experiences de M. Francklin sur l'impression des objets lumineux sur les nerfs optiques*," which more closely resembles the Spanish translation. Furthermore, the lengthy extract in the *GLM* contains a short paragraph at the end that does not seem to be written by Alzate (p. 64) and yet does not appear in the *JP* text where the article ends "&c, &c, &c." It is possible, given these facts, that this text was taken from the *Esprit des Journaux*; this has yet to be verified.

16 Also appears in the *JS*, September 1780, pp. 383–91. Alzate does not translate the text but refers to the work as an introduction to his criticism of Beccaria who had claimed to be the first to undertake such an observation. Alzate points out that he, as well as Ulloa, had published on a similar theme before Beccaria.

17 Refers directly to the *JP*, January 1787, p. 47. Translates the original title and the complete text of pp. 47–48, providing ample footnotes criticizing the work as the translation progresses. This text also appears in the *JS*, January 1787, pp. 119–20.

18 Alzate states that this article has been read at the Paris Royal Academy of Sciences in March 1787. He highlights three isolated points extracted from across the original text (pp. 290 and 292), providing the statements translated and in italics interspersed with his own criticism. The article is linked to the previous article criticizing Ignaz von Born as well as to Alzate's earlier periodical publication, the *Observaciones sobre la Física*, which he now defends using the present arguments.

19 According to Alzate, this report was read at the Paris Royal Academy of Sciences in June 1785. Provides a direct translation of the title but translates only a small section of the French report.

20 Direct translation of the full text and title, including the footnotes from the French translation.

21 States that it is a report from the *JP*. Direct translation of the complete text and title.

22 Reminds the reader that this is the continuation of the previous report. Direct translation of the complete text and title.

23 Informs the reader that he is translating the extract but does not state where it originates. Translates the title and all but the last paragraph of the French text; adds extensive footnotes with his own observations.

24 Translation of one short paragraph that is found on p. 111 of the subsection "De l'Indigo." He uses this text to show how advanced this particular area of technology is in Mexico compared with Europe.

25 This is a translation of only a small opening section of the text due, Alzate states, to the fact that the remainder of the report provides dubious information. Alzate states only that it is a work recently published in Paris in 1789.

26 Alzate translates the complete title but only sections of these letters, although they are presented in the Spanish text as complete without reference to omissions. He includes the original footnotes to all the correspondence.

27 Alzate again translates only sections of the text, regrouping the phrases and even, at times, adding his own thoughts as part of the translation.

28 Only four lines are translated. The third Michaelis letter is not included, as Alzate does not consider it pertinent to the present subject of the *GLM*.

29 A short postscript statement that these articles by Saussure came into his possession only after he had written the preceding report. Alzate references these articles to support his criticism of Lavoisier's system but does not quote from them.

30 Translates the original title but only part of the first paragraph of Willemet's text (pp. 145–46) before concluding the article with his own thoughts.

(continued)

165

TABLE 8.I.
(continued)

31 A complete translation of the text, including a translation of Pingré's corrections to the Collection. He states that Pingré published in 1774 and provides a title that is almost identical to the original. He includes the original footnotes as well as his own comments—the former numbered while his own lettered.

32 Refers directly in a footnote to the prize set by the Academy of Tolouse in France demonstrating that the Indians of Tonalá were more skilled in inventing non-toxic varnish than the Europeans.

33 The Pallas text is mentioned in passing as an introduction to the article in the *GLM*, describing it as referring to a viagero físico in Siberia in the year 1777 or 1778.

34 The Stehlin letter is mentioned in passing as an introduction to the article in the *GLM*. The reader simply is told that it is a "Carta dirigida por Stehlin . . .al Dr. Mary." Both this and the above texts are used to show that had information been published on the natural wealth available in New Spain many of the disputes between Naturalists would have been avoided.

35 Brief mention of this work in the footnotes of the *GLM* with a translation of three lines of the text (he states that it is on p. 83 of the *JP* but it is found on p. 162).

36 Brief announcement of the publication of the work stating that it is from *JP April* 1788, p. 304. He translates the title and provides a commentary on it.

37 Although Alzate sources this article to *BPE* 1783, p. 261, it has been included under the *JP* entries as this has not yet been verified and also to show the cross-appearance of texts in the French periodicals. Appears as postscript within the main text, drawing the reader's attention to the main contents of Magellan's argument and tying it to the subject of the previous article.

38 Translates the complete article, at times rewording sections; his own comments appear in the footnotes. A supplement is included on pp. 68–70, *GLM*, to show the already established Indian practices in Mexico.

39 Translates the title and the entire text and adds footnotes. Observations appear in the footnotes.

40 Translates the title and the entire text, including French translator's note. His observations are added in expansive footnotes.

41 Translates almost the entire review. Explains that he is translating an extract published by *JP* in 1776, and adds footnotes and a note by the Spanish translator at the end of the piece.

42 Translates the entire text, including the title. Alzate tells the reader that it is from *JP*, June 1782. The French footnotes are maintained, as is the original formatting of the text, including internal quotations in quotation marks.

43 Translates the majority of the review, including the original title. Extensive commentary is added in the footnotes as well as his own statements within the main body of the text, often in brackets. The differentiation between the original and Alzate's work is not evident from the format without comparing with the French text.

44 Translates directly all of the text from pp. 229–34, including the original title. After this point Alzate provides only an overview of the main details of the text, explaining that he does not want to tire the reader.

FIGURE 8.2. *(Left) Article by Pingré in the* Journal de Physique, *1774. (Right) Translation published in the* Gazeta de Literatura, *1791*

SOURCES: *(Left)* Journal de Physique *(May 1774), 317; The Bakken Library, Minneapolis. (Right)* Gazeta de Literatura *2 (November 1791), 235; Nettie Lee Benson Library, The University of Texas at Austin.*

167

TABLE 8.2

Articles from the Bibliothèque Physico-Économique *published in the* Gazeta de Literatura de México, *1789–1792*[1]

GLM	Bibliothèque Physico-Économique	Article Title
Volume 1, number 3b (8 October 1789), p. 23	1788, p. 237	Climent ou Mortier impermeable (by Volney)[2]
Volume 1, number 3b (8 October 1789), p. 24; Volume 1, number 4b (24 October 1794), p. 25	1788, pp. 348–52	Guérison d'une affection paralytique des extrêmites inférieures, par l'usage de la teinture de Cantharides[3]
Volume 1, number 7b (9 December 1789), p. 53; Volume 1, number 8b (23 December 1789), p. 89	1788, pp. 300–4	Observations sur les hommes de la province de Terre-Neuve en Amérique; par M. l'Abbé Gilli[4]
Volume 2, number 21 (14 June 1791), p. 165	1785, pp. 184–87	Manière de M. de Francklin pour imprimer aussi vîte que l'on écrit ; extradite des Mémoires sur la Méchanique & la Physique, par M. l'abbé Rochon . . .[5]
Volume 2, number 39 (15 May 1792), pp. 312–13	1785, pp. 310–12	Remèdes contre la piqure des Cousins, & precautions pour s'en garantir; publiés par M. l'abbé Rozier[6]
Volume 2, number 45 (28 August 1792), pp. 356–60	1785, pp. 285–93	Moyen de conserver les Enfants, sur-tout à l'époque de la Dentition[7]

1 The *Bibliothèque Physico-Économique* is referred to as *BPE* in the notes, although Alzate always calls it the *Biblioteca Económica*.
2 States that the work is taken from the *BPE*, published in Paris in 1788, vol. 2, p. 259. He includes a short paragraph and expands on how useful this idea could be in Mexico. It is possible that the difference in pagination here is due to the difference in publication houses between the cited text and the text used for this study. This has still to be verified for the 1788 volume of the *BPE*.
3 Reference is provided as "Ibid.," 1788, p. 247, as it continues directly on from the previous entry. It is a direct translation of three separate paragraphs from the original and includes the footnotes. See above reference for difference in pagination.
4 Reference again is stated as being the *BPE* 1788. Alzate focuses on sections of the text, translating short paragraphs and providing a general summary of other sections in order to criticize the content.
5 Referenced as found in the *BPE* 1785; this is a direct translation of the text.
6 Complete translation of the text, which is referenced as found in the *BPE* IV, p. 310. (IV = quatrième année). We can tell from the pagination cited by Alzate that he used the publication originating at Hôtel de Mesgrigny and not that of Hôtel Serpente.
7 Referenced as *BPE* 1785. Complete translation of the text.

TABLE 8.3

Articles from the Journal des Sçavans *published in the*
Gazeta de Literatura de México, 1789–1792[1]

GLM	Journal des Sçavans	Article Title
Volume 1, number 22 (25 June 1789), p. 97	October 1785, pp. 1689–92	De Vienne—Programme d'un prix propose—L'objet du Foundateur de ce Prix est de diminuer le nombre des Procès, sans resserrer la liberté des Paildeurs . . .[2]
Volume 2, number 8 (13 December 1790), p. 66	October 1771, pp. 117–37	Observaciones Meteorológicas de los últimos nueve meses de el año 1769, hechas en esta Ciudad de Mexico, por D. Joseph Antonio de Alzate y Ramírez. Eclypse de Luna del 12 de Diciembre de 1769 observado en la Impérial ciudad de Mexico. Impresso en Mexico en la imprenta del Lic. D. Joseph de Jauregui en la Calle de S. Bernardo. 1770.[3] See repeated reference below.
Volume 2, number 46 (11 September 1792), pp. 369–70	October 1771, pp. 117–37	Observaciones Meteorológicas de los últimos nueve meses de el año 1769, hechas en esta Ciudad de Mexico, por D. Joseph Antonio de Alzate y Ramírez[4]

1 For conciseness the *Journal des Sçavans* is referred to as *JS* in the notes, although Alzate always made reference to *Diario de los Sabios*.

2 Referenced as coming from the *JS*, October 1785. It is a direct translation of the short problem posed by the Journal and offers a solution as presented by Mariano Castillejo—the serious nature with which problems of this type are posed is ridiculed at the end of the article.

3 Makes reference to the appearance of this article in the *JS* and to the *Memorias de la Academia de Ciencias de París*. Uses this proof of publication in the *JS* to counter Beccaria's claim to be the first to have undertaken such observations in 1772.

4 Mentions that he had published this article in the *Diario Literario* in 1768 and that now it has come to his attention that the editors of the *JS* have printed it in 1771, adding comments (he states that this can be found on p. 559 but it is on pp. 125–26). He proceeds to translate the commentary for the readers, *GLM* p. 370, complaining that European publications only arrive in New Spain after a long delay.

TABLE 8.4

Articles from the Journal Encyclopédique published in the Gazeta de Literatura de México, 1788–1791[1]

GLM	Journal Encyclopédique	Article Title
Volume 1, number 11 (4 August 1788), pp. 93–94	September 1785, pp. 346–47	Institutions de médecine pratique, traduits sur la quatrieme cerniere edition de l'ouvrage anglois de Cullen . . . par M. Pinel[2]
Volume 2, number 14 (21 November 1788), p. 32	August 1786, p. 548	L'art des arpenteurs rendue facile . . . par M. L. -A. Didier[3]
Volume 2, number 18 (3 May 1791), p. 142	June 1785, pp. 312–18	Fin des observations sur la phthise de naissance par M. Portal[4]

1 The Journal Encyclopédique is referred to as the JE in the notes, although Alzate always used the term Diario de Bouillon.
2 Provides a title to this short review of Cullen's work that is a direct translation of the French and translates the entire French text. States that it originates in the JE, September 1786, p. 136.
3 Refers directly to the JS, August 1786, p. 548, giving the exact title of the work in Spanish as further reference for the reader. Does not include any of the text.
4 States that the extract is taken from the JE, June 1785, but gives the author's name as Poltal, instead of Portal. Mentions the title only in the footnotes.

(text continued from page 158)

one to two years between the appearance of articles in France and then in Mexico. In fact, when dealing with the Spanish periodical press we find a gap of only five months between the previously mentioned publication of Alzate's criticism of de la Porte's *Voyageur François*, in January 1788, and its republication in the *Memorial Literario* in May of the same year. Unfortunately, the question of the availability of these particular French periodicals is one that cannot be resolved in this particular study but remains as an important and interesting aspect of the research still to be undertaken. Other practical details regarding the editions of the periodicals in Alzate's use, however, do come to light within the information at hand. An examination, for instance, of the page references provided by Alzate for articles from the 1785 *Bibliothèque Physico-Économique*, when compared with other available editions, indicate that he had at his disposal the periodical printed at the Hôtel de Mesgrigny and not the edition originating at the Hôtel Serpente, both of which differ by about twenty pages in the pagination. Similar patterns have yet to be found in relation to the other publications.

Turning now to the Spanish and Spanish American articles, we can see clearly that this forms a much less significant body of material than the French, with only six articles included (see Table 8.5).[48] Of the works printed in Spain, we find *Memorial Literario, Instructivo y Curioso* (1788), *Diario Curioso, Erudito, Económico y Comercial* (1787), and *Memorias Instructivas, Útiles y Curiosas* (1785), although the latter is not strictly speaking a periodical publication.[49] Closer to home, Alzate includes articles from the *Mercurio Peruano* and the *Papel Periódico de la Havana*, both of which are the latest publications to be presented in the *Gazeta de Literatura* appearing in 1791 in each case.[50] The question of availability with regard to the Spanish-language periodicals is slightly easier to ascertain through announcements made in the *Gazeta de México*. All three of the Spanish publications are mentioned at various times, suggesting that either the works were already accessible or were entering the country at that time. The announcements take the form of subscription opportunities,[51] availability for purchase at certain bookshops,[52] or information regarding decrees by the Inquisition banning work in total or in part.[53]

Overall, we can tell from the information gathered in these tables that the four areas that most interested Alzate were articles relating to public health/medicine, meteorology, mining and minerals, and technology. It also becomes clear that, with thirty-five references, the third and fourth subscription runs that comprise volume 2 (1790–92) saw the greatest increase in extracts from the French periodical press. The

TABLE 8.5

Articles from Spanish works published in the Gazeta de Literatura de México, 1788–1795

GLM	Periodical Title	Periodical	Article Title
Volume 1, number 6 (24 February 1788), pp. 51–53	Diario Curioso, Erudito, Económico y Comercial	5 November 1787, number 503, pp. 553–57	Padrón de vecinos existentes en la Corte de Madrid[1]
Volume 1, number 7b (9 December 1789), p. 51	Memorial Literario, Instructivo y Curioso	May 1788, pp. 87–98	Historia de la Nueva España, por el Viajero Francés, alias el Abate Delaporte[2]
Volume 1, number 13 (6 November 1788), pp. 1–8	Memorias Instructivas, Útiles y Curiosas	Volume 10 (October 1785), pp. 249–73	Memoria XC: Sobre los diversos métodos inventados hasta hoy para precaver de incendio los edificios[3]
Volume 2, number 33 (17 January 1792), pp. 260–66	Papel Periódico de la Havana	4 August 1791	Jueves 4 de agosto de 1791 al Editor[4]
Volume 3, number 41 (17 June 1795), pp. 319–21	Mercurio Peruano	Volume 1, number 21 (March 1791), pp. 190–92	Carta sobre la profesión de abogado[5]
Volume 3, number 41 (17 June 1795), pp. 321–23	Mercurio Peruano	Volume 1, number 3 (January 1791), pp. 21–23	Desagravio de los mineros. Señores de la Sociedad de Amantes del País

1 The actual table is not included in the article, but the statistics are referred to within the text and a brief comparative table with Mexican statistics is added at the end of the article. Alzate states that he has taken it from the GM, 15 March 1788. This fact can easily be verified with the announcement and table appearing in vol. 3, no. 5, pp. 39–40, "En el Diario Curioso, Erudito, Económico y Comercial de Madrid del Jueves 15 de Noviembre de 1787 se lee la siguiente noticia que por ser bastante Curiosa nos ha parecido insertarla en nuestra Gazeta."

2 Makes reference to the republication of his article on de la Porte in the Memorial Literario as part of his introduction to the criticism of Abate Gilli's work in the BPE.

3 These were also published in the JP, October 1778, pp. 249–73, and April 1779, pp. 306–22, but Alzate references them directly to the Memorias, vol. 10. Does not include the text but uses it as a backdrop to his own observations.

4 I have not yet been able to verify this extract from the PPH. It has been included due to the fact that Alzate, five years later, writes an article in the GM (vol. 8, no. 42 supplement, 21 October 1797) in which he again provides an extract from the PPH from the same year, 1 January 1791, pp. 345–48. This suggests that Alzate did at least have access to copies from that year of publication.

5 Both texts from the MP are published in their entirety, including the footnotes, and are referenced as coming from the MP.

first two subscription runs making up volume 1 hold twenty extracts (1788–89), while the fifth and sixth subscription runs, volume 3 (1793–95), decrease to only eleven. This drop in number by volume 3 reflects a similar situation with regard to the number of active contributors to the *Gazeta de Literatura*. This falls from twenty-four individuals in volume 1 and nineteen in volume 2, to thirteen contributors in volume 3. Unlike volume 1, in which both a variety of authors and a sustained correspondence are in evidence, by volume 3 there is little example of such variety and Alzate seems to have turned increasingly to the publication of prolonged debates and reports, divided over several issues, most notably his report on the cochineal beetle that ran uninterrupted from February to September 1794. What, then, does Alzate tell us about his own reasons for taking this material and presenting it to the Mexican reading public? We turn now to consider this in our final section.

ACQUIRING AND ATTACHING:
THE EDITOR'S VIEW

In the Prologue to the *Gazeta de Literatura*, Alzate lays out both his purpose and the means by which he aims to attain his stated goals. As mentioned earlier, he sets the work in a clear geographical and national context, yet at the same time he does not limit himself to these boundaries in terms of intellectual content. Providing the voice for New Spain does not entail an insular or closed pursuit of knowledge limited to that found within the Americas.[54] The translation of foreign articles was to play a significant part in this process:

I will attempt through it [the Gazeta] to present reports and dissertations relating to progress being made in commerce and in navigation, either in the form of extracts, copies, or translating that which is useful: progress in the Arts will not be the least of those areas at which my ideas are aimed: Natural history, which provides so many wonders in our America, will be one of the objects of predilection.[55]

Certain key elements lie behind Alzate's translation and adaptation of articles in the *Gazeta*. First and foremost is his perception of his role in the Republic of Letters as a promoter of knowledge and defender of the truth. The fact that this is an international community and that Alzate has direct links with both Spanish American and European institutions of authority creates a platform that he feels justifies the validity of his own opinions. From this standpoint, he can then work to promote and defend the established knowledge and practices in the Americas; to disseminate information on science and the "useful arts"; and to defend his own observations and arguments through reference to his European counterparts. A few examples have been highlighted in this section in an attempt to create some idea of the tone adopted by Alzate in his treatment of these issues. To provide a wider overview, where possible, the various uses of the texts have also been indicated in the list of journal articles presented in the notes to Tables 8.1 to 8.5.

From the start, we must be aware that the *Gazeta* cannot be considered to be a neutral space for the gathering of knowledge. The highly personalized comments and criticism in the main text of the articles, as well as in the footnotes and end summaries, clearly indicate that every article formed an important part of Alzate's personal agenda based on his understanding of the needs of the Mexican nation. In addition, Alzate is aware of the growing need for a published record that would serve, on the one hand, as concrete evidence against the critics of the Spanish nation as a whole and, on the other, as a means of claiming back rights of ownership to what might now be termed "intellectual property" within the international community, that is, records of first discoveries and naming practices as well as established medical, technological, and botanical practices that predate European inventions or discoveries. In one memorable footnote to an article that had already provided extracts and criticisms of the work on amalgamation by Ignaz von Born, whom he declaimed as espousing ideas that were technologically lagging behind those of New Spain,[56] Alzate links the deception of the German's work to that of the outrages against the Spanish nation brought about by botanist Joseph Dombey: "What did Dombey do? Was it not necessary for the nation to publicly make his insolence and his actions known to the world? And even after this attack, did not the Spaniard, Cavanilles, sacrifice the name of several plants to him, now known as *Dombeya*, when as a lesson they should have been given the titles *Dombeya maliciosa* (malicious Dombey), *Dombeya ingrata* (ungrateful Dombey), etc.?"[57]

The concern to record and defend Mexican practices leads Alzate to consider the fact that valuable knowledge is not necessarily attained only through officially recognized institutions. These thoughts are reflected in his treatment of the account published on the construction of the monument by Marin Carburi in honor of Czar Peter, in which through extensive footnotes he observes how advanced Indian knowledge was in these matters of technology.[58] Yet his admiration of the Indian practices, especially when dealing with agriculture, would later lead Alzate to defend himself against the label of *entusiasta* (enthusiast) in his introduction to a report on indigo by Quatremer d´Isjonval.[59] Alzate is simultaneously attempting to awaken the Mexican national consciousness in two ways: first, by printing extracts so that his readers can see firsthand what is being published about them in Europe, demonstrating that the arguments are unfounded; and second, by encouraging his readers to become more aware of and involved with their surroundings, to own their natural resources and the traditional knowledge available to them, and to look for ways to exploit these riches that are beyond the reach of the average European. This he attempts to achieve not only through criticism but also through the publication of the problems set by the scientific academies in Europe, although he does not offer an equivalent reward in monetary terms. Thus, in the matter of testing whether wine sold in the taverns contained alum, as promoted by the academy of Lyon, he points out:

A question of such interest presented to the world in print does not exclude Americans from taking part in the search for a resolution to the problem. Are our hands tied? . . . Let us try, then, for our part to cooperate in finding a resolution that is of such great interest to the Europeans, in gaining the greatest advantage possible in their commerce, and to the inhabitants of America, in order to create an active market for so many and such rare natural produce.[60]

In this way also, new ideas on science and technology are opened up for the Mexican reader to allow for testing and response. One such example can be found in Alzate's attempts to encourage the installation of lightning rods on Mexico City's most prominent buildings. Despite having published lengthy translations of articles by Franklin, as well as correspondence between Lichtenberg and Michaelis, Alzate eventually expresses his exasperation by the lack of response from his readers and once again turns to the authority of the texts he has quoted:

Don't give credit to the words of the author of the Gazeta: but let the doubters examine those authors he cites, and if, after this, they remain obstinately set against what is written and what has been proven, they will be held responsible for the death of many and for the costs that are an inevitable part of restoring buildings damaged by such a powerful meteor.[61]

As in many cases, Alzate is concerned not so much with the theory behind the instruments themselves as with the practical outcome and the alleviation of a threat against public health.[62] In fact, his treatment of articles related to health in general indicate that, in the majority, they are used in a positive light to promote new practices. Two approaches are generally followed if a medical work was not available, either Alzate translates the reviews taken from the *Nouvelles Litteraires* in the hope that a book trader will purchase the work, as is the case with William Cullen's *Institutions of Medicine*,[63] or if the text is available in one of the French periodicals, and is not of too great a length, he will translate it himself.[64] At the same time, Alzate encourages the physicians and surgeons to adopt the techniques advocated in the translations, indicating that these texts will provide them with a *norte seguro* (true guide) in their patient care.[65] As well as promoting the use of new medical techniques, Alzate does at times offer the *Gazeta* as a space for criticism of the translated material if the ideas prove to be unfounded, but only by a *verdadero médico* (true doctor), the evaluation of which lay in the editor's hands.[66]

Before concluding this brief review of Alzate's presentation of translated material in the *Gazeta de Literatura*, it is necessary to state that there were inherent dangers involved in using articles taken from the European journals, whether in defense of an idea or as a basis for criticizing an author. This is especially true of those articles that had already undergone a process of translation in order to be used in the French journals. One clear example of this problem is Alzate's treatment of the *Saggio di Storia Americana* by the Italian Jesuit, Filippo Salvatore Gilij.[67] Taking the work as found in the *Bibliothèque Physico-Économique* (1788), Alzate launches a point-by-point acerbic attack on the facts and ideas presented by the Jesuit, accusing Gilij of being, at best, an inexact observer and, at worst, an outright liar.[68] However, when the French text and the Italian original are compared, we find that the very ideas most criticized by Alzate are those that have been mistranslated or badly adapted in the *Bibliothèque*. As a result, Alzate needlessly attacks a work that supports many of the ideas he espoused, and fails in the thorough observation and assessment that he demands of others.[69]

CONCLUSION

It is clear that the European periodical press played an important role throughout Alzate's career as an editor in Mexico. From the outset of the *Gazeta de Literatura* he states directly that much of his inspiration to undertake the publication of a fourth periodical is due to his appreciation of the work of Manuel Antonio Valdés and the *Gazeta de México*, but also due to his reading of the European journals. The additional fact that Alzate was directly linked to the Paris Royal Academy of Sciences as corresponding member highlights his desire for active participation in the Republic of Letters on an international level, but especially with reference to France. The particular attention paid to the French scientific periodical press is apparent when we consider the number of references directly linked to the *Journal de Physique*, over and above the Spanish press. Given the fact that Alzate clearly felt himself to be an equal, again within the Republic of Letters, to his European contemporaries, it is hardly surprising that such a "universal storehouse of knowledge" should serve as both an inspiration and a source of material for him. It also reinforces the fact that, in the area of scientific knowledge, Alzate privileged information originating in France over the Spanish domains. That said, it was not a process of passive adoption, but a matter of testing new ideas in the fires of the Mexican experience, disseminating what was judged valuable, and fomenting discourse in a public sphere that would become *la voz México* (the voice of Mexico).

The Indies of Knowledge, or the Imaginary Geography of the Discoveries of Gold in Brazil

JÚNIA FERREIRA FURTADO

A few years ago, on a research visit to the Evora Archives in Portugal, I had in my hands an old book by J. C. Pinto de Sousa, titled *Biblioteca histórica de Portugal e seus domínios ultramarinos* (Historical Library of Portugal and Its Overseas Colonies),[1] which cited several historical books and manuscripts on Brazilian history. One of them caught my attention: *Relação das minas brasílicas* (Report of the Brazilian Mines), written by José Rodrigues Abreu (see Figure 9.1). Pinto de Sousa pointed out that Rodrigues Abreu's work was the first known description of the Minas Gerais gold region of Brazil (see Maps 9.1 and 9.2).[2] It was an intriguing statement, as this book and its author are unknown in Brazil today. The discovery began an unexpected search for both of them.

This essay investigates both who José Rodrigues Abreu was and his report on the Minas Gerais region at the beginning of the eighteenth century. Although the cited manuscript has not been found, Rodrigues Abreu left his impressions of the gold region in another text, which informs us of his impressions about the area.[3] The vision evoked by his observations of the geography of that region is at once rational, magical, and mythological. Rodrigues Abreu's text was one of the first written on Minas Gerais, allowing him to distinguish between direct observation and that which had been told to him by others, as he would point out in his writings about the area. The preponderance of *seeing* over *hearsay* formed the basis for any empirical study of the period, and in his essay about Minas Gerais, Rodrigues Abreu describes the region as paradise while at the same time trying to analyze it rationally. This chapter will

FIGURE 9.1. *Portrait of José Rodrigues Abreu*
SOURCE: José Rodrigues Abreu, *Historiologia médica fundada e estabelecida nos princípios de George Ernesto Stahl*, 4 vols. (Lisbon: Oficina de Antonio de Sousa da Silva, 1739). Biblioteca da Ajuda.

focus on the apparent paradox of these two images—one marked by wonder and the other by reason.

José Rodrigues Abreu was a physician who accompanied Governor Antonio de Albuquerque on his expedition to Minas Gerais for the resolution of the Emboabas War—a dispute over control of the mines between recent immigrants to the area, called *Emboabas*, and the Brazilian-born explorers from São Paulo, called *Paulistas*.[4] The etymology of the word *Emboaba* is uncertain, and its meaning is rich in disparate connotations. Sometimes it was used to refer to the Portuguese alone; at other times it referred to everyone except the Paulistas—such as those from other parts of the colony, like the *Baianos* (born in Bahia) and *Pernambucanos* (born in Pernambuco)—as well as the Portuguese. It was used to distinguish the Paulistas, who had discovered and opened up the mines, from newcomers to the area who were reaping its profits. The term also distinguished those who had arrived in Minas from the north, via Bahia, by the *caminho da Bahia* (Bahia route), from those who had come via São Paulo, by the *caminho velho* (old route).[5] The

MAP 9.1. *Present-day state of Minas Gerais, Brazil*

Paulistas called themselves "sons of the soil" and "sons of the land" and opposed the intruders, called Emboabas, which in the "Brazilica" language meant "chicken with paint," in an obscure reference to the boots worn by the Emboabas, as the Paulistas themselves preferred to walk barefoot like the Indians.[6]

As the king of Portugal had promised the Paulistas possession and control of the mining area, they were naturally opposed to the invasion by Portuguese settlers and people from other regions of Brazil. During the conflict, insubordination was at an extreme, and the revolutionaries severed liaisons with the Portuguese crown. Manoel Nunes Viana, a local rebel, was proclaimed governor during the uprising in an open challenge to royal authority. The Portuguese crown did not sanction the actions of actual governor Fernando de Lancastre, who led a failed military expedition against the rebels, thus a new governor, Antônio de Albuquerque, was sent to replace Lancastre in 1709 and to negotiate an end to the Emboabas War. Unlike his predecessor, Albuquerque was a negotiator. His career in overseas administration had prepared him for his new position: he had governed the Maranhão captaincy between 1691 and 1701 and had demonstrated his negotiating skills upon returning to the area in 1705 as the king's representative, in his work with the French in the demarcation of the border with French Guyana.[7] Albuquerque's profile as a negotiator was improved by his ability to speak the "Brazilica" language spoken by the Paulistas at that time.

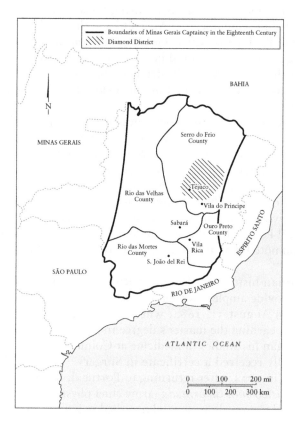

MAP 9.2. *Areas of colonization in Minas Gerais, 1700s*

Following his successful negotiations with France, Albuquerque was nominated governor of the São Paulo and Minas do Ouro captaincy. As was the custom, an entourage accompanied the governor to the mining region for his own safety and as a demonstration of his power. Author José Joaquim da Rocha mentions the "captain José de Souza, who had come together with his guard."[8] André Gomes Ferreira appointed "a captain, two foot soldiers, one assistant, and four servants," and José Álvares de Oliveira describes Albuquerque as having "one assistant, one lieutenant, and three soldiers."[9]

In addition to the officers and foot soldiers, an anonymous figure formed part of the governor's party, probably as his assistant: it was the doctor, José Rodrigues Abreu, who accompanied Albuquerque during his entire second stay in Brazil between 1705 and 1713. Albuquerque was sick when he first returned from Brazil to Portugal in 1701.[10] When he went back four years later, probably still weak, he demanded the presence of a doctor by his side. During these eight years, Rodrigues Abreu visited the captaincies of Maranhão in the northwest, and Rio de Janeiro, São Paulo, and Minas Gerais in the southeast. His presence in Albuquerque's entourage could be attested by the *sesmaria*, or piece of land, that the latter granted Rodrigues Abreu in Minas in April 1711. The land was located in the first mining fields of the Vila Rica village on the way to Ouro Branco.[11]

While in Minas Gerais, Rodrigues Abreu wrote the *Relação das minas brasílicas*, though at some point after 1747 the manuscript was lost. Yet, in his medical book *Historiologia médica* (Medical Historiology), published in four volumes between 1733 and 1752,[12] Abreu incorporated part of his notes about Minas Gerais in the entry on gold, in his dictionary of the "simples" in volume two.[13] Diogo Barbosa Machado, who wrote the valuable *Biblioteca lusitana* in 1747, an account of all Portuguese writers, supplied some information about the author and noted that at that time four copies of the *Relação* were still circulating.[14]

If Brazilian history has forgotten Rodrigues Abreu, the Portuguese archives provide ample information about his active life. He was born in Évora on August 31, 1682, where he studied Human Science and Philosophy, earning the master's degree in August 1699. The following year, he began his studies in medicine at Coimbra University, for which he eventually received a certificate in Surgery.[15] In 1705, he sailed to Brazil, and in 1714, after returning to Portugal, Rodrigues Abreu was appointed *físico-mor da armada* (army chief physician). It was his job to examine and license all doctors who wished to join the army and to punish those who attempted to do so without his license.[16] Eleven years later, he headed for the island of Corfu on an expedition against the Turks. In 1729, he accompanied King Dom João V on the mission of bringing about the marriages of the princes of Brazil and Asturias.[17] The exchange of the princesses was performed on the banks of the Rio Caia, near the border of Portugal and Spain, in the presence of both kings, Dom João V and Philip II, as well as several anonymous observers, among whom was the discreet doctor. The ceremony was majestic and included the building of a castle on the riverbank near a stone bridge to host the entourage (see Figure 9.2).[18]

José Rodrigues de Abreu's loyalty and zeal to the crown were royally rewarded. As was common in the early eighteenth century among important Portuguese doctors, especially those who served the royal family, he received several patents of nobility.[19] Rodrigues Abreu ended his days among the nobility, living on royal income and favors. By 1724, he had been appointed doctor of the royal house and a knight in the Order of Christ,[20] he was a *fidalgo*, or nobleman, of the royal house, and he was a family member of the Holy Inquisition and physician to the king himself.[21] In 1750, he was elected member of the recently created Academia Médica Ibérica (Iberian Medical Academy) in the city of Porto. Two years later, he lived on Rua Parreiras, where his books were sold when he died in 1755.[22]

Rodrigues Abreu did not waste the journeys, and as he suggests, he "wandered across all these lands with wise observations, learning new things about the medicinal virtues of herbs and plants produced by those vast lands."[23] The expansion of the borders of the Portuguese empire and the movement of the overseas voyages meant new additions to the practice and circulation of empirical knowledge—the result of observations by sailors, traders, clergymen, administrators, doctors, and others. (For references to this process, see the essay by Fontes da Costa and Leitão in Chapter 2.) In his works, Rodrigues Abreu tried to systematize the knowledge he had acquired in his travels, as was common among Portuguese men during the Enlightenment of the eighteenth century.

Like Rodrigues Abreu, other Portuguese doctors sought to renew medical knowledge that was marked by Galenism until the first half of the eighteenth century, as in the rest of Europe. The journeys of these doctors to other lands and the study of Portuguese students at other European universities were fundamental to the exchange of ideas and transformation of the sciences being promoted by scholars in other courts, but also by the Portuguese, with their practical experience and empirical observation on sea voyages. Rodrigues Abreu's travels with the Portuguese fleet and into Brazil allowed him to experience new worlds. He used his observations to reinterpret traditional concepts about medicine

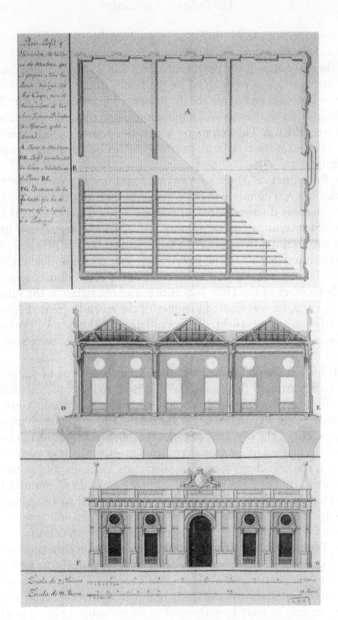

FIGURE 9.2. *Castle near Rio Caia at the Portuguese-Spanish border*

NOTE: The castle was built to host the Spanish and Portuguese wedding
parties in a 1729 double royal marriage between the two countries.
SOURCE: Archivo Histórico Nacional, Estado, Mapas, planos
y dibujos, fol. 754. Ministerio de Cultura, Spain.

and began studying nature on an empirical basis. (On the relationship between empirical knowledge and Portuguese science, see Walker in Chapter 13.)

Rodrigues Abreu also participated in a group in Portugal named the *estrangeirados*, or "foreign-made," those who sought to contribute to the political, economic, and intellectual development of the kingdom through travel to Europe or the overseas empire. This small group—made up of overseas representatives, ministers of the government, and intellectuals—widely discussed the issues of the kingdom.[24] Notable members of this informal group were Martinho de Mendonça Pina e Proença and the Cardinal of Mota, both of whom, not by mere chance, took on the charge of writing the introduction to Rodrigues Abreu's book. The *estrangeirados*—some of whom were in Brazil as royal civil servants or as part of the Overseas Council[25]—by and large belonged to the intellectual elite promoted by Dom João V.

THE BOOK OF NATURE

The disappearance of Rodrigues Abreu's report does not prevent us from reconstituting at least part of his impressions of Minas Gerais. A few years after his return to Portugal, around 1730, he began to write the *Historiologia médica*, an account that attempted to join his practical experience with his theories of modern medicine. On February 25, 1734, King Dom João V granted the necessary license for a ten-year circulation of the book, which the doctor himself "had composed, written, and printed."[26] The book was published in stages between 1733 and 1752, comprising four volumes in all. The first part was dedicated to the medical professors, and Martinho de Mendonça in his introduction noted the author's ability to stand apart from traditional doctrines by summoning the principle of authority, the grounds on which Scholasticism was founded. He exalted the author as "a new [Vasco da] Gama, who leads the Portuguese across seas never before navigated in order to discover the wealthy Indies of knowledge."[27] It was a clear allusion to the poet Luís de Camões, who described the journeys of the Portuguese navigators, also suggesting links between the Portuguese journeys and the improvement of knowledge (see Almeida, Chapter 4). Volume 2, written as a dictionary of "simples," was dedicated to Cardinal João da Mota, who was the king's confidant when this volume first reached the public in 1739. Da Mota and Martinho de Mendonça were both important men of the Enlightenment, connected with the world of books and science, and organizers of the king's library—one of the biggest of the time but unfortunately destroyed in the Lisbon earthquake of 1755. Martinho de Mendonça also published several Portuguese authors abroad. Volumes 3 and 4 came out in 1745 and 1752, respectively, and were dedicated to the medical practice. Both were addressed to Dom Pedro, prior of Crato, Dom João V's son.[28]

All of the censors agreed that the author had based his writings on the texts of the most disreputable writers, especially those of "George Ernest Stahl, infamous writer of our times."[29] Nevertheless, they affirmed that he had done so in such a way that it had neither offended the laws of God nor the kingdom, and he wrote with such clarity that the text had much to teach to both beginning and experienced doctors.[30] Because of Rodrigues Abreu's innovative instruction, the censors decided that the work deserved to be published. What did the censors mean by the assertion that the book brought together two contradictory philosophies? How did Rodrigues Abreu bring together two opposing concepts, one according to the Catholic church and the other according to the heretical thoughts of Stahl? On the one hand, it is important to point out that, by taking Stahl's studies and those of other heretical philosophers as a starting point, Rodrigues Abreu placed himself among the vanguard of that time, favoring the principle of rationalism over authority and pointing to the excellence of empiricism and experimentation. On the other hand, it must be stressed that these changes happened within the context of traditional thought; new ideas mingled with previous ones with complex results—often refusing the interpretations of a linear evolution of the history of science and ideas.

Since the sixteenth century, a number of authors began arguing against aspects of Galenism, as empiricism gathered strength as a basis of medical practice. Experimentalism and anatomy put several of its concepts in check, but at the same time, alchemy and astrology were still important auxiliaries to medicine.[31] Paracelsus (1493–1541) created one of these paradoxical theories typical of the study of medicine at this time. Against Galenism, he argued that each disease was specific, had an independent existence, and should be treated with chemical remedies. He said that medicine should be practiced by reading the Book of Nature.[32] (For a further discussion of the concept of the Book of Nature, see Pimentel in Chapter 5.) Founder of iatrochemistry, Paracelsus called for a more rational explanation of the world, based on mechanical and mathematical principles.[33] He maintained several points in common with the alchemists, calling for a knowledge based on experimentation and consistency; but he also defended the existence of a spiritual essence, suggesting that God was in everything, contrary to the naturalism of Galenism. He also believed that the human and celestial bodies obeyed the same laws and that the stars influenced the human body.[34] Alchemists and astrologers had similar points of view. Alchemy held that a spirit was behind all things, alive or not, a primary element that was separated into the four lesser elements: earth, air, fire, and water. Paracelsus and the alchemists held the same animistic view of nature.[35] Both looked for a fifth element that could not be destroyed, had the power to cure all diseases, and went against the rules of decay.[36] Paracelsus and the astrologers believed in the influence of the planets and stars on bodies, human morality, and weather.

As we can see in Paracelsus thought, naturalism and animism were not always in opposition, as he proclaimed that chemical elements could be used to treat specific illnesses and at the same time recognized the existence of the spirit behind all things, influencing all creatures. The same kind of apparent contradictory thoughts can be seen in Rodrigues Abreu's text and in the books by Stahl, which inspired him. With no linear scientific development, opposite principles could be sustained and could, in fact, reinforce each other.[37] For example, a search for a more rational explanation of the world could at the same time reinforce the idea of a God above all things, in accordance with the thinking of Paracelsus, Stahl, or Rodrigues Abreu. As a good Catholic, José Rodrigues Abreu believed in God as the supreme mover behind all living creatures, but as a Stahl follower he proclaimed that each disease had a specific treatment that should be tested empirically.

To understand how Rodrigues Abreu described Minas Gerais, it is necessary to examine his and Stahl's ideas. Stahl was a German physician and chemist. As a man of his time, his ideas were influenced by alchemy and astrology as well as new theories based on experimentation. He called himself a philosopher, and as such he tried to read the Book of Nature, saying that the whole of nature was regulated by chemical principles; however, in argument with iatrochemistry, and closer to the animist position, he asserted the existence of an *anima sensitiva*, or God-created soul, but only in living creatures. He divided the material world in two: the organic and the inorganic, with only the former containing this vital force.

The idea of all living creatures being subordinate to souls was the essence of Stahl's theory. The body existed because of the soul, where the body was the instrument of the *anima*.[38] As Paul Hoffman suggests, "L'âme, chez Stahl, est un être actif, intelligent, autonome, architectonique, capable de varier les mouvements vitaux, qu'il produit conformément aux fins de la vie. . . . L'âme a une réalité ontologique,"[39] and according to Stahl the soul was "the Energy or constant action of God in sustaining his creatures."[40] If these ideas drew Stahl close to the principles of the Catholic church, the similarity stopped there. His idea of a soul in all the organic elements was instead much closer to the alchemists' idea of a primary element.

Stahl's chemical principles, while intended to achieve "the foundation of a scientific manner of inquiring and preparing the natural and artificial bodies for the uses of life," were filled with heretical thoughts. In section two, he defends an empirical way to produce the philosopher's stone, the basis of alchemy.[41] He states that this art came from the Egyptians and Arabians and was propagated among Europeans by the *Declamations* of Paracelsus. As a good philosopher, Stahl asserted his opinion only after observing "the better operation will succeed," and he concluded that "the most commodious method of all promises to be that which proceeds with running Mercuries, and reguline sulphurs." Stahl believed in the possibility of producing a philosopher's gold, which had the property of being able to be multiplied or increased in quantity, and thus improved in quality, defying the rules of decay and putrefaction.[42]

In the introduction to his book, Rodrigues Abreu affirmed Stahl's ideas. But he found a loophole with which he could express his ideas without angering church authorities. Stahl declared that a spirit was behind all creatures—animate or inanimate. He explained combustion as a burning of the spirit principle of the material. But employing Stahl's division between organic and inorganic matter, José Rodrigues Abreu argued for the existence of a "vital agent, intern[al] spirit, soul, *archeo*, plastic virtue . . . an inorganic spiritual principle, the first thing that moved and ruled all moral and rational actions, but also the animals." In addition, he insisted that this spiritual principle was present only in living creatures, and that "God govern[ed] the world." In this way, the stars and celestial bodies could not influence human judgment or their moral conscience, although they could influence human bodies. With these statements, he was closer than Stahl to the animism of the Catholic church, which denied the possibility of inorganic material possessing a spirit. In closer accordance with these principles, José Rodrigues Abreu declared that God was behind all processes of Creation and behind all living creatures, and only these could have an anima, or soul.

Rodrigues Abreu tried to clarify his disagreements with the alchemists and stressed them in several statements. When speaking about the therapeutic virtues of gold, he tried to stay away from the legends and doubtful beliefs of the alchemists. They claimed that gold "receive[s] the influences of the best and most brilliant of all stars, the sun," that it was produced by art, and that a Muslim could make gold from silver and another kind from red oil. Rodrigues Abreu discarded all ancient and traditional knowledge of gold. Most important, he assured that it was impossible to make gold more perfect by heating it. The main principle of alchemy was to find the perfect gold or quintessence, a substance that did not follow the principle of decay. He also criticized Pico della Mirandola's statement that gold was found in the gizzard of the partridge, and the belief in the existence of a Hebrew manuscript that taught one how to make gold from an herb such as *mangerona* (marjoram), which grows by the influence of the moon.[43]

While Rodrigues Abreu disagreed with ancient precepts of alchemy, he also accepted the idea that gold was a perfect mineral. The influence of alchemy was partly hidden in his text. Agreeing with Stahl, Rodrigues Abreu believed that gold was "the most perfect metal of them all"[44] but also affirmed that alchemy was full of "many enigmatical traditions, contradictory accounts, and mutual confutation."[45] Since Rodrigues Abreu intended to construct knowledge with empirical methods, he did not discard the alchemist's idea that gold was a perfect substance. He rejected only the beliefs that came from ancient alchemy that maintained that it was possible to create a perfect gold by artificial methods, because experimentation had not validated these claims.

Although Rodrigues Abreu suggested that gold became purer when put in fire and melted, he rejected the notion of its containing any miraculous property. He denied its use as a medicine for most diseases—as was recommended in "Solar Medicine," as though gold were the fifth element.[46] He argued that "the Royal Academy of Sciences in Paris applied gold to several patients' bodies internally and externally, and none of them became better or worse."[47] Like Stahl, Rodrigues Abreu had no doubt that empirical tests would always outweigh the evidence of mythological beliefs, but as we will see in his descriptions of Minas Gerais, perfection was related to gold, and presence of the metal produced this same flawless quality in nature.

Like his master, Stahl, who also read the Book of Nature with its experimental and rational principles, Rodrigues Abreu wondered where gold originated. Disagreeing with contemporary myths, he intended to report what he "observed in the mines of Brazil . . . of the climate of those lands, of the situation of the hills, abundance of rivers, and the wealth of its treasures." For Rodrigues Abreu, only experience could serve as a basis for the construction of conclusive thinking, thus establishing the transitive relationship between reality and speculation, as "it was [still] not certain how gold was produced."[48]

Following these principles, in his report he tried to differentiate what he had learned through his own accurate observations from that which was generally believed based on other people's reports. Empirical and rational, he favored experience over hearsay, as was becoming common at that time among the naturalists in Europe and Latin America. This mechanism of direct observation can be noted in his description of boa constrictors present in the Minas Gerais region. Although Rodrigues Abreu saw them as true monsters and aberrations of the ever-prodigal nature of the area, there was nonetheless strong evidence of their existence because "we *have seen* the skin of one of them, which was ordered to be hung in the palace room where the governors were gathering in the village of Nossa Senhora Carmo" (emphasis mine); however, having not gathered enough information to positively proclaim their existence, he left it to the opinion of the practical people of the interior, who said of the boas, "one is born from another's bones."[49]

As a reader of the Book of Nature, Rodrigues Abreu could only accept knowledge based on empirical experience. A good reader could find the rules that the Creator had established and that governed the world, but when this was not possible, and because he was a good Christian, he also looked for the explanations in the Biblical tradition. Minas Gerais would represent for him one of these paradoxes. There, gold was fused to the earth itself and all of God's rules triumphed, making Minas Gerais the earthly paradise itself. In that wonderful land, the imagination had no limits, connecting Biblical, traditional, and empirical theories.

PARADISE REGAINED:
REPEATED LINES THAT CROSS ONE ANOTHER
IN THE THEATER OF THE WORLD

Rodrigues Abreu's geographical observations on Portuguese America were scattered among his medical notes, his descriptions of diseases, and the natural medicines found throughout his journeys. In his report about Minas Gerais, however, his practical and rational spirit was overcome by wonder. In his imagination, that place corresponded to the idyllic, the paradisiacal, although he said he would describe only "what was observed in Brazilian mines . . . while he lived there." It was the land of "benign and healthy airs . . . due to the conjunction of some benevolent star."[50] As the land itself was inorganic matter, it was not heretical to accept the influence of celestial bodies over it. For Rodrigues Abreu, Minas Gerais symbolized the opposite of the captaincy of São Paulo, where goiter was abundant. When describing São Paulo, he adopted an Emboaba discourse, defining the Paulistas as savages, ugly, and uncivilized. On the other hand, those born in Minas Gerais had a marvelous nature and well-formed human bodies.[51] From an Emboaba point of view, Minas Gerais was paradise, but only for the Portuguese: the rich land was revealed to be the redemption of the Portuguese empire, as had been predicted in millenarian prophecies important to Portuguese culture at that time. (On earlier millenarian prophecies in relation to Spanish Pacific exploration, see Sheehan's essay in Chapter 12.)

In Minas Gerais, everything, even the stars, combined in an exuberant and rich nature, crowned by gold, silver, and other metals. Hunting and fishing were abundant, the waters of the torrential rivers were crystalline, and the forests supplied wood of every kind. The brush was high and virgin, and the superb mountains reached the clouds. As Rodrigues Abreu suggests, there were infinite diamonds, "excellent emeralds, daughters of the same land, very fine marble, very clean and clear jaspers, transparent crystals, amethysts, rubies, topazes, square stones, etc."[52] No longer the cold scientist, Rodrigues Abreu described Minas Gerais as paradise on earth, joining wonder, prodigal nature, and mineral wealth, which were part of the early modern European collective imagination.

We can be sorry for the loss of the original manuscript of the *Relação das minas brasílicas*, as it would have provided detailed information about the colonization of Minas Gerais in the first decades of the eighteenth century. Yet the wealth of Rodrigues Abreu's existing account, incorporated in his book *Historiologia médica*, is found not only in its documenting of important empirical data on the region but also in the possibility for historians of culture to dissect the mental mechanism that led Rodrigues Abreu to this construction of reality. In this sense, the loss of the report has generated more fertile commentary on the natural world of eighteenth-century Minas Gerais. The text of *Historiologia médica* reveals thought-provoking challenges to those who believe that history is made not only of facts but also from the impressions that men have of the real, even when they convert that reality into wonders.

The persistence of thought marked by a theological and nonrational reading of reality, present in Rodrigues Abreu's description of the geographical conformation of Minas Gerais, can be understood only when one perceives the role that the Brazilian mining region represented for the author. Its wealth, guarded in the entrails and on the surface of the earth, drew it close to paradise, close to something that escaped human comprehension and had its own logic, given by the Creator and not revealed to man. Insofar as these complex events could not be explained by reason, it was necessary to seek other sources, and working in this way, Rodrigues Abreu paradoxically defended the existence of a terrestrial paradise, a concept that was no longer accepted even by the Catholic church in the eighteenth century, and he located this paradise in Minas Gerais.[53] How can it be that a man of such rational spirit, a reader of heretical authors, could describe Minas Gerais based on the belief in the existence of an earthly paradise?

Part of the answer can be found precisely in Rodrigues Abreu's effort to be empirical. The data he collected in Minas Gerais were inexplicable: the wonder of nature, the perfection of the bodies, the uncountable wealth, and the gold appearing anywhere with no order or rationality to its origins. When the Book of Nature could not provide a rational explanation for the earth's events, these answers had to be found in another source of thought, the place where God revealed himself, thus it was impossible to use rationalism to understand the inexplicable. The only explanation he could find was that it was there where God had placed paradise. As the Bible and theological tradition became its source, an apparent paradox emerged in Rodrigues Abreu's system of ideas. When he took from Genesis the concrete existence of an earthly paradise, an outdated idea by that era, and placed it in Minas Gerais, a second apparent paradox emerged. His writings, however, disclosed no such paradox, as his conclusions came from direct observation of the area and allowed the presence of two seemingly contradictory statements in his system of thought: the rational and the paradisiacal descriptions of the land.

Rodrigues Abreu's geographical presentation of the area describes precisely the location of the mines, which began "in the area of Ibitipoca, a six-day journey before arriving at the village of Rio das Mortes, and ended in Tocambira, beyond Serro do Frio."[54] He was meticulous about the distances that separated Minas Gerais from the captaincies of Rio de Janeiro, Bahia, Pernambuco, and São Paulo, not failing to note, like a good empiricist, that "there may be mistakes, since these measurements are more a result of reflection and the computation of days."[55] As in the rest of the discourse, the tradition of rustic men, mainly those from São Paulo traveling in that area, stood in contrast to that of the erudite and the experts, and Rodrigues Abreu invariably chose knowledge from the latter sources. He described the three ways to Minas Gerais: the old route that departed from São Paulo, the Bahian route that crossed the cattle ranches, and the new route that departed from Rio de Janeiro. He could state confidently that the mines were located on the same latitude as the captaincy of Espírito Santo (Holy Spirit), twenty degrees and five minutes, sixty yards from the coast, "according to the observation of some pilots who came to those lands and took measures of the sun."[56]

Once in Minas Gerais, however, the account of the geographical position of the mines abandoned precision and began to suggest the wonderful and the dreamlike. This process of disconnection from reality was reflected in an imaginary cartography, in which the mines were at the center of America. This displacement from reality and abandonment of the excellence of the empirical and observable can be noted in the emphasis Rodrigues Abreu places on local hydrography in his geographical description. For him, "in this place, as the center of repeated [or parallel] lines, the most torrential rivers of America emanate." This phenomenon occurred because "all the rivers that are contained in and are distilled from the mountains run inland."[57] Here, Rodrigues Abreu printed the first of a succession of images that placed Minas Gerais in the place of paradise. In opposition to the rules of nature, the Minas Gerais rivers run inland and not to the sea. Also, he named four rivers—the same number as the rivers of paradise—which had their sources in the center of the captaincy.

Following the course of his imaginary understanding, Rodrigues Abreu swiftly imploded the limits he himself had established between reality and the imaginary, between myth and what could be observed. The interior of Minas Gerais became the center of this "corporal American illustration of many parts" in whose center the "notable Gold Lake, or *Xarais* (as it was called by the natives), was born at the heart of this body, located almost in its center; it divides the lands into two arms, or two rivers, working as a barrier: in the North, the Amazon [River]; and in the South, the Plate [River]."[58] This lake distinguished and demarcated Minas Gerais from Peru, where the Spanish had found silver. In this way, the sources of the metallic wealth of America were close to each other, but put apart. Born from this inner lake, spreading to the north and south, the Amazon and Plate Rivers divided the Portuguese possessions from those of the Spanish, to the north as well as to the south. Minas Gerais became Portuguese America itself, paradoxically constituting its center and its external limits.

From Minas Gerais came four other important rivers that would define its inner borders. This chain of rivers enclosing the mines also recalled the four rivers that originated in paradise. That was the same image that the heretic, Pedro de Rates Henequim, used before the Inquisitors to justify his claim that Minas Gerais was "the earthly paradise in which Adam was raised."[59] By joining these two images, Abreu's cartographic vision of the area placed Minas Gerais in the pulsing center of two concentric circles whose limits, on both sides, were established by the rivers that surrounded, embraced, and protected it: the inner circle isolated the area from the rest of Portuguese America, with the outer ring separating off the Spanish lands; in the center, Gold Lake was like a heart, the source of life, in whose boundaries an abundance of wealth could be found.

This geographical image was not an entirely new one at the time, since the outer part of Rodrigues Abreu's hydrographic circle had been part of the cartography in Portugal during the seventeenth century, when the Portuguese first mapped the interior of Brazil—still seen as wholly virgin and unknown. Several maps of the sixteenth and seventeenth centuries represented Brazil in this way, among which are notably André Homem's *Planisphere* of 1559 and the anonymous *Teatro mundi*, from about 1600. The *Atlantic Letter*, attributed to João Teixeira Albernaz I, drawn in about 1640, is the last-known Portuguese cartographic representation that still represented the mythic image of Brazil as an island. From Portuguese mapmaking, the image spread to Dutch, Italian, English, and other European countries with mapmaking traditions (see Map 9.3).

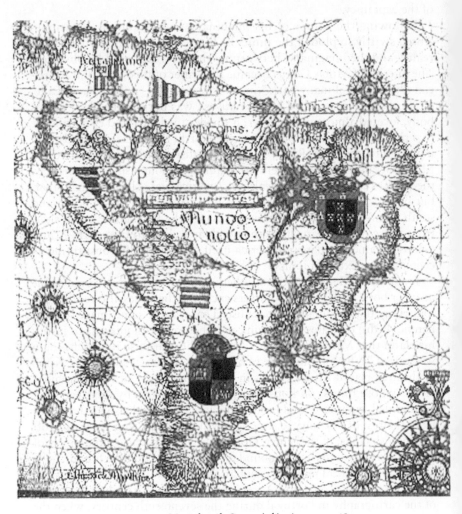

MAP 9.3. *Map detail*, Carta Atlântica, *ca. 1681*

SOURCE: João Teixeira Albernaz I, in Gilberto Costa, Júnia Ferreira Furtado, et al., *Cartografia da conquista do território das Minas* (Belo Horizonte: Editora UFMG, 2004), 14.

These images bore no correspondence to reality. The rivers' sources were in basins and were not connected; furthermore, the Amazon and Plate Rivers ran very far away from Minas Gerais. In part, this image stemmed from the Indian legend of an inner lake from which sprang two big rivers forming the natural borders of the interior of Brazil. The first travelers that wrote about Brazil in the sixteenth century perpetuate many of these indigenous legends. The French priest and cosmographer André Thevet, who was in Rio de Janeiro in 1555 during the French invasion, and Gabriel Soares de Souza, a Portuguese explorer and naturalist who wrote a book in 1587, both described several of the natives' myths, such as the belief in the immortality of the soul that goes to idyllic gardens after death, or the existence of a universal flood.[60] When speaking about the São Francisco River, Rodrigues Abreu also invoked another indigenous legend, one that described the circular course of the river and its disappearance within the earth.[61] Abandoning the discourse of the erudite, he reported that "they say there is a drain in which that sea of water pours through and that at some distance the river will appear again."[62] He also stated that its course traveled a circular route along a great part of its length. These beliefs derived not only from indigenous mythology but also from Christian eschatology associated with the existence of a terrestrial paradise. French theologian and philosopher Pierre d'Ailly (1350–ca. 1420) described the courses of the Nile or Gion Rivers after they left Eden as circular and subterranean as far the Indian Ocean, until they reappeared under the hills of the Moon.[63]

One of the reasons for this interpretation of Brazilian hydrography was the need to find natural limits between the Portuguese and the Spanish possessions. Sergio Buarque de Holanda has said that "at this moment, the myth of Brazil as an island was born: the idea that Brazil was an island defined by the ocean and by the communicating hydrographic systems of the Amazon and the Plate Rivers."[64] The misunderstanding of the South American inner geography, the acceptance of Indian information and myths, the impact of the inland spread of the Portuguese settlement, and the Portuguese' fear of losing control over their colony to the Spanish crown all enforce the desirability of a natural border between the Spanish and Portuguese territories in America. The Portuguese and Spanish crowns were joined from 1580 to 1640, in a period known as the Iberian Union. Rodrigues Abreu also believed the rivers were useful for separating the lands of both crowns, as he wrote that sometimes the Paulistas, looking for Indians to enslave, crossed the inner river and entered the Spanish lands.[65]

Jaime Cortesão had another explanation. For him, the image was not a fantastic interpretation of Brazilian geography, but one derived from a misunderstanding: Gold Lake was in fact the Pantanal. During the rainy season, the Paraguay River in Brazilian lands flooded and spread to the lowlands, which appeared to be a lake. The misunderstanding of local hydrography fused the Paraguay and Plate Rivers as one, and created an image of an inner lake that could form the common basins. For Cortesão, this interpretation was completely unrelated to political interests, which could be proved by explaining how the Dutch, Italians, and Germans had represented Brazil in the same way on their maps of the time without political ends.[66]

By the beginning of the eighteenth century, the Paulistas were entering the area from São Paulo, going up the rivers in several expeditions. They soon realized there was not an inner lake but a seasonal flood of the Paraguay River. They built up an image of an enormous swamp and called it the Pantanal.[67] Wisely using the region's network of rivers, the Paulistas discovered the entire Brazilian central-west interior by trespassing over the imaginary line of Tordesillas.[68] The Portuguese maps that relied on these sources in the eighteenth century did not describe a golden lake, or the image of Brazil as an island, but rather the Paraguay River and the Pantanal itself.

By the time Rodrigues Abreu arrived in Minas Gerais, the Paulistas had already described the Pantanal, dismissed the image of an inner lake, and become very familiar with the hydrographic network. Rodrigues Abreu used neither their knowledge nor their methods of direct observation, however, when he described the geographical conformation of the land. Two reasons explain this. First, since the Paulistas were rustic men and he himself was an Emboaba, he dismissed them as untrustworthy sources. Second, his main sources in the report were sixteenth-century Portuguese maps, Spanish navigation charts, and the studies of other Europeans such as Hans Staden from Germany,[69] Jacobo Cartier from France,[70] and Jeronymo Benzoni from Italy.[71] Most of these sources gave imprecise information about the land, and many also incorrectly suggested that Peru was fairly close to the Portuguese settlements.

What is curious about the description of Minas Gerais by the rational and empirical author, José Rodrigues Abreu, was the persistence of a geographical representation of Brazil as an island, long out of use, having been substituted by more realistic representation. A broader, concrete knowledge of inland Brazil was decisive in ending this myth, accomplished through Paulista expeditions that explored the interior and through occupation of the central captaincies of Minas Gerais, Mato Grosso, and Goiás in the first half of the eighteenth century. Rodrigues Abreu was among those who traveled across the country, and the chronological gap separating his journey and the publication of his book on medicine was sufficient for him to have incorporated a more precise knowledge of the area. During this period, a well-developed Portuguese mapmaking tradition existed in Brazil, and Rodrigues Abreu, as part of the *estrangeirados*' circle, was probably well aware of it.

Because Rodrigues Abreu hated and dismissed the Paulistas, who constituted an important source for building this new image of the land, as an Emboaba he ultimately chose to keep the paradisiacal images of his old books, as they were better able to illustrate what he saw. Vigorous, savage, rich, and exuberant, in his view the mines were protected from human greed, safeguarded by Portuguese redemption. The two hydrographical circles that protected the mines were completed in his description by a third circle formed by a chain of mountains. The rivers and mountains completely enclosed the land.[72] Like Ptolemy, who claimed that the earth was at the center of the universe, Rodrigues Abreu inaugurated an earthly cosmography centered not merely on Brazil, but on Minas Gerais, an almost unassailable land that "hides, keeps, and never faces men's greed in vain," where nature has "deposited one of its largest treasures, defended by almost unassailable walls of hills upon hills, which hinders the step of the ambitious travelers, so that with effort they have to discount what interests them."[73] Very much like paradise, redemption in Minas Gerais could be achieved only through great effort. A man of his time who fused reason and myth, José Rodrigues Abreu was located at the intersection of two systems of thought, one marked by wonder, another by science.

CHAPTER TEN

Space Production and
Spanish Imperial Geopolitics

NURIA VALVERDE AND ANTONIO LAFUENTE

When speaking of maps it is necessary to distinguish between the concepts of territory and of space in order to differentiate the ways of experiencing, remembering, or thinking of places. The concept of space indicates the management of territory, or a raw extent of land. It is thus a human production with historical and cultural connotations, whereas territory denotes the land itself. Hence the importance of maps in representing space, especially those that are to be validated before a court or in other corridors of power. Like any other kind of representation, maps are not neutral. Their makers select the items believed to be worth depicting on a flat surface, and design them to seem unequivocal and objective. But design also implies human expectations, values, and hopes, hence the construction of a symbolic space upon which past, present, and future political aspirations can be reworked. Rather than showing what a territory is like, maps show what the space that they create may become by using different strategic approaches. Thus, in an international context, maps are designed to stabilize the most advantageous way for a country to define its political weight, economic interest, and field of influence. A map is a document and also a monument to history: to its value as a record we have to add a diplomatic function, the efficacy of which depends on the stability and timelessness of the map. (For the relationship between cartography and courtly politics, see Portuondo in Chapter 3; for cartography as a project of elites, see Pimentel in Chapter 5.)

This point of view transcends the traditional interpretation of maps as tools that serve to show us the way to places, and emphasizes that they represent an original and efficient form of managing time as well as space. Thus, paradoxically, geographers would be experts at translating spaces into time: how long a journey might take (roads), the traces of administrators (estates or towns), the memory of conflicts (borders), the existence of sponsors (the names of things), or the signs of progress (ports or mines).

The production of maps is not as uniform as we might think, and differing practices record the different ways that empires sought to establish territorial claims and aspirations. The data, assumptions, and uncertainties contained in maps and the corresponding management of time have a political purpose: they establish borders and fix territorial ownership, and in this way maps have geopolitical value. For that reason, in the eighteenth century, the new technologies applied to space production—scientific expeditions, geodesic cartography, the algebra of populations and resources, and the engineering of borders and forts—made maps particularly significant.

Most of the geographical information produced by travelers and merchants as well as missionaries during the late seventeenth and early eighteenth centuries gathered data both on the physical environment and on ethnography. Fostered by commercial interest or by the purpose of evangelization, their investigations led to the production of chorographies that reinforced a holistic approach to the landscape based on the accumulation of empirical data.[1] The maps they produced were thus topographic repositories ordered within both the territorial and jurisdictional fields, whose intention was to show ownership and establish political hegemony.[2] Two things happened to this type of representation in the mid-eighteenth century. In Europe, the geographic paradigm promoted by Jean-Baptiste Bourguignon d'Anville helped to perfect the protocols of documentary criticism in cartography, especially concerning the location of towns. His method also emphasized the importance of purging "formal" or "modern" maps of descriptive ethnographic information.[3] In the case of Spain, this effort to replace the rhetoric of description with that of formalization was coupled by an imperialist desire to form a homogenous image of the totality of their American possessions.[4]

With these efforts, the information traditionally present in maps disappeared from the new cartography, replaced by empty, blank spaces. These spaces, which Anne Godlewska has attributed to a "sense of geographical ignorance,"[5] actually gained progressive strategic importance in the interpretation of maps over the course of the eighteenth century. A reading of such empty spaces is needed to understand the development of maps as historical documents rather than as repositories of true and accurate information. Brian Harley has stated that unintentional (not explicitly commanded) cartographic silences are due to the elimination of uniqueness imposed by scientific standardization, and that the affirmative support they give to the political status quo is related to an unconscious rejection of social groups by the powerful upper classes. Our approach, however, seeks to explore a more subtle role for the empty spaces in maps, one that is not so closely related to an exclusionary ethos, but rather one that illustrates imperial desires to establish separate legislative and managerial orders (biopolitics and geopolitics) that called for different degrees of stability.[6] To do so, we shall explore the different cartographical practices deployed during the Spanish Enlightenment to produce inland and coastal spaces in America, and the symbiotic connection between geographical information and political context.

ATTEMPTING THE BIG PICTURE

In 1750 no one in the Spanish court would have been able to understand the disparate terms of reference of different geographical sources, to compare them critically, or to translate all this information into a single work. That is, there was a lack of know-how of the technologies used to unify geographical knowledge provided by different traditions and to replicate it accurately. Yet by 1775, having sent people abroad for training, the crown had at its disposal one of the most important achievements of eighteenth-century Spanish cartography: a general map of southern South America. Yet it was heavily criticized upon its release, mainly by influential naval hydrographers and military engineers. We shall explore the way this map was produced in order to understand how the emerging role of cartographic "silences" revealed changes in the political use of geographical information that were incompatible with the all-encompassing representation, the rather academic ideal underlying this long-awaited map.

In 1775, Minister of State Grimaldi received the results of the work that had been under way for ten years from the geographer Juan de la Cruz Cano y Olmedilla (1734–1790): the engraving of a modern, detailed map of the southern cone. This map gave a uniform representation of an area characterized by increasing political tension, historically mythical, rich in natural resources, and unexplored in many of its regions. Cruz Cano, trained in Paris in the studio of d'Anville, was originally meant to copy and oversee the engraving of plates of the map drawn by Francisco Millau y Maravall (1728–1805), a sailor who had taken part in the expeditions emanating from the Treaty of Limits (1750) on the boundaries between the American territories of the Spanish and Portuguese empires. It was to be a relatively easy task, but in 1766 Cruz Cano decided not to copy Millau y Maravall's work but instead to create a new map based on all those available.[7]

Taking data coming from an extraordinary variety of sources, Cruz Cano compiled a rough draft of such great dimensions that could only be managed when the map was laid flat on the ground, and on which he verified longitudes "with respect to all the nations which have established their meridians."[8] This work would bear fruitful results. In 1769, for example, the Spanish edition of *Byron's Journal of His Circumnavigation* (London, 1764–66) appeared as *Viage del comandante Byron alrededor del mundo* (A Voyage Around the World Undertaken and Performed by the Hon. Commodore Byron), translated and annotated by Casimiro Gómez Ortega (1740–1818) and provided with a map by Cruz Cano of the Magellan Strait, which consisted of a dialogue over two centuries between explorers, administrators, and sailors, and containing sources ranging from the accounts of Pedro Sarmiento de Gamboa (1555–1620) to the latest map of Milhau (1768). Such a critical dialogue was possible because Cruz Cano worked with four systems of reference for longitude, which included— along with the Pico de Tenerife, still the most used, and the older one of Isla de Hierro—those of Paris and London, which were by then becoming dominant. A simple glance at the map shows the importance of the toponymic differences and the war of names maintained with the English. The result is a toponymic and topographic encyclopedia, a sort of museum, which allowed the hoarding of geographical space and created a guide to the management of immense flows of colonial information.

If we compare this academic map of the Strait of Magellan of 1769 with the same area in Cruz Cano's map of South America of 1775, we immediately notice that the information that is eliminated is as important as that which it contains. In the later map, the longitudes refer to the "east of Teide," although the other references for establishing longitudes remain (these are the meridians of Paris, London, Madrid, and Hierro) and, of course, the toponymy is unified to Spanish, or as the author explains, the old names are recovered. Also, coastal measurements and routes (naval information) disappear, announcing what was already known but was not yet visible: that the coasts were going to be managed independently from the continental masses. As well as recovering the original names (which made it easier to compare old documents), the internal political borders are introduced—essential information for knowing "where Royal Taxes are collected, where there are Viceroyalties, where there is government, and where there is a Corregidor (local magistrate)"[9]—as well as the highways and postal stations. Three objects (jurisdictions, communications, and staging posts) "are not to be found in any geographer, and have only now been possible to obtain by dint of much time and good original information."[10] Indeed, he had made a tremendous effort to flesh out the American skeleton with abundant colonial meat that, as a whole, included different forms of authority, represented political and cultural relations, defined areas of influence, and recuperated documentary sources. Nevertheless, the map would soon become a source of argument, and not only for its errors or the persistence of some geographical myths.[11]

Even before it left the presses, the map already had its enemies. When Admiral of the Fleet Manuel Antonio de Flores appeared in 1769 at the home of Juan Cruz Cano with his fellow naval officer Fernando Senra to supervise the development of the map, "they regretted that it was not a flat projection, [a map in] which only the coasts were included."[12] The sailors did not merely hate bureaucracy or distrust land-bound geographers; rather, they rejected the disproportionate character of that style of producing documents, based on technologies that were neither mathematical nor mechanical, and the usefulness of which they found hard to discern. As navigators they had been trained to produce and to read another type of map, other forms of representing and codifying information. And their way of incorporating geographical information into the political universe was also different.

Cruz Cano's attempt, instead of representing space, unites and accumulates space depending on its appearance in time or, rather, history. The purpose of his map was not to show distances but rather to fix the univocality of the colonial world, unifying visual codes so that the map could defend metropolitan sovereignty over a homogenous whole, without instability or differences. Cruz Cano was not aware that local information could have strategic value and might produce certain political discrepancies. The viceroy of Río de la Plata, D. Pedro de Ceballos (1715–1778) had the map before him while he recommended his successor to use the roads from San Juan and the Diamante River to muster troops and to improve or raise the fortifications in Pergamino and Esquina de la Cruz Alta, Melinque, and Punta del Sauce.[13] But if we project onto Cruz Cano's plan the troop movements and forts with which Ceballos intended to stabilize the province of Cordoba and stimulate its colonization from the coast (Map 10.1), and compare it with the plan of 1794 (Map 10.2) of the road between Valparaiso and Buenos Aires drawn up by Felipe Bauzá (1764–1834) and José Espinosa y Tello (1763–1815), we can see that Cruz Cano's map described the relationship of the people to their surroundings (biopolitics) and not relationships of force on the borders (geopolitics).

By contrast, in the map drawn by the sailors, trained as hydrographers in the Astronomical School of Higher Studies in the observatory of Cadiz, such things as administrative borders, areas dominated by different indigenous nations, and information on resources (the carob

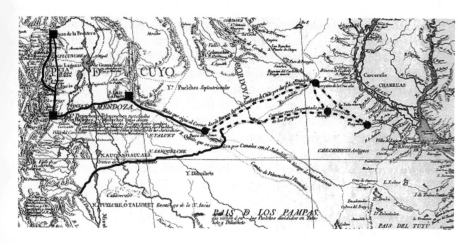

MAP 10.1. *Map detail showing troop movements and forts, province of Cordoba, Argentina, 1775*

NOTE: Viceroy Pedro de Ceballos's plans to stabilize the province of Cordoba: squares depict suggested villages for recruitment, circles depict forts, troop movements are shown in solid lines, and suggested routes between forts in dotted lines (all added by the authors).
SOURCE: Juan de la Cruz Cano, *Mapa geográfico de América meridional dispuesto y gravado por D. Juan de la Cruz Cano y Olmedilla . . . teniendo presentes varios mapas y noticias originales con arreglo a observaciones astronómicas* (Madrid, 1775). Archivo del Palacio Real. © Patrimonio Nacional, Spain.

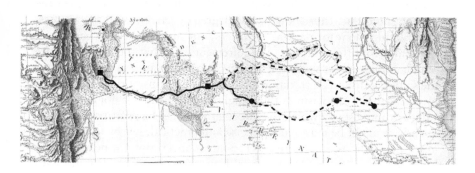

MAP 10.2. *Map detail of the road between Valparaiso and Buenos Aires, 1794*

NOTE: Ceballos's same plans to stabilize the province of Cordoba as depicted on the 1794 map. Squares, circles, and lines have the same meaning as above.
SOURCE: Felipe Bauzá and José Espinosa y Tello, *Carta esférica de la parte interior de la América meridional para manifestar el camino que conduce desde Valparaíso a Buenos Ayres; Por observaciones astronómicas que hicieron en estos parages en 1794 . . .* (Madrid: Dirección de Hidrografía, 1810). Reproduced in Julio Guillén y Tato, *Monumenta Chartographica Indiana*, vol. 1 (Madrid: Ministerio de Asuntos Exteriores, 1942), map 96.

plantations of the Pampas, for example) all disappeared. There remained the rivers, with their courses properly corrected, a denser and more carefully selected network of roads, and above all, broad spaces labeled "unknown land." The information was suppressed in order to better calibrate the territory and to suggest the direction in which new spaces could be created. Thus, the blanks were functional because they avoided improvisations or the design of unreliable strategies. The unknown territory, as long as it did not change its status, counted only as a possibility and never as a political object. Hence in that interface of territory that we call a map, the "silences" served to represent areas without history or, in other words, places whose resources could not be mobilized and from which, as a consequence, no advantage could be obtained.

The fact is that the versatility of Cruz Cano's map as a strategic tool depended on what, at the end of century, was already a great defect. The military engineer Francisco Requena, admirer of the Aragonese Cruz Cano, put his finger on it: "he should print," he wrote in his report in 1802, "with lighter and less strong strokes everything that is put in the center of America purely for [geographical] reference, and not well examined nor organized by celestial observations, in order to differentiate, as all Geographers do, what is well-known from what is doubtful."[14] Requena seems to be suggesting that the manufacturer of maps should take more care not to give equal weight to all the sources or methods of gathering information. The less certain the information, the weaker the outline should be, thus showing the degree of reliability of the sources, formulating the information "by layers," in such a way that each layer responded to different levels of requirements and that the disappearance of one did not affect the whole. Cruz Cano's map, on the contrary, gave the same validity to all traditions (within the margins of rigor established by the historical criticism of the Enlightenment), and did not explain the values or interests by which the information could be prioritized. Consequently, it might be reasonably precise and practical, but by leaving open many possibilities, it gave rise to too much uncertainty: it neither stabilized an object (colonial space) nor eliminated a conflict (disputes with Portugal), nor did it set a trend; that is to say, it did not permit the identification of a line of progress in the management of the colonial territory.[15]

DEPICTING LANDS AS ACHIEVEMENT

To trace a line of progress in the management of territory had also been one of the main aims of the chorographic cartography produced by religious orders, and by the Creoles educated by them. Deeply committed to a moral dimension comprising the reformation of native beliefs and behavior, the integration—however minimal—of natives into a socio-political order, and the preservation of natural wealth, chorographic maps produced by religious orders provided a basis for interpreting inland territory in a dynamic way, even if they depicted it in a rather stable fashion. As we shall see, the new regional maps commissioned by the viceroyalties inherited many of the features of the old cartography, but faced with domestic political tensions, they stressed the signs of territorial instability. Once the new maps had put aside the moral dimension developed by the ecclesiastical tradition, they were able to make room for exhausted resources in their descriptions: clearing the way, as we shall argue, for a more technological understanding of empty spaces. By paying attention to this kind of map, we hope to highlight the process by which inland spaces, even in the most arid areas, were depicted in a fashion that, while emphasizing the management of resources, only made sense if the maps were expected to be improved by changes controlled by local actors.

The division of America into viceroyalties favored ways of approaching territory that created affinities within these administrative spaces, but it also led to clashes with other viceroyalties. Nevertheless, the general cartographic representation of these spaces was not carried out at the same time, which meant that such representations were not a collective answer to the needs of the court in Madrid. Let us consider, for example, New Granada, whose first modern small-scale general map arrived in 1790, after being commissioned by the Viceroy José Ezpeleta (1740–1823). At the beginning of the nineteenth century another map appeared, the *Carta esférica que comprehende parte del Nuevo Reyno de Granada* (Cosmological Chart Comprising Part of the New Kingdom of Granada; 1804–5) of the engineers Joaquín Fidalgo and Carlos Cabrer. In 1808, by order of Viceroy Antonio Amat y Borbón, the lieutenant-colonel of engineers Vicente Talledo y Rivera finished a third map, his famous *Mapa corográfico del Nuevo Reyno de Granada* (Chorographic Map of the New Kingdom of Granada), drawn, as its full title indicates, according to "the best astronomical observations, up-to-date news, and trigonometrical operations."[16]

In New Spain, the demands for a general cartographic representation came earlier.[17] The first map of 1746 we owe to José Antonio Villaseñor y Sánchez (fl. 1733–1756). Two decades later, it would be Father José de Alzate y Ramírez (1737–1799) who drew the *Nuevo mapa geográfico de la América septentrional española, dividida en obispados y provincias* (New Geographical Map of Northern Spanish America, Divided into Bishoprics and Provinces; 1767). Alzate also created in 1772 his other great map of New Spain, the *Plano geográfico de la mayor parte de la América septentrional española* (Geographic Map of the Greater Part of Northern Spanish America), a correction of the previous one in order to "bring it into line with new astronomical observations." The same year, Joaquín Velázquez Cárdenas y León (1732–1786) would draw his *Mapa manuscrito de toda la Nueva España* (Manuscript Map of All of New Spain; 1772). In 1779, Miguel Constanzó (fl. 1764–1790) compiled by order of Viceroy Bucareli (1717–1779) the map of the internal provinces, and some years later, Antonio Forcada drew up the *Mapa manuscrito de todo el reino de Nueva España, desde los 16° a los 40° de latitud* (Manuscript Map of the Entire Kingdom of New Spain, from the Sixteenth to the Fortieth Parallel Latitude; 1787).[18] Finally, in 1794 we already have that prepared by Carlos de Urrutia (1750–1825), *Plano geográfico de la mayor parte del virreynato de Nueva España* (1793), included in the *Noticia geográfica del reyno de Nueva España* (1794), a text of statistical and demographic character that the viceroy and second Count of Revilla Gigedo, Juan Vicente de Güemes Pacheco (1740–1799) had asked him to produce.[19]

In fact, as we see, the administrative reforms of the colonies undertaken by the Bourbons implied a remarkable increase in cartographic activity. For example, in 1741 when topographic accounts were made of the 129 jurisdictions encompassed by New Spain, only 5 included maps in their replies,[20] a deficiency that Alzate attributed to the instability of the *alcaldes* (town council officials), for "since it is so laid down by law, an *alcalde* resides in the same territory just a short time and, therefore, cannot have that topographic instruction that the priests have."[21] And, indeed, the interest of the different religious orders in ethnographic, linguistic, or anthropological studies, as well as economic reasons, had given these orders a considerable advantage in designing and establishing a network of such social instruments as schools, hospitals, churches, and craftsmen's workshops. No wonder then that the priests, as Alzate calls them, together with the Creoles, tended to see space as an object of great plasticity subject to commercial, meteorological, or sanitary vicissitudes and which, therefore, had frequently to be reviewed. The 1767 map drawn by Alzate mentioned above is a classic example. Its decorative border is made up of small boxes showing the country as a cornucopia of vegetable and animal life, inhabited by people occupied in a diversity of work.[22] The territory is always seen in relation to these two poles: resources and people. When the map is unfolded, what its small symbols show is always the relation between both their absence and presence.

But this border would be omitted in the second version of the map, producing another type of cartographic blank, one that silenced the moral history, which had been so successfully developed by Jesuit or Franciscan literature (see Figure 10.1). The cancellation of those resources of epic rhetoric highlighted the renewed importance of the settlements and strategic posts, and hence the dialogue with other New Spanish geographers, especially with Costanzó, Luis Surville,[23] and the anonymous author of the map of the province of Sonora of 1770. Thus Alzate would record forts (for example: "Our Lady of the Pilar is the prison of the Adaes, created in 1717"), mines, and indigenous establishments in his maps. For his part, Costanzó would also record forts (existing or reformed) and detachments, in addition to urban nuclei, including missions, estates, royal mines, indigenous settlements, staging posts,

FIGURE 10.1. *Detail of Miguel Venegas's*
Mapa geográfico de la California, *1757*

NOTE: At the borders of the map the ethnographic or "moral" information typical of Jesuit accounts completed and shaped the territorial approach.
SOURCE: Miguel Venegas S.I., *Noticia de la California y de su conquista temporal, y spiritual hasta el tiempo presente*, vol. 1 (Madrid, 1757). Biblioteca Nacional de España, Madrid.

and stopping places. He included abandoned sites and derelict mines as well—signs not of what is, but of what was; in other words, that show us a territory under construction.[24]

In the case of the Internal Provinces—of which the three most important (New Biscay, New Mexico, and Sonora) totaled 226,600 inhabitants and were mostly dedicated to trading and mining[25]—these were the object of mapmaking that is of particular interest, in the midst of extraordinary tensions between the indigenous populations and the metropolis, in verifying how maps contributed to the redefinition of strategies of territorial management. Maps were required in order to sanction or correct measures affecting the consolidation of new spaces. Their influence was particularly clear in the policies of the "line of forts."[26] Captain of Engineers Nicolás de Lafora, who with the regimental draughtsman Second Lieutenant José Urrutia would accompany the Marquis de Rubí in his commission to visit the forts on the northern frontier of the viceroyalty and to propose the reforms that he considered advisable,[27] would submit his "Report of Captain of Engineers Don Nicolás de Lafora concerning the securing of the borders of New Biscay. Based on what he has seen of them, the reports of the most practical people and on the most correct maps of this country" (1766). Later, he prepared the *Mapa de la frontera de Nueva España* (1771), in which he showed how forts should be redistributed: the plan and report would become the essential reference for later reforms.[28] The "depopulation" symbol might mean the exhaustion of a resource or the victory of the insurgent Indians.[29] But at the same time, the discovery of new deposits or the foundation of other colonies confirmed the prospects of continuity in the fight against depopulation, for Governor Riperdá himself foresaw that the province of Texas "will only have its name left."[30] The maps were still descriptive, but they also sought to be prescriptive—and this in spite of the fact that their references could be extremely unstable and fragile. If settlements, particularly of miners, were short lived,[31] the "line of forts" would face the threat of becoming isolated from each other.[32]

The construction of the territory as achievement, as the gradual attainment of stability, not only established local actors as important political agents—as in the case of the earliest chorographic production[33]—but also basically redefined the concept of territory as a fundamentally technological and economic enterprise, toward which all social and natural actors are drawn. This participation took different forms, from consultation,[34] to individual or collective economic contribution,[35] to the creation of establishments and infrastructures.[36] The identity of places is not defined exclusively by their plants or animals, but by their mines and water supply.[37] Maps were evidence that only the proper use of suitable technologies could replace the immense emptiness depicted in the abandoned provinces with prospects of new colonies.

THE EMPIRE AS A STRATEGY OF ACCESS:
FACING DETERRITORIALIZATION

The domestic concern for visualizing successful and unsuccessful establishments on a biopolitical basis was the reverse of the strategy behind the cartographic work on the coast area. The so-called "Nootka crisis" shows to what extent the work of hydrographers mirrored the pressure of international conflict and was shaped by the need to reach a political consensus on coastal rights. We shall argue that, from the point of view of geopolitics, the absence of inland areas from hydrographic charts helped to establish the legal possession of the sea in terms of a stability that was unattainable and of little interest for inland politics.

The multifarious corrections that the Cruz Cano map underwent, beginning with the erasure of the demarcation line of the Spanish-Portuguese border, contribute to the metropolitan authorities' obsession with having fixed, almost immutable, cartography. This did not rule out modifications, but map entries could no longer be removed. For this reason, time and caution were required when filling in a map. Thus, in Bauzá's map of Patagonia of 1798 (Map 10.3),[38] the interruptions of the coastline were evidence of a lack of certain knowledge, and similarly, all the information about the area that was considered doubtful or inadmissible also disappears—such as the communication routes that Cruz Cano drew in Tierra del Fuego (Map 10.4). We have before us, then, a very different map from those mentioned before. In contrast to the profound historicism and overload of data on the *intendencia* or regional maps, naval hydrographers created a style of mapmaking, following guidelines from the metropolis, that aimed for a radical level of stability. This stability was, of course, restricted almost exclusively to the principal target of these maps, the outline of the coast.

We are talking here about a change that affected all imperial powers in the same way, and which needs to be explained. There is nothing obvious about a process that turned the memory of places—all the information compiled by Cruz Cano from the most varied documentary sources—into irrelevant information: extraneous curiosities, trivialities for scholars. Harley interprets this displacement, prompted by a scientific theory, as a landmark in the process of dehumanization of the landscape, concealed beneath the inability of hydrographic engineers to include the descriptions of peculiar and local features.[39] We respect this approach,

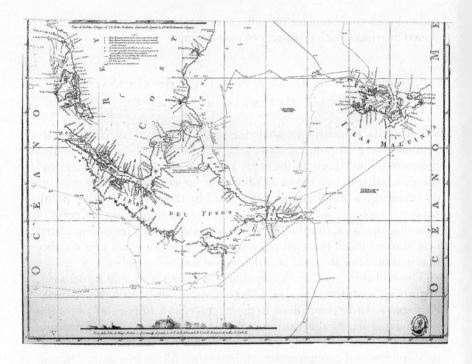

MAP 10.3. *Detail of Bauzá's* Carta esférica, *1798*

NOTE: Depiction of missing and imperfect knowledge along the Patagonia coastline and inland.
SOURCE: Felipe Bauzá, *Carta esférica de las costas de la América meridional
desde el paralelo de 36°30' de latitude S. hasta Cabo de Hornos levantada de ordern
del Rey en 1789, 90, 94 y 95 por varios Oficiales de la Marina* (Madrid: Depósito
Hidrográfico, 1798), map 95. Biblioteca Nacional de España, Madrid.

but there must be something more to be said about the proliferation of
empty space, the desire for silence imposed by imperial management.
Indeed, at the end of the eighteenth century Spanish cartographic production was extraordinarily developed, while at the same time means
were created to reunite dispersed historical material referring principally
to the colonies.[40] Not all maps came into public view, but many were
published or copied and interchanged at the request of other European
countries. Suddenly, maps acquired an enormous diplomatic value.

At stake were the limits of empire; maps were no longer exclusively
scientific and increasingly became documents in an international dialogue of markedly legal character. Strategies employed by the old empires for structuring territory show to what extent the construction of
imperial limits, in contrast with an empire's empty spaces, has not been

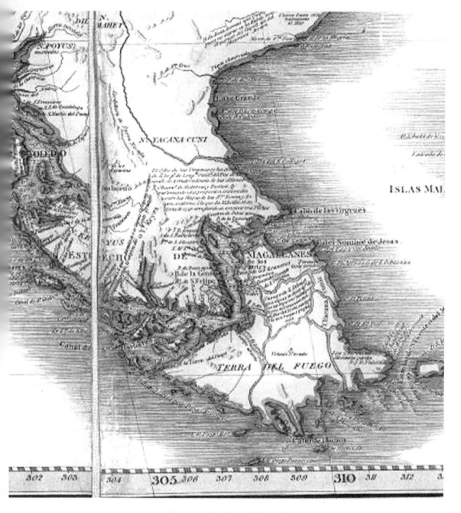

MAP 10.4. *Detail of Cruz Cano's*
Mapa geográfico de América meridional, *1775*

NOTE: Cruz Cano drew the navigable canal of San Sebastian showing small torrents and
passages suitable for canopies and people he had documented. Tierra de Fuego thus appears
as an already manageable territory with several established communication routes.

SOURCE: Juan de la Cruz Cano, *Mapa geográfico de América meridional
dispuesto y gravado por D. Juan de la Cruz Cano y Olmedilla . . . teniendo presentes
varios mapas y noticias originales con arreglo a observaciones astronómicas*
(Madrid, 1775). Archivo del Palacio Real. © Patrimonio Nacional, Spain.

a constant.[41] The preoccupation with accurately drawing the coastal perimeter was the consequence of a deep conceptual transformation of ocean space that had been developing since the end of the seventeenth century. If the increase in marine traffic raised the question of whether ocean space should be of a public nature, a commons,[42] then the political and legal analysis of the insular condition opened by Great Britain encouraged new geostrategic balances.[43] The proposal to establish contact with the enemy on the basis of regional economic criteria, and not the balance of European powers, lent political robustness to the trading enclaves—even giving preference to the islands, subject to limits of supply.[44] Great Britain had sufficient military arguments to guarantee commercial privileges for its colonies.

What comes later is easy to explain because the development of naval technology reduced distances and increased contacts between the different colonies and with the metropolis. Space was contracted and deterritorialized, and in just a few years European colonial policies changed dramatically. The main strategic option was no longer the search for a point on the coast from which to reach the interior, or to depict the perimeter shaping the continental mass, as was the intention of late-fifteenth- and early-sixteenth-century navigators.[45] (On the search for continental land in the South Pacific during this period, see Sheehan in Chapter 12.) Now the focus was on the coast itself, the main area of claims for jurisdictional rights, even when there were no claims of ownership.[46] But the spectacular development of trade gave constant cause for tension that, by opening bitter disputes on the legitimacy of certain commercial privileges, ended in the revision of criteria for the acceptance of claims of territorial possession.

At the end of the century, a British company established a fur trading settlement in Nootka, an island to the west of Vancouver that had been incorporated as a part of the Spanish empire in 1774 by an act of symbolic appropriation based on the Treaty of Tordesillas of 1494. When Spanish authorities seized British ships in 1789 for encroaching upon the Spanish crown's rights, the British claimed their right to enjoy free navigation and to possess those establishments not previously occupied, formed with the consent of natives. Such a hostile response brought Spain and Great Britain to the brink of war,[47] making it clear that the intricate jurisdictional network that maintained such a fragile equilibrium was sustainable only so long as the principle of property was not put into question. Great Britain, who until then had been guided by the classic criterion of discovery and symbolic possession, upheld occupation as a basic principle to justify these rights.[48] The signature of the El Escorial agreement (1792), that would end such tensions, as well as the difficulties of the later commission of limits to make the treaty effective, would show that the battle for the islands dealt a death blow to traditional colonial balances. A large part of the Spanish possessions in America had not been ratified in any treaty, and in Spain there was suspicion of the danger that the English would "populate and fortify the immense coasts and islands of our America from the River Plate, going round Cape Horn to Valdivia and Chiloé; nor are there sufficient forces nor strength to prevent it nor yet any means of knowing the places where they are until it is too late to cast them out."[49] It is no wonder then that José Francisco Bodega y Quadra (1743–1794), to avoid the threat of disintegration posed by the English demand for the return of their possessions at Nootka, insisted that George Vancouver (1757–1798) recognize that occupation did not imply property,[50] a tailor-made formula to counter the new British arguments for defining unconquered lands. In any case, the Spanish crown had to recognize that it needed to reformulate its policy of production of coastal space.

The treaty contemplated a moratorium on the creation of new settlements on the islands and coasts of southern America; but, according to a secret article, the pact would only be valid as long as no power settled in that area.[51] This situation forced the Spanish crown to keep constant watch, which made the possession of the coast and adjacent islands in some way a geostrategic disadvantage. Bodega y Quadra must have trembled at the possibility that a large part of what had been taken for a continental mass might turn out to be an archipelago[52]: hence his hurry, and that of Vancouver, each for opposite reasons, to examine the coast "at whatever risk" and "from 47 degrees southwards with such precision that not one river or bay be overlooked as far as 41 degrees."[53] This is also the reasoning behind the decision to leave settlements like San Blas, with a terrible climate and insufficient depth for large frigates, or to occupy places strategic for trade, like Nootka, and to make the already existing settlements to the north of California profitable. All of these suggestions were directed toward the construction of a coastline that was independent from the inland territories—one that, after being recognized and translated into an international language (that is to say, mathematical), would permit the localization of a handful of sites suitable for marine trade. This strategy had the advantage of assuming the commoditization of the coastal strip, which was favored by the new navigational facilities while still restricting occupation to the objectively occupiable areas. In addition, the coasts were marked as the limits of new space. In other words, the densely filled "inland" area on the map, in relation with which the coastal strip established references and, so to speak, communicated, was located in the oceanic space.

Although the question of property, jurisdiction, and sovereignty would be a lasting problem, maps achieved diplomatic status because, as they accumulated the different courses on the same plan, new discoveries and successive reconnaissance were introduced directly and fully into international legislation.[54] At the same time, the territorial vacuum of the inland space turned the coasts into a legal watershed. And as it did so, the hydrographic charts identified as the only valid interlocutors in the international panorama their indirect author, the Spanish crown, and those powers in a position to construct a territory on the ocean.

The enclaves, in short, gained value insofar as they came to form part of transoceanic trade, which meant that the policies of space production had to be projected onto the surface of the seas. In practice, the oceans then ceased to belong to the public domain, and those "silences" that hydrographers had depicted inland of the continental masses were offset by these claims offshore. Thus, the "deterritorialization" implied by the purge of the Cruz Cano map ran parallel to this "reterritorialization" of the empire of the seas, which the hydrographers depicted in order to be able to take part in the new struggles of geopolitics.

CONCLUSION

The erudite commander Alexander Malaspina once defined the British empire as an amphibious body with a trading head and military body.[55] It is also possible to invent an ad hoc chimera for the Spanish empire, another monster whose body was half inland and half a shoreline empire. Cartography, indeed, echoed this duality, because what was valid at the level of biopolitics (the cross between geographic features and the tracks of human activity) was inadmissible on the scale of geopolitics (the cross between two types of documents, scientific and diplomatic).

However, strategic considerations are not sufficient to justify the differences between what lay on either side of the shoreline of the empire. We also need to consider other arguments of an epistemic nature. The internal argument emphasized the location of resources and populations, and it was restricted to showing the existence of certain territorial bonds that, besides creating a hierarchy of places, also worked as lines of force that hinted at future mobilizations. The territory at that time was constructed like an organization in a permanent process of growth, which implied the rejection of earlier cartographic traditions. On the other hand, the oceanic representation of the empire rejected any idea of change, whatever the situation in the continental interiors, trying to turn the coasts into an ontologically stable object, which would in consequence be manageable by a lasting international treaty.

Within this context, the failure of Cruz Cano's map indicated the end of a world whose actors had lost political relevance and which, together with the archaeological, anthropological, or ethnographic knowledge to which they were intimately bound, had to migrate from the world of wonders and curiosities to that of treasures and thesauruses. And just as maps became technical documents, objects too underwent a process of thesaurization (according to the new conventions produced by the scientists) before being put away into the reduced space of the museums.[56] So, as display cabinets filled, the maps emptied. Cartographic voids, nevertheless, behaved differently according to whether they referred to the inland or the shoreline empire. The "silences" that came from the interior preached the end of exceptionality and the beginning of the fact that the whole territory could be treated with the same yardstick: that is, by means of biopolitical tools (demographic tables, balances of payments, inventories of resources, medical geographies, and botanical classifications) and management projects consistent with territorial evolution. On the contrary, using silences in coastal maps contributed to the illusion that the hydrographic solution was universal and, being thus free of political connotations, could be the basis for a global solution to geopolitical conflicts. Using the voids in such a different way, the Spanish crown disconnected two worlds, which allowed it to become an important actor on two stages at once, competing with the British in the maritime theater without, in another arena, jeopardizing its own colonial policies.

PART IV

Commerce, Curiosities, and the Circulation of Knowledge

CHAPTER ELEVEN

Knowledge and Empiricism in the Sixteenth-Century Spanish Atlantic World

ANTONIO BARRERA-OSORIO

This chapter discusses the emergence of empirical practices in the Spanish American empire during the sixteenth century. Neither Spain nor Latin America figure in any significant way in our history of science books.[1] Yet Spain and its American empire were at the center of events leading to the development of the new science in the seventeenth century. I argue that the emergence of personal experience and empirical information as a source of knowledge—a phenomenon traditionally ascribed to the rise of the new sciences (particularly astronomy and mathematics) in the seventeenth century—was rather the result of the commercial and imperial expansion of Europe during the sixteenth and seventeenth centuries. I discuss here the case of Spain as representative of epistemological changes taking place within commercial and long-distance empires, and I examine the type of empirical activities both institutions and people deployed as they extended the commercial and political structure of the empire. In a first section, I discuss some of the nonnavigational activities taking place at the House of Trade (Casa de la Contratación, est. 1503) and Council of Indies (est. 1524) in Spain, and in viceroyal courts. The House of Trade institutionalized protocols for the verification of information, instruments, and technologies related to the Atlantic World, and these protocols were implemented at the Council of Indies as well as in viceroyal courts.

Many thanks, *muchas gracias*, to the editors Daniela Bleichmar, Paula De Vos, Kristin Huffine, and Kevin Sheehan for putting together this volume. In particular, thank you to Daniela Bleichmar for her comments.

In a second part of this essay, I examine the origin of Francisco Hernández's expedition to the New World (1571–1577). I suggest that it traces back to Seville and its traffic of people, curiosities, commodities, and ideas. In other words, the expedition began in the streets of Seville, not at the court of Philip II (1556–1598).[2] The context for the circulation of ideas, people, commodities, and curiosities was the Spanish empire, which "was equal or indeed superior to other leading states for transport, transfer, and communications."[3] My thesis is simple: the commercial and imperial context of the Atlantic World created conditions for, and fostered the emergence and institutionalization of, empirical practices in Europe. This article shows how artisans, technologies, and commodities brought changes in sixteenth-century epistemological practices in the context of long-distance empires.[4]

INSTITUTIONS:
HOUSE OF TRADE, COUNCIL OF INDIES,
AND VICEROYAL COURTS

The Casa de la Contratación constituted a pioneering model for the scientific academies of the seventeenth century. Information in the form of reports arrived to the *Casa* to be studied by a group of experts (pilots and cosmographers). (On the role of the Casa de la Contratación, see Portuondo in Chapter 3.) These experts used this information to produce charts, navigational instruments, and professional pilots.[5] The *Casa's* activities have been studied by a series of historians, usually in the context of Spain's commercial activities in the New World.[6] Its navigational and mapping activities are well known, but other empirical activities are less so; for instance, its role as a center for testing instruments to be used in the New World and for collecting reports about the New World. A prime example is the case of the cosmographer Diego Ribeiro (or Ribero, as the Spaniards called him), and his metal bilge pump.

In 1524, Ribeiro received a conditional license (confirmed in 1526) to develop and sell a metal bilge pump for extracting water from wooden ships.[7] The conditional license explained that, taking into consideration Ribeiro's service to the crown

and the benefit and universal good expected from [the pumps] to come to us, our subjects, and residents of our kingdoms, we say that once you make those pumps as you bound yourself to do, and conduct a trial with them in Coruña or Seville, where they would be examined and seen by the people we would designate for the purpose of that experience, and if they approve and consider them useful for navigation and if we want to use them, we would grant you . . . sixty thousand *maravedís*.[8]

The decree also established other prerogatives. In 1531, the year Francisco Pizarro landed on the western coast of the Inca empire, Ribeiro had his first pumps ready; the crown ordered the *Casa* to look into this matter and to name appropriate people to test the pumps.[9] A few months later, the crown once more reminded the officials of the *Casa* "to appoint shipmasters, sailors, and people with knowledge and experience in navigation to test and experience the aforementioned pumps in a vessel."[10] The crown also ordered the *Casa* to send reports of these trials to the Council.

The pump's first trial took place on November 25, 1531, in the ship *Santa María del Espinar.* Three royal officials from the *Casa* and five experts in navigational matters tested the pump and found it to be more efficient and easier to handle than wooden pumps.[11] However, the Council was not convinced by the results. In 1532, the Council ordered the *Casa* to pay Ribeiro sixty thousand *maravedís* as a preliminary payment; meanwhile, the Council would organize more extensive tests.

On May 4, 1532, Ribeiro delivered a bilge pump weighing 303 pounds to the captain of the ship *Alta Mar.* The vessel sailed for Santo Domingo and, before it could reach its destination, had to head back to Seville because it was taking in too much water. According to reports from the captain, three pilots, and five sailors, the pump's great efficiency was the reason the ship made it back to Spain. The *Casa* sent the *Alta Mar*'s report to the Council with its own positive opinion about Ribeiro's pumps. Ribeiro, alas, never enjoyed the benefits of his successful invention: he died on August 16, 1533. His heirs and legal representative, Diego de Oliver, continued Ribeiro's process before the crown.[12]

The tests conducted by experts and royal officials on Ribeiro's bilge pumps provide an example of a practice later institutionalized at the *Casa* for validating new instruments and technologies. The crown expanded empirical practices already in use among artisans, and incorporated these testing practices into the state's regulation of instruments, technology, and inventors. The establishment of testing rules by the late 1520s had been, in part, the result of Ribeiro's own suggestion to conduct tests. These rules compelled artisans to take their instruments before royal officials, where experts would test the actual performance of the instrument. Inconclusive tests led to more tests under better conditions. The results of the tests provided royal officials with the information necessary to validate the claims made by the inventor.[13] By the 1570s, the Council of Indies already had a "rule that those who have [inventions], appear before the Council for examining."[14]

In 1535, Vicente Barreros, a carpenter, offered wooden pumps that he claimed were cheaper and better than Diego Ribeiro's. Barreros asked for a license similar to the one granted to Ribeiro. The crown had created a legal precedent for this request by stipulating in 1526 that if someone "wanted to compete and challenge" Ribeiro's claims about his pumps, Ribeiro "would allow him to do it," and if "within two years someone came with a better invention, he would have" the same prerogatives granted to Ribeiro.[15] The crown, following the established procedure, ordered the *Casa* to investigate whether Barreros's pumps were "more useful, better, and cheaper" than the metal pumps designed by Ribeiro.[16]

After conducting the necessary tests, *Casa* officials issued a positive recommendation and the Council of Indies granted Barreros a license to make and sell his pumps for five years.[17] In this way, the crown stimulated competition between inventors for improving the efficiency of bilge pumps. Royal licenses secured the space for inventing and improving instruments and technology.

The crown granted a license once an instrument had been tested and its efficiency determined by a group of experts. Personal experience in the form of reports was the criteria for determining the validity of information, knowledge, instruments, and technology. Furthermore, the proven efficiency of new technologies and instruments became the criteria for validating them. To test for efficiency, the crown established a system of verification based on trials and rewards for individuals who succeeded in their inventions.

These procedures to control experience were not implemented only at the *Casa*. The Council of Indies, formally established in 1524, became a center with similar empirical procedures to determine the validity of reports and instruments. The case of Antonio de Villasante shows a parallel pattern to the case of Diego Ribeiro.[18] Villasante was a resident of Santo Domingo who, in 1528, presented the Council of Indies with a report describing a new medicine, a balsam useful for healing wounds and abrasions and relieving stomach and tooth pain. Villasante performed tests with this balsam and presented the results in his report to the Council. The crown was very interested in this commodity for its commercial possibilities. Given that Old World balsam was a very popular but scarce medicine in this period, the Santo Domingo balsam could provide a good source of income not only for Villasante and his partners but also for the crown. Yet in 1530, a competing report from a physician in Santo Domingo, a certain Dr. Barreda, challenged Villasante's findings. He was not the only one; in Spain, other physicians had also challenged the healing properties of the Santo Domingo balsam. The Spanish crown alludes to these dissident physicians when ordering some twenty-two physicians and surgeons in different cities of Spain to conduct tests with Villasante's balsam.[19] The issue at stake was that the Santo Domingo balsam, although different from the Old World balsam, was still a worthy medicine, especially for treating wounds.

The crown, interested in the commercial possibilities of this balsam, then sought to control dissident physicians—that is, those who denied its "balsam" properties (for these physicians, an issue of identity)—by ordering them to speak or publish only after they had performed experiments with the balsam, as the crown and Villasante termed it (for them, an issue of its worthiness). Furthermore, the dissident physicians had to bring their experimental findings before local magistrates, who would send these results to the crown.[20] By asking physicians to experiment with the balsam and then to show their reports to royal officials, the crown established empirical procedures to validate claims about New World medicines and about information concerning the New World in general. Meanwhile, local magistrates were instructed to try to foster the sale of balsam, "in the best way they see fit."[21]

The crown, however, not only attempted to discipline dissident physicians into experimenting with the balsam, it also ordered physicians, surgeons, and hospitals to do so. Having listened to the dissidents, the crown sought to produce accurate knowledge—in one case sending a sample of the balsam, useful to "cure injuries and many illnesses," to the cardinal in Toledo for use on patients chosen by the physicians and surgeons of the hospital there. The crown requested that hospital administrators "be attentive to inform us of the cures and experiences realized in the hospital with this balsam."[22] Hospitals in Seville, Burgos, Galicia, and Granada received similar orders.[23]

The crown approached physicians, too. The physician Andrés de Jodar, for instance, a resident in Baeza (in Jaen, Andalusia), received the order to use balsam for those "cures and experiences" that he would deem appropriate.[24] Moreover, whatever he may find "certain and true" by means of "art" and experience, he should "put in writing," "sign" his report, and send it to Villasante's partners in Spain. Twenty-two physicians and surgeons in different cities of Spain received similar orders.[25] Villasante's partners, Franco Leardo and Pedro Benito de Basniana, used these reports to commercialize balsam in Spain, and they hired physicians and surgeons to help with this task.[26]

By 1532, information was already arriving at court. A certain Juan de Vargas had been using the "balsam from the Indies" to heal the sick in the area of Cuellar (Segovia).[27] He seemed to have been quite successful, for the crown ordered the officials of Cuellar to collect information from patients who had been healed with Santo Domingo balsam. The scribe of Cuellar, Melchor de Angulo, sent this information to the crown and received 108 *reales* for the eighteen days he worked on this assignment.[28]

The crown also asked Juan de Vargas to come to court, which he did, in late 1532 or early 1533.[29] During his stay in Madrid, he tested the balsam and received monetary compensation for his work.[30] Despite royal support for Vargas's position, some practitioners in the medical community still opposed the use of his balsam, as they maintained it did not have the same properties as Old World balsam. In 1539, the physician and apothecary of the village of Amusco (Palencia) denounced Vargas for using the New World balsam. The authorities there arrested him and confiscated the balsam, though he was later released and the crown demanded an explanation of the matter by Amusco authorities.[31]

In the end, the crown could not dismiss Barreda's contention that the Santo Domingo balsam was different from Old World balsam. The crown's only concern, however, was the fact that American balsam, as both Villasante and Barreda had argued, was especially good at treating wounds. The crown sought, first, to develop the right method to use it; second, to end the confusion between New World balsam and classical balsam; and, finally, to convince other physicians that it was a worthy medicine. This situation shows the interplay between the production of medical knowledge regarding New World medicines, and the political and economic interests of the crown in controlling this knowledge and its products. In this particular case, controlling knowledge about balsam amounted to controlling its commercialization.

Ribeiro, Villasante, and Barreda offer but a few cases among many in which entrepreneurs approached the crown with reports and inventions, asking for protection and licenses. Crown officials and artisans developed empirical procedures to determine the validity of these reports and inventions, usually through the establishment of committees of experts who had to perform tests and report back to crown officials.[32] In the cases of Villasante and Ribeiro, both men presented reports to crown officials (at the House of Trade and the Council of Indies). The crown ordered Villasante to send samples of balsam to physicians and hospitals in Castile; these physicians in turn were asked to report on their tests. A similar situation took place with Ribeiro, who had to provide pumps for experts to test at the House of Trade. In both cases the crown brought together experts, merchants, and artisans to examine artifacts and commodities and asked them to produce reports based on their own personal experience and tests.

Similar practices emerged across the Atlantic. In the Americas, viceroyal courts became centers for the validation of knowledge based on empirical practices, as was the case with the development of mining technologies. In the early 1540s, German and Spanish miners began testing new methods for exploiting silver in New Spain. Silver had been discovered there in the mid-1520s, yet mines at that point were only small enterprises.[33] The discovery of silver in Taxco in 1534 marked the beginning of New Spain's large mining activity.[34] Years later, Juan Velázquez de Salazar, in his report to Juan de Ovando (president of the Council of Indies, 1571–1575), explained that the production of silver entered a period of crisis in the early 1540s when working the exhausted mines became expensive and unproductive. Around 1542, Juan Alemán, after receiving a "report sent to him from Germany," proposed a new method (based on bone ash and lead) to mine poor silver ore.[35] On June 8, 1550, the German miner Gaspar Loman obtained a license to exploit silver for six years using Juan Alemán's method.[36] The method worked well for a few years, and then the production of silver declined again—but, fortunately for royal coffers, the culture of inventions created incentives to artisans and professionals for developing new instruments and technology. In 1553, the tailor Bartolomé de Medina arrived in New Spain with the purpose of developing a cheaper and more efficient method of mining New Spain's silver.[37]

Medina (ca. 1497/1504–1585) had traveled with his own means, expecting to receive official protection, and thus economic advantages, once he had found a more efficient method for exploiting silver.[38] In the early 1550s, a scribe at the court of Viceroy Don Luis de Velasco recorded the story as told by Medina himself:

Being in Spain, he had news of the method employed to exploit gold and silver [in New Spain], and of the great cost . . . of it; and to know if this was so, he came to this New Spain to see it with his eyes and to endeavor how to exploit those metals at a lower cost, and thus with great industry and care and work of his person and expense of his patrimony he understood, on account of the experiences mentioned, a method to exploit those metals by using quicksilver and obtain from them all their ore . . . with less cost of people and horses.[39]

After great work and costly experiments, Medina found a method based on mercury for smelting silver: the amalgamation method. This method became known as the "patio system," because it took place in small yards or patios. Medina's method consisted of preparing a mix of ore, water, large amounts of salt, and mercury. This mix was prepared in small boxes or enclosures on a patio divided into shallow rectangular pools. The process took about twenty-two days and was done outdoors. This system prompted the construction of a system of patios surrounded by administrative and working buildings.[40]

Medina's innovation consisted of the adaptation of the amalgamation process—already known to alchemists and German miners, and discussed in the work of Vannoccio Biringucci (1480–ca. 1539)—to the industrial production of silver.[41] Perhaps Medina, as a tailor, already had experience smelting small quantities of silver and gold to use in dress-making.[42] Before leaving Spain, it is known that he had the help in his shop of a German expert known as Maestro Lorenzo, with whom he probably discussed the possibility of using mercury.[43] By late 1554 or early 1555, Medina had developed the amalgamation process in New Spain. His method was implemented quite successfully: by the late 1550s, over 120 people were already using it; by 1562, in Zacatecas alone there were thirty-five schemes based on the amalgamation process.[44] Medina's technology transformed social relations in New Spain and Peru by mak-ing the exploitation of silver viable again and accessible to new entre-preneurs. The method reached Spain in 1557, Peru in 1572, and Central Europe not until the 1780s—two and a half centuries later.[45]

In Peru, Medina's system not only produced an explosion of improve-ments and royal licenses, but more important, it also transformed the social situation of the indigenous people. The production of silver in Peru had fallen from 379,244 marks in 1550 to 114,878 in 1572. After 1572, however, production rose again (to a peak of 887,448 marks in 1592).[46] Indigenous labor was organized into a draft-labor system (called *mita* in Peru and *repartimiento* in New Spain) to meet the new levels of produc-tion. The *mita* (Quechua for "turn") was "a draft labor scheme designed to bring to Potosí some 13,400 male Indian workers annually."[47]

In 1569, Cardinal Espinosa (president of the Council of Castile), Francisco de Toledo (viceroy of Peru), and Juan de Ovando (soon to be appointed president of the Council of Indies) discussed, among other American policies, the implementation of the new refining process and the possibility of intensifying the draft labor system in Peru.[48] Viceroy Toledo subsequently received instructions from Philip II to implement both systems in Peru, which he did in late 1572.[49] The area designated for supplying the labor force extended 800 miles, from Cuzco in the north to Tarija in the south, and around 250 miles across the width of the Andes. Workers (males between the ages of eighteen and fifty) had to go to Potosí for one year; this provided around 13,500 men a year, work-ing in groups by turn, one week on and two off.[50] Their wages were set daily, according to the job.

Medina's method not only transformed social relations, it also gen-erated competition for developing more efficient and cheaper ways to implement the amalgamation process. Once the patio system was es-tablished in New Spain, a series of other inventors sought to improve it, mostly by providing more efficient and cost-effective uses of quicksilver. In 1559 or 1560, for example, Alfonso Martínez de Leiva requested a license to carry out some experiments in the Zacatecas mines to reduce the time of the amalgamation process. The viceroy granted him a license for three months and, if these experiments were successful, exclusive rights to his method for six years.[51] Other miners sought ways to reduce the use of quicksilver, and thus to reduce the cost of the process. In 1559, Pedro González and Alonso de León improved sieves to reduce the ore to a fine powder, thus making the mercury reaction more efficient.[52] In 1562, Pedro Díaz de Baesa used fluted earthen jars to save quicksilver in the last step of the amalgamation process. On July 10, 1563, Juan de Plascencia licensed a new sieve for metals and an instrument for spinning the mesh used in his improved sieve. In 1567, Leonardo Fragoso and Cristobal García developed a new metal washer for reducing the loss of quicksilver. Gaspar Herrera developed a method for reducing the amount of quicksilver in 1566.

Some of these improvements required great effort and expertise. In the context of competition generated by the legal protection granted to inventions and technology, innovators carried out their experiments and work in secret and at great personal expense. The viceroy Martín Enríquez commented that a certain Juan Capellín worked for "nine years in many inventions and buildings in his house and some secret places to learn and understand" the method for reducing the time of the amalgamation process and, thus, the amount of quicksilver used in it. By 1576, he had designed a mill with metal mallets and an instrument to retrieve quicksilver used during the amalgamation process.[53] Bernardino Santacruz developed a "wooden box" and a "small tank" with a "secret" to save quicksilver. The viceroy found his invention useful and granted him a license on September 3, 1580.[54]

Not only entrepreneurs became interested in the amalgamation process. It also piqued the interest of natural historians Juan de Cárdenas and José de Acosta, for example.[55] As physicians and cosmographers, they became brokers between the empirical culture of inventors, with its tests and informal circuits of information, and the textual culture of natural history. Where did this empirical culture begin? It began in the streets of Seville and the roads of the New World. The next section discusses the emergence of this culture outside royal institutions.

FROM INDIOS MÉDICOS TO THE STREETS OF
SEVILLE AND SCIENTIFIC EXPEDITIONS

Similar empirical practices were established outside royal institutions on both sides of the Atlantic: in Franciscan monasteries in New Spain and medical gardens in Seville. To give but one example, in the second half of the 1520s, a group of Franciscan missionaries arrived to an area later called Michoacan, New Spain. When they arrived, they fell sick—as often happened when Europeans traveled to and around the New World. The leader of the Tascaroras, Cazonci Tzintzincha Tangaxoán II (d. 1530), offered to one of the Franciscans the services of his physician, an Indian doctor. The Franciscans "seeing the lack of medical services that he had there," accepted his offer. The *indio médico* (indigenous physician) gave him the powder from a root, and purged him so well that on the same day he was already feeling better.[56]

The rest of the Franciscans and other Spaniards followed this example and took the medicine, which they called *mechoacan*, and all of them healed from their maladies. The Franciscans sent a report about this medicine to their counterparts in Mexico City, who consulted with physicians there, and offered it to those who were sick. "Many began using it," and it became very famous in the New World—with people planting it in their gardens all the way to Tierra Firme.[57] For example, Bernardino del Castillo, a mill owner in Mexico, planted *mechoacan* as well as many other plants from the New and Old Worlds.[58] From such experiences, the Franciscans produced a book for Charles V about medicinal plants from Mexico, *Libellus de medicinalibus indorum herbis* (Little Book of the Medicinal Herbs of the Indians), also known as *Codex badianus* (1552). The Franciscans brought *indios médicos* to their Colegio de Santa Cruz in Tlatelolco and asked two of them, Martín de la Cruz and Juan Badiano, to write the book.[59]

Soon after *mechoacan* became popular in New Spain, a certain Pascual Cataño brought it to Seville. Cataño arrived sick to Seville and sought the advice of Nicolás Monardes (ca. 1512–1588), then a young physician just starting his practice in Seville in the 1530s.[60] When the doctor advised him to use a certain purgative, Cataño suggested using *mechoacan* instead. The doctor looked at him in disbelief: how could he use a medicine for which there was "nothing written or known"? Cataño, perhaps amused by his doctor's reaction, answered that he "had the experience" of using it before with excellent results, and that he was going to take it. The doctor insisted that under no circumstances would Cataño do so, and his authority ultimately won out over Cataño's personal experience.[61]

Cataño took the doctor's medicine, and it did not work. Monardes proposed to give him a different dose. But this time Cataño held his ground and took the new medicine. It worked. The doctor started to look into this new medicine and others coming from the New World. He found other people who had came lately from New Spain who had used it, and who had been restored to health as well. The doctor kept asking questions. Upon hearing that a Franciscan had brought a plant of *mechoacan* from New Spain in a barrel, he ran down the street of Seville, found the monastery, and at the entrance of the infirmary saw a new plant; he realized immediately that it was *mechoacan*.[62] Later, Monardes wrote about the plant, and sent reports in Latin and Spanish about it to doctors and friends in Europe.

Monardes continued to write about *mechoacan* and many other medicinal herbs coming from the New World. He kept asking soldiers, merchants, Franciscans, royal officials, and women about new medicines and plants: many brought medicines from New Spain in their luggage, just as modern-day travelers might carry pain-relief and allergy medicines. The doctor asked travelers to provide the names and uses of the substances they brought, and to describe their own experiences using them. Most referred back to Indian uses and Indian names of the various herbs. At some point, he started receiving his own samples of medicines with accompanying reports. Monardes established a botanical garden—and he was not the only one to do so in Seville. He so fully trusted the empirical reports accompanying his samples that he used them without any of the original concerns he had had with *mechoacan*. On one occasion, he received a medicine said to induce abortions. Knowing only that the Indians used it for that purpose, he tried it in Seville and it worked.[63]

Monardes published three reports on these medicines in 1565, 1569, and 1575. The first report ended with these words:

We are certainly worth of great reprehension for nobody is writing anything about all the herbs and plants and other medicinal plants from New Spain. Nobody knows anything about their characteristics and shapes to compare them with our own medicines. If someone had the desire to investigate and experiment with so many medicines as the Indians sell in their markets, that would be a thing of great utility and benefit.[64]

The doctor added that the Indians had experience and knowledge about those plants, suggesting that it is just a matter of asking them about their medicines.

Contemporary to the doctor's inquiries in Seville about medicines, the royal cosmographer Alonso de Santa Cruz (ca. 1500–1572), a leading cosmographer in Philip II's court, proposed in the late 1550s to ask specific questions to explorers and colonists arriving to Seville—just as Monardes had done regarding medicines from the New World—rather than asking for general information about the land and its natural things.[65]

Almost simultaneously, in New Spain (between 1558 and 1569) the Franciscan Bernardino de Sahagún (ca. 1499–1590) elaborated a questionnaire to write his *Historia general de las cosas de la Nueva España* (General History of the Things of New Spain). Sahagún sent questionnaires, reconstructed from his book, to Indian villages, asking for information about aspects of their culture, society, and land. On natural history, the questionnaire included questions about names of animals; the history of the name; a physical description of the animal; its environment, activities, and food; ways of hunting or catching it, popular histories about it; and sayings related to it.[66] Both Sahagún and Santa Cruz collected information intensely during the 1560s.

They were not alone in this search for information and natural products. In the late 1550s, Marcos de Ayala, a resident in the town of Valladolid (Yucatan) who had been living in New Spain and Yucatan for over twenty-five years, presented a report on the *campeche* wood, a dark-blue dye from Yucatan to Viceroy Don Luis de Velasco. Ayala wanted to exploit this commodity, and in order to validate his knowledge about it he performed the experiment with this dye at the viceroyal court, and was granted a license to exploit the *campeche* wood.[67] In 1565, when Monardes published his first book on medicines, the crown dispatched a royal decree to the governor of Yucatan requesting samples of Campeche wood, as stated in the *real cedula*: "because We want to know the qualities of this wood and seeds and understand their effects and uses in these kingdoms . . . and to test [*el ensaye*] them as it is appropriate, We order you to send a reasonable amount of campeche wood, indigo, and cochineal to Our officials of the Casa de la Contratación in the first ships leaving those provinces."[68]

Similarly, a soldier in Peru, a certain Pedro de Osma, who had read Monardes's book, was searching Peru for medicines and natural products. "As a reward for the benefit that I have received [from your book]," Osma wrote to Monardes, "I sent you a dozen [bezoar] stones [from llamas], by way of Juan Antonio Corso, rich merchant. If they arrive, you must test them in many diseases and you will see what great effects they have."[69] Monardes included Osma's report in his second book on medicines, and probably added Osma's samples to his collection in Seville.[70]

Research activities in Mexico and Peru such as those undertaken by Pedro de Osma and Marco de Ayalas responded not only to the crown's project for searching the land for medicines and commodities but also to the larger goal of collecting. The New World yielded strange new things for study and research. Curiosities or marvels, and not only commodities, attracted the attention of Europeans.[71] Marvels from the New World, ranging from maguey to armadillos, constituted singular entities subject to empirical investigation because they did not figure in the books of the "ancients"—as the first natural historian of the New World, Gonzalo Fernández de Oviedo (1478–1557), called classical texts.[72] In this way, curiosities as well as commodities shaped the emerging strategies for the empirical study of nature. Curiosities were in themselves a sort of commodity: a natural product with either commercial or ornamental value. As the destination of numerous *curiosidades* and *maravillas*, as they were called in Spain, Seville was transformed into a center for collecting items from the New World.[73]

The second half of the sixteenth century witnessed an increased interest in curiosities that resulted in the institutionalization of empirical practices—from experimental gardens and collections of oddities to the first natural history voyage of exploration to the New World. As shown with the Santo Domingo balsam, new commodities and curiosities legitimized the empirical approach to nature occurring in gardens, in display cabinets, and in explorations. During this period, the trafficking of curiosities became more systematic. In 1554, Philip II ordered his officials in Hispaniola to send "all the seeds and plants found" in the island for his gardens in Spain. They sent him some "plants of pineapples," among others.[74] Between 1563 and 1566, Philip received for his gardens at Aranjuez, Casa de Campo, and Valsain seeds from Flanders and Seville—these perhaps coming from the New World.[75] In 1578 two trees from Chile, one of them a balsam tree, arrived in Seville for the king's garden.[76] As mentioned at the beginning of this chapter, the Spanish American case shows that this institutionalization began in the sixteenth century, in the context of commercial and long-distance empires, rather than in the seventeenth century.

I finish this essay with Francisco Hernández's expedition. Philip II's interest in natural history, Monardes's publications that had reached the court by 1565, and the growing commercial interest in New World commodities prompted Philip to send the first natural history expedition to the New World in 1570.[77] Monardes's book and his plea for researching the New World were probably instrumental in the organization of the expedition.[78] Philip II sent Dr. Francisco Hernández to the New World to "gather information generally about herbs, trees, and medicinal plants."[79] Hernández was ordered to "find out how the abovementioned things are applied, what their uses are in practice, their powers, and in what quantities the said medicines are given, as well as the places in which they grow and their manner of cultivation."[80]

Four years later, Hernández sent "seeds, plants, and other natural things" to the king.[81] New World reports and curiosities had highlighted the empirical approach to nature through gardens, explorations, and protocols at royal institutions.[82] In 1573, after two years in Mexico, Hernández explained the magnitude of the American task:

What I may now advise Your Majesty is that four volumes of paintings of plants have been completed recently, in which there are 1,100, and another in which there are 200 animals, all exotic and native to this region, and scripts in draft and almost half of the descriptions in fair copy, of the nature, climate of the places to which they are native, the sounds they make, and their characteristics, according to the Indians, whose experience stretches over hundreds of years here. I have relied both on the evidence of other curious persons and of the doctors of this land and my own experiences, beyond what can be deduced using the rules of medicine. In all this, great care has been taken that no plant is painted unless I have seen it ten or more times in different seasons, smelled and tasted all its parts and asked more than twenty Indian doctors, each one individually, and considered how they agree and differ, and unless I have subjected it to the rigorous methods of identification and examination that I have developed here for this project.[83]

From the roads of New Spain to the streets of Seville emerged the empirical culture characteristic of the sixteenth- and seventeenth-centuries' long-distance empires. Artisans, merchants, and physicians provided ideas for new protocols, and the crown quickly adopted them. The cases of Ribeiro and Villasante show how the crown institutionalized their suggestions into protocols and practices at the Casa de la Contratación and Council of Indies; Monardes illustrates how medicines from New Spain arriving in Seville shaped new approaches to validate and produce knowledge. Monardes understood that someone with expertise had to travel to the New World to collect samples and information, and the crown translated his call into the Hernández expedition of 1571. Spain provides a case of how a new culture emerged around the circulation of commodities, people, and ideas during the sixteenth century. By institutionalizing artisans' practices at the Casa de la Contratación, Council of Indies, and viceroyal courts, the Spanish crown launched an empirical program without parallel in the sixteenth century.

CHAPTER TWELVE

Voyaging in the Spanish Baroque

Science and Patronage in the Pacific Voyage of
Pedro Fernández de Quirós, 1605–1606

KEVIN SHEEHAN

On November 23, 1606, having traversed more than thirteen thousand nautical miles of the South and North Pacific, the ship *San Pedro y San Pablo* dropped anchor in the harbor of Acapulco. As the former expeditionaries went their respective ways, the royal comptroller of the port, Francisco de la Carrera Güemes, meticulously noted each and every item the ship carried.[1] Three hundred and thirty empty water pitchers, three cables of Chile cordage, two copper spoons, twenty Peruvian gunpowder jars—the list went on interminably. Precisely and indiscriminately, her contents were catalogued and sent to the warehouses of the viceroyalty of New Spain. One wonders what Carrera Güemes made of the item he blandly described as "one copper apparatus, in two pieces, for extracting fresh water from seawater"? Captain Pedro Fernández de Quirós, the self-proclaimed inventor of the apparatus and erstwhile commander of the expedition—which had set sail from the port of Callao on December 21, 1605, in search of unknown islands and a continent in the South Pacific—was under no illusion as to the voyage's significance. (On this search for continental land, see also Valverde and Lafuente in Chapter 10.) In his account of the voyage he describes how in early February 1606, after nearly a month and a half at sea and with the stock of water declining rapidly, he had cut the daily ration and taken personal possession of the keys to the hatch. He then ordered the *instrumento de cobre* (copper instrument), which he had had fashioned in Peru, to be set atop a brazier. According to Quirós, some two to three *botijas* (bottles) of "very fresh and healthy" water were converted from seawater on a daily basis. With certain refinements, he later wrote, this could be increased to a rate of one *botija* every one and a half hours with the use of very little firewood.[2] If such an experiment had been conducted by James Cook in his Pacific voyages, it undoubtedly would have been recorded as yet further evidence of these voyages' enlightened nature. That such a technique was employed centuries earlier by a Portuguese navigator in the service of the Habsburg monarchs of Spain has gone largely unnoticed by the English-language narratives of exploration and discovery in the Pacific Ocean, hinting at an underlying set of biases in the assessment of Iberian scientific voyaging in this era.

The history of Pacific exploration has generally been presented in terms of a canon of principal voyages and explorers. Beginning with the Portuguese Ferdinand Magellan and his trans-Pacific voyage of 1520, maritime historians have focused upon successive waves of European navigators, culminating with the so-called Enlightened voyages of the eighteenth century. In the historical literature on the subject, there is a noticeable tendency to focus on pathfinding, landfall, and cartographic precision as the hallmarks of exploratory achievement. In this trajectory, initiatives taken by Spain and Portugal are overtaken first by the Dutch, then later the French and the English. This narrative parallels that of the march of Western domination, the rise of the nation-state, and the victory of enlightened thought over ignorance. These histories inevitably conclude with the crowning achievements of the Englishman James Cook. The Iberians may well have been the protagonists of a "heroic age";[3] nevertheless, it is Enlightened English voyaging that produced the map of the Pacific as we know it and results in what has been termed "the enthronement of fact on fantasy."[4] In short, this is a story of winners and losers.

Is there indeed a place for Spanish exploration in a narrative that stresses the prominence of empirical observation, and the employment of scientific instruments over a reliance on nonempirically based textual authority? In many respects, the voyage of Pedro Fernández de Quirós from Peru across the South Sea in 1605–1606 in search of a southern continent is representative yet also contradictory of these tendencies. Quirós spent years both before and after this expedition seeking support for his project to confirm the existence of a landmass stretching from the Strait of Magellan to the Island of New Guinea.[5] In the process, he became a familiar figure at the residences of the viceroys of New Spain and Peru, as well as the papal court of Clement VIII and the court of Philip III in Madrid. As a means of persuasion, he wrote more than fifty memorials outlining his proposals, giving details of the cosmology and hydrography upon which his claims were made for the existence of another New

World on the far side of the Pacific. An accomplished cartographer, he is believed to have drafted more than two hundred maps, charts, and globes.[6] He also designed a series of scientific instruments to ascertain latitude and longitude, and proposed a voyage of circumnavigation to test them. Evidence of the power of his ideas is found in the eclectic group of influential persons that came to support his project, as well as the broad dissemination of his ideas in early-seventeenth-century Europe and the Americas via printed memorials. Ironically, while seeming to embody the hallmarks of the beginnings of early modern science, Quirós is more frequently remembered as an individual "deeply imbued with the superstitions of his time and nation."[7]

A VOYAGE IN THE BAROQUE

To understand the origins of this seeming contradiction, we need to place ourselves on the shores of the Western Pacific island of Espíritu Santo in the New Hebrides on May 14, 1606. That morning, the members of the Quirós expedition waded ashore after more than five months at sea. The beach that Sunday witnessed a complex baroque pageant of word and gesture. The words of the ceremony pronounced by Quirós expressed the profound optimism of this moment. Clasping a cross made from the wood of an orange tree, he intoned an act of possession heavily influenced by verses taken directly from the Apocalypse. On the eve of the possession ceremony Quirós had founded a new chivalrous order, the Knights of the Holy Spirit, admitting all members of the expedition—even those of mixed blood and the black slaves who had accompanied their adventurer masters. The city traced out along the banks of the nearby River Jordan was named New Jerusalem, and Quirós appointed its municipal officers from among the expeditionaries.[8] A region that had for generations of Spanish seafarers been known as the Islas de Salomón in the dream of finding the origin of the gold of Solomon's temple was now transformed by another vision. The new land was formally named the Parte Austral del Espíritu Santo—later diplomatically changed to La Austrialia del Espíritu Santo, in honor of the ruling Habsburg House of Austria.

The event and its surrounding circumstances were layered with manifold meanings. When taken together, they suggest the profound influence of utopian currents of thought. Here combined were the aspirations of religious and political elites in the Americas and on the peninsula for a renewal of Spanish imperial endeavor through the founding of another New World. Here was the great event hoped for by the new king

Philip III, an antidote to the growing sense of *desengaño* (disillusionment) perceptible in Spanish society. Here was the opportunity to enact the spirit and letter of the laws governing discoveries set in place by Philip II some thirty years earlier.[9] Here, too, in the shape and structure of the New Jerusalem were the dreams and aspirations of the architects of Renaissance urban renewal. Perceptible also are undertones of a belief in the emergence of a new spiritual age characterized by a time of ecclesial renewal through the evangelization of a people more amenable to the message of the Gospel. In this combination of events and their interpretation, it is not far fetched to perceive faint echoes of Joachim of Fiore's dream of a new age, and the approaching end of time.[10] (For the relationship of millenarianism and Spanish imperial ambitions, see Ferreira Furtado, Chapter 9.)

The settlement of New Jerusalem was short lived, for within a few weeks the vessels of the expedition became separated and Quirós was forced to return across the Pacific before enacting a detailed survey of the coastline of La Austrialia. The other vessels in his fleet under the command of Luis Vaez de Torres eventually found their way to Manila, but not before demonstrating both the lack of any mainland in the vicinity of Espíritu Santo and the insularity of New Guinea. These discoveries would be lost to European memory for centuries. Quirós's discovery would, in contrast, catch the imaginations of generations of European seafarers.[11]

KNOWLEDGE AND EMPIRE

The story thus far contains fascinating elements of cultural and intellectual history. This grandiose plan for discovery, exploration, and settlement was derived from a complex set of ideas drawn from contemporary cosmography, natural history, nautical science, political tracts, and theology. This image was constructed from exchanges across a broad community of scholars and patrons. In short, the Quirós voyage, however unique in terms of its goals or achievements, was very much the product of contemporary aspirations in early-seventeenth-century Spain.

To what extent might this voyage and its surrounding negotiations be attributed to emerging baroque scientific sensibilities? To answer this question, we must refocus our understanding of the term "science" to place it within the context of the intellectual, social, and cultural history of early modern Iberia. Science as presented in this essay refers to cosmography, nautical science, cartography, and natural history.[12] In the Spain of the sixteenth and seventeenth centuries, these disciplines were not considered simply the preserve of intellectuals. They were actively cultivated by the ruling elite and co-opted in service of the empire. For much of the sixteenth century, Spain had actively sought new frontiers and peoples. Vacant regions and abruptly terminating coastlines on contemporary charts were not so much signs of the limits of knowledge and dominion as they were challenges and invitations for further discovery and settlement.

Initially tentative in the role of patron of voyages of exploration, the Spanish crown increasingly followed the example that had been set by the monarchs of Portugal, through wholeheartedly embracing maritime endeavor as a crucial factor in securing profitable imperial possessions. As David Goodman has argued persuasively in his seminal work on the crown sponsorship of scientific endeavor during the reign of Philip II, the Spanish monarchy instituted a series of crucial reforms and innovations during the second half of the sixteenth century aimed at harnessing science for the construction of empire.[13] This activity extended from the founding of learned institutions like the Royal Academy of Mathematics to specific reforms aimed at improving the functioning of Spain's maritime activity.[14] The motives were thoroughly pragmatic. The process employed was based on a systematic review of information that had been gathered by subjects of the crown at the distant corners of its empire.

Increasingly in the sixteenth century, the crown sought to guide and control the gathering and processing of information obtained from the far frontiers of its expanding empire. This project could never have been contemplated without the cooperation of Spanish subjects working on the fringes of that empire.[15] On July 25, 1574, writing a memorial to Philip II from his convent in Peru, Fray Gerónimo de Villacarrillo, O.F.M., began his petition with an opening statement concerning the nature of the obligation between subject and lord within a Christian commonwealth: "We, Your Majesty's vassals, have the obligation to advise you concerning your royal service, and particularly concerning matters dealing with the discharging of Your Majesty's conscience."[16] This was not an isolated perspective. From one end of the Habsburg empire to the other, royal officials, churchmen, soldiers, scholars, navigators, and merchants each saw themselves as somehow participating in and contributing to the fulfillment of the corporate obligations of empire.

The first decade of the seventeenth century saw a series of important policy changes for the Spanish crown under a new king, Philip III, and the influence of his favorite, the Duke of Lerma.[17] Diplomatic efforts secured a period of unprecedented peace for Spain after decades of European conflict.[18] While Philip may well have lent only begrudging support to a truce with the Low Countries, the short-term economic benefits of peace in Europe were undeniable. Within a broader global imperial strategy, however, the benefits would be debated. Two factors emerged repeatedly in discussions among the king's councilors during this period. The terms of the Twelve Years' Truce had left Philip's Portuguese possessions in Asia vulnerable to Dutch ambitions. The truce also created a dilemma for Spain in terms of continuing the program of frontier expansion that had resulted in the creation of a vast overseas empire in the sixteenth century.

The question of the Pacific now emerged as a crucial series of options for Spain. New expansion might well bring renewed wealth as well as the conversion of countless peoples to the Catholic faith. In this, Spain would be continuing her providential mission. However, such expansion might result in a further depopulation of the Iberian Peninsula, and an overextension of Spanish financial and military resources. Moreover, the discoveries of new lands in the Pacific were bound to come to the attention of the English and particularly the Dutch, who were now threatening Portuguese possessions in Asia. Discovery was thus as much a liability as a potential blessing.[19]

Confronted by a reticent imperial policy, expansionists at court lobbied passionately for the king's favor and approval. In many respects they appear to have won it, for the new king was intent upon distancing himself from the legacy of his father and renewing Spain's fortunes in the face of strident English and Dutch maritime activity.[20] The Pacific emerges in the Spanish mind as a type of social and technical laboratory, its settlement proposed as a panacea for Spain's troubled empire in the Americas. These factors help to explain how a skilled but little-known Portuguese coastal pilot became one of the most influential writers on Pacific navigation in the first half of the seventeenth century. The process by which this occurred reveals something of the intimate connection between elite patronage, cosmological theory, applied nautical science, and imperial ambition that occurred in the shadow of Philip III's court during the first decade of the seventeenth century.

THE PACIFIC PROJECT

Here then was the crux of the spiritual and temporal challenge presented to Spain in the latter half of the sixteenth century. This challenge would take on a particular urgency as Spanish navigators and adventurers gazed westward across the Pacific from the shores of the viceroyalty of Peru. For, somewhere beyond the horizon, rumor had it that the place from whence the fleet of the biblical King Solomon had journeyed—islands laden with gold and pearls—now awaited rediscovery. Ever since his first voyage in search of the mysterious islands rumored to lie to the west of Peru, and the resulting discovery of the Solomons in February 1568, Álvaro de Mendaña had been involved in a series of negotiations with a view to the eventual settlement of these lands. By early 1595 the negotiations and preparations were virtually complete, and at this point for the first time we find specific mention of Quirós in documentary sources.[21] On April 9, 1595, the fleet set sail from Callao and headed north along the Peruvian coast picking up supplies, before choosing a westerly course in search of the Solomons. Mendaña would eventually anchor at Graciosa Bay at the island of Santa Cruz on September 21, 1595. Within two months, with Mendaña dead, and the settlement plagued by mutiny and sickness, Quirós—now under the orders of Doña Isabel Barreto, Mendaña's widow and successor—pioneered a route from Santa Cruz to Manila. The survivors of the voyage eventually reached Manila on February 11, 1596.

Quirós might well have ended his days in the company of explorers like Juan Fernández, the *piloto mayor* of the South Sea, who pioneered the southerly route to Chilean waters in the sixteenth century—that is, as a navigator and pilot with considerable local knowledge and expertise in the coastal navigation of the Americas but little more. However, participation in Mendaña's ill-fated voyage in 1595 catapulted Quirós into the spotlight. Determined to pursue Mendaña's goal, Quirós journeyed to Spain in early 1600, and then on to Rome for the celebration of the holy year decreed by Pope Clement VIII.

THE POWER OF PATRONAGE

During this Roman sojourn, Quirós was introduced to an influential group of scholars, diplomats, and ecclesiastics. The support of this seemingly disparate group in turn gained him access to the court of Clement VIII and then to that of Philip III of Spain. It was also during this time that Quirós had access to the library of the Spanish ambassador and further proceeded to refine his understanding of cosmology and nautical science through conversations with Roman academics.

Even before reaching the Eternal City, it appears that his understanding of the Pacific had evolved significantly beyond an interest in the discovery of the Islands of Solomon. In a 1597 memorial addressed to the viceroy of Peru, Don Luis de Velasco, Quirós had requested approval for another voyage of exploration in search of a mainland believed to extend between the Cape of Good Hope and the Strait of Magellan as far south as ninety degrees.[22] This opinion approximates the charts produced by some Portuguese cartographers—most notably those of Fernando Oliveira in his *Ars nautica* (Art of Navigation) of 1570.[23] Here, the ill-defined outline of a landmass extending eastward from the Strait of Magellan below the Cape of Good Hope and South Asia was depicted based primarily upon the authority of the ancient author Diodorus Siculus. With the majority of the landmass described by Quirós falling within the Portuguese lines of demarcation, this southern continent was essentially the product of a Portuguese mind-set.[24] Not surprisingly, given the loss of life suffered on the Mendaña voyage and the ambiguous geographical focus of Quirós's proposals, Velasco referred the matter to the aged Philip II for a decision. Reflecting what had been a Peruvian perspective on exploration in the South Pacific for the previous forty years, Velasco's letter to the king referred to the proposals primarily in terms of the Islas de Salomón.[25]

Quirós arrived in Rome fortunate to have made the acquaintance en route of Fray Diego de Soria. Soria had met Quirós in Manila in the aftermath of the 1595–1596 voyage. He had traveled to Spain and Rome as procurator of the Dominican province in the Philippines and would eventually be nominated bishop of Nueva Segovia in the same islands.[26] Versed in Scholastic theology, Soria also had impeccable connections. He was representative of a school of missionaries that had actively advocated the expansion of Christianity in Asia by force of arms if necessary.[27] It is highly likely, therefore, that Quirós's nascent missionary zeal for the peoples of the undiscovered lands had in part been fired by conversations with this erudite ecclesiastic. What is certain is that Soria was instrumental in introducing Quirós to one of the most influential Spaniards in early-seventeenth-century Rome, Don Antonio Cardona y Córdova, duke of Sesa and Spanish ambassador to the Holy See.[28]

The Portuguese pilot clearly made an impression on Sesa and was invited at length to become a member of the duke's household.[29] In return, Quirós agreed to tutor Sesa's son in the art of navigation and cosmography.[30] What followed was a period of intense activity over some eighteen months, facilitated in no small part by the duke's growing passion for Quirós's proposals. Quirós was introduced by Sesa to a select group of Rome's intellectuals. Among these were the Jesuits Christopher Clavius and Juan Bautista Villalpando, and a number of mathematicians, geographers, and clerics from the Roman curia and papal household. (On the role of Jesuit mathematics in Spanish imperial science, see Pimentel in Chapter 5.) Sesa described one of these encounters in a letter of recommendation to Philip III: "I informed Father Clavio and other mathematicians and geographers of the matter. All of them are now persuaded that there cannot fail to be a great tract of mainland or a number islands stretching from the Strait of Magellan to New Guinea and Java Major and other islands of that great archipelago."[31] A radical shift had occurred in Quirós's cosmology. No longer were the Austral regions concentrated within the portion of the globe granted to monarchs of Portugal. Instead, they now lay securely within the potential dominion of the king of Spain. What had been Portuguese was now thoroughly Castilianized.

While accepting the hospitality of the Spanish ambassador, Quirós also busied himself with the preparation of a text on navigation. Although never published, this treatise was clearly intended to establish his credentials in the scientific circles cultivated by the duke. In devising a number of instruments to aid in the determination of longitude, it is likely that he consulted works such as Luis de Fonseca Coutinho's *Arte da agulha fixa e do modo de saber por ella a longitud* (Art of the Fixed Compass and the Method of Ascertaining Longitude by It). Determining a way of calculating the variation of the compass needle was in fact a recurrent obsession among contemporary Iberian cosmographers. (See Portuondo's essay in Chapter 3; note also Barrera-Osorio's references to the role of instruments in exploration in Chapter 11.) Some of Quirós's closest associates in the years following the 1605–1606 voyage—among them Lorenzo Ferrer Maldonado and Dr. Juan Luis Arias de Loyola—had themselves devoted much energy to the question.[32] Both of these individuals incidentally would eventually volunteer to go to La Austrialia.

Clavius and Villalpando appear to have been impressed with Quirós's technical expertise, as the text of the *cédula* (royal order) that would eventually accompany him to Peru suggests:

He demonstrated there two [instruments] of his invention, one to be used while navigating to show the difference that the compass needle makes between northeast and true north, and the other to measure latitude with greater ease and certainty: and both of these have been praised by Fathers Clavius and Villalpando, of the Society of Jesus, and the Doctors Turibio Pérez and Mesa, who have lectured in mathematics in Salamanca, and by other famous geographers.[33]

Conversations with Christopher Clavius undoubtedly provided a broader intellectual basis for Quirós's plans. Possibly the most relevant of these works is that on the sphere.[34] It is likely that Quirós had access to this when he wrote his own treatise on navigation, for the arguments used by Clavius and Quirós for the spherical nature of the earth are virtually identical.[35]

While the influence of the learned Jesuit mathematician and cosmographer Clavius may seem natural for a mariner attempting to craft a thesis on navigation and hydrography, that of Juan Bautista Villalpando is more mysterious. Villalpando, also a member of the Society of Jesus, was a client of Sesa and the Spanish crown. The duke had been instrumental in providing the necessary funds for the completion of Villalpando's monumental exegesis of the Book of Ezekiel.[36] Villalpando's work was not simply a piece of erudite biblical scholarship. It was also an elaborate reconstruction of the Temple of Solomon.

There were also fundamental political implications to this work. In a letter written to Philip II some years earlier, Villalpando outlined the ideology underlying his mathematical analysis of Ezekiel and the theological meaning of the temple architecture. In its perfect proportions, Solomon's temple becomes a symbol of the terrestrial Church and the heavenly Jerusalem. Villalpando's praise of the builder of the biblical temple becomes, by association, praise of Philip II, a new Solomon and builder of El Escorial. The commentary is dedicated not only to a monarch who is "obeyed by the greater part of Europe and the East and West Indies," but also one who could look forward to the future dominion over other kingdoms, thus procuring a reign over the entire circumference of the globe.[37] (On the relationship between Spanish expansion and eschatological ideas, see Pimentel's essay in Chapter 5.) Villalpando was in fact more interested in the creation of a perfect mathematical model, an identification between the temple and cosmic perfection, and an exercise in harmony and proportion than in any literal interpretation of Ezekiel or the exercise of a type of exegetical archaeology.[38] His aim was a reconciliation and harmonization of revelation with humanism, concretized in an architectural form that was both a prophetic statement of Habsburg universalism and a catechesis in Tridentine theology.

In this, Villalpando exhibited something of the intellectual inheritance he had received from his master, the architect of El Escorial, Juan de Herrera. Herrera's *Discourse on the Cube*, thoroughly influenced by the philosophy of Raymond Lull, stresses the intrinsic relationship between harmony, order, and truth—values concretized in the edifice of the Escorial.[39] (For further references to the work of Juan de Herrera, see Portuondo in Chapter 3.) Herrera was an undoubted influence in the development of Philip II's literary collectionism, manifest in the king's efforts to fill the royal library in the monastery of San Lorenzo with as many of Lull's works as could be found throughout Europe.[40] Herrera was not the only follower of Lull at the Spanish court. In 1591 Philip II appointed Dr. Juan Luis Arias de Loyola, another defender of Lull, to replace Juan López de Velasco as chronicler of the Indies.[41] The relevance of these intellectuals for the story of Spanish expansionism in the South Pacific becomes apparent when noted that both Arias and Villalpando

were avid supporters of Pedro Fernández de Quirós, providing him with
the mathematical and cosmological "proofs" for the existence of a continent situated somewhere to the west of Peru.[42]

THE COURT OF PHILIP III

Armed with a series of papal briefs approving his mission and, most
important, Sesa's letter of recommendation, Quirós finally arrived in
Madrid on the eve of the octave of the feast of Corpus Christi, 1602.
He traveled to El Escorial, where the king was in residence. Not surprisingly, given the documentation he now had at his disposal, Quirós was
promptly granted a royal audience and delivered his first memorial to
Philip III on Monday, June 17, 1602.[43]

Events transpired quickly. Unlike the protracted series of negotiations
that would follow the 1605–1606 voyage, we do not find a series of arguments and responses between Quirós's *memoriales* (letters of petition)
and the Council's *consultas* (consulations). Instead, Quirós's narrative
account of his voyages provides details of negotiations at this point. The
Relación suggests that Quirós had contact with individuals at the highest levels of the court.[44] Naturally, the members of the Councils of State
and Indies were numbered among these.[45] It does seem that some were
not confident in his ability to lead an expedition, or held reservations
concerning the viability of further discoveries and conquests. In the face
of doubts and criticisms, Quirós was forced to address another memorial
to the king later in 1602.[46]

When compared with later memorials, we are struck by the relative
brevity of these two prevoyage memorials. The first bases its petition on
a cosmology and an ethnology gained through experiences on Mendaña's
1595–96 voyage in search of the Solomons. The fundamental petition—
originally placed before Velasco and now brought to the king—is focused on the discovery of new lands that experience suggests exist in
the vicinity of the islands discovered by Mendaña. The benefits of their
discovery were twofold: "that the undiscovered dwellers in the Antarctic
region be brought to pasture, benefited and sustained in the evangelic
doctrine, and that your Majesty may be the lord, known, obeyed and
served from one pole to the other, as you are from East to West."[47] The
second memorial takes up similar themes, and also appears to answer
criticisms: "That I can count on many years of experience at the very
outset, whereas those appointed to command voyages of discovery in
the past had but the slightest experience in such enterprises; and I am an
eyewitness and come from those very parts to these to give intelligence

of them."[48] On Saturday, April 5, 1603, Quirós received his long-awaited dispatches. Thus within the space of a year he had managed to procure a series of favorable *cédulas* that would enable him to enlist the assistance of the viceroy in Peru. The *cédulas* commanding the discovery and exploration of the Austral lands reproduced, almost word for word, the Duke of Sesa's original recommendation.

CONCLUSION

Quirós's obsession with the founding of a New Jerusalem in the land of the Holy Spirit, his idyllic descriptions of the region and its inhabitants, his colorful possession ceremony, and his elaborate articulation of his project through repeated memorializing have tended to reduce his contribution to the story of Spanish maritime endeavor. Was his behavior on the shores of La Austrialia any more bizarre than that of Joseph Banks declaring the island of Tahiti "an Arcadia of which we are going to be Kings,"[49] or his stripping naked and smearing himself with charcoal on the same island?[50] Moving beyond a historiography that has judged these voyages in terms of their lack of success in discovering a southern continent, or the apparent millennial delusions of their principal protagonist, this essay offers a revision of baroque Spanish voyaging in the Pacific by looking at its intellectual context.

How, then, did the Quirós memorials, negotiations, inventions, experimentation, and voyaging mesh with a trajectory of reporting and information gathering that had been established for a century in the Iberian world? In some cases, the links between Quirós's vivid descriptions of *La Austrialia* and that of contemporary and later authors and cartographers are explicit. As news of the "discoveries" in the South Pacific became disseminated throughout the Iberian world in the first two decades of the seventeenth century, cartographers such as Manoel Godinho de Erédia reproduced the contours of La Australia in accord with their own plans for settlement of the region.[51] (On Erédia, see Walker's essay in Chapter 13.) Moreover, this movement of information was not restricted to the Iberian world. The concerns of Philip III's councils were indeed proven warranted, as Quirós's writings became translated and well known among navigators, geographers, and cartographers beyond the borders of Spain. On October 25, 1615, while in the Atlantic in search of a westward route to the Pacific, the Dutch navigator Jacob Le Maire roused the spirits of his despondent crew by reading them Quirós's *Eighth Memorial* and informing them that the purpose of their voyage would be the further exploration of the region recently discovered by Quirós.[52]

At other times, a relationship is more implicit. Such is the case with Roger Bacon, and the literary work that represents the final evolution of his thoughts on epistemology.[53] (On Bacon, see also the chapters by Almeida and De Vos in this volume.) In his *New Atlantis*, written in 1624 and published in 1627, Bacon describes a mysterious land found in the South Sea opposite the coast of Peru. In Bacon's tale, a group of travelers attempt to navigate across the Pacific in search of Japan and China. Blown off course, they arrive unexpectedly on the shores of a hitherto unknown land whose inhabitants speak Spanish. They encounter the local nobility who have instituted a chivalrous order bearing the title "The House of Solomon." They also receive a lengthy discourse on the social structure and institutions of what amounts to a utopian society.[54]

The similarities between the underlying inspiration for Bacon's narrative and crucial details found in the accounts of Spanish exploration and settlement in the Western Pacific seem more than coincidental. The template for the *New Atlantis* may well be found in one of the most widely circulated of Quirós's memorials—the so-called *Eighth*. Written in late 1608 or early 1609, and printed in Seville and Pamplona in 1610, the *Eighth Memorial* became widely circulated not only at court but also beyond the borders of Spain. By 1617 this memorial had been reprinted a number of times in Spanish. It had also been printed in English, French, German, and Latin.[55] The year after Bacon composed his *New Atlantis*, Samuel Purchase published another English translation.[56]

The baroque voyage of Fernández de Quirós displays a curious blend of motives and methods, ranging from the pragmatic to the utopian. To characterize the voyage in terms of a closing of the "heroic" age of Spanish seafaring is an oversimplification. In its focus on the implementation of cosmological theories, the use of experimental instruments, and the testing of new maritime routes, there is something decidedly modern and innovative here. Yet it must also be stressed that the voyage was part of a long trajectory in the Iberian experience of applying science to the expansion of empire. In this it becomes emblematic of the mentalities of a time in which the Spanish sought pragmatic solutions to their mounting problems of empire. One of the reasons why the Spanish imperial system was grudgingly admired by its nascent rivals, the English and Dutch, was its capacity to effectively govern vast regions, not simply through gunpowder and pike, but more crucially through the gathering and processing of information on a scale never before witnessed.

Rather than stress the manifest failure of the Quirós voyage to achieve its stated goals, or of the crown to respond effectively to repeated requests for new initiatives, this essay has focused on the processes involved in the gathering and dissemination of knowledge in early-seventeenth-century

Spain. From this perspective then, the encounters with indigenous knowledge in the Pacific, the exchange of ideas among groups of intellectuals and technical experts, and the dissemination of new discoveries—all found within the context of this voyage and its negotiations—reveal a vibrant and creative Spanish interest focused on a search for solutions to the problems of empire building. Within this story, the sciences of cosmology, navigation, mathematics, and the various branches of theology, which had been carefully cultivated during the reign of Philip II, continued to be considered essential tools to the maintenance of effective imperial hegemony.

Acquisition and Circulation of Medical Knowledge within the Early Modern Portuguese Colonial Empire

TIMOTHY WALKER

Portuguese colonial expansion during the sixteenth, seventeenth, and eighteenth centuries had a profoundly important scientific dimension, the impact of which has far outlasted the economic ascendancy of the Lusophone maritime empire. Portuguese exploration added extensively to the human understanding of global navigation and geography, but contributions to the fields of pharmacological botany and medicine were no less significant. (On the work of Portuguese navigators in this global exchange, see Almeida in Chapter 4.) Descriptive works about Asian, African, and South American medicinal plants by Portuguese observers during the early modern period introduced Europeans to many of the efficacious drugs commonly employed in extra-European healing traditions. In the Portuguese colonies, medical practitioners encountered a radically different sphere of healing knowledge, one that they would explore, exploit, expropriate, and export for more than three and a half centuries. In an unmatched feat of scientific acquisition and dissemination, Portuguese colonial officials spread indigenous drugs and information about various native healing methods to European territories on four continents.

This chapter will examine the role and influence within Portugal's maritime dominions of medical techniques, remedies, and drugs originating

I wish to thank the American Institute of Indian Studies and the National Endowment of the Humanities; this research was completed with a grant provided through these organizations. For logistical support in Goa, I am grateful to the Fundação Oriente and the Xavier Centre for Historical Research. I also wish to thank Professor Luís Felipe Thomaz for his assistance.

in colonial holdings in Asia, Africa, and the Americas. It will focus particular attention on the collaborative interaction of indigenous healers and European physicians, surgeons, and pharmacists in medical facilities located in the Portuguese enclaves, where indigenous techniques were often employed to the edification of colonial agents (missionaries, colonial administrative officials, marine commanders, and state-licensed medical practitioners), who then disseminated those techniques to Europe and other European colonial locations. A second focus will be on the trade of medicinal plants to the *metrópole* and between disparate Portuguese colonial regions. Indigenous peoples of the Portuguese colonies thus made important contributions to "Western medicine" during the early modern period, but did so through European intermediaries, who often altered the original application of native medicines (or the philosophy of indigenous healing techniques) to meet their own ends.

Throughout this period of dynamic cultural blending and information exchange, Portuguese military and missionary considerations fueled official interest in indigenous medical practices. Initially commercial motivations were secondary, but grew as the colonial system expanded and matured. In the colonial military sphere, the exigencies of personnel survival drove European inquiries into native medicine. Typically during the sixteenth to eighteenth centuries, new conscript arrivals to the eastern colonies from continental Portugal numbered from a few hundred to as many as three thousand annually,[1] but their ranks shrank rapidly due to tropical diseases. The sixteenth-century Dutch traveler John Huyghen van Linschoten observed that, in the Royal Hospital of Goa, "every yeare at the least there entered 500 live men, [who] never come forth till they are dead."[2] By the early seventeenth century, Portuguese India was known for its terribly high mortality rate. According to one contemporary estimate, in the three decades between 1604 and 1634, Portuguese military deaths exceeded 25,000 men in the Royal Military Hospital of Goa alone (see Figure 13.1).[3] Until the early nineteenth century, yearly mortality rates of 25 to 50 percent were common for newly disembarked Portuguese soldiers.[4]

Portuguese commanders, appalled at the ineffectiveness of European remedies against previously unknown tropical diseases and desperate to preserve the fighting effectiveness of their garrisons, turned to native healing practitioners to treat their men.[5] Information about the efficacy of any local remedy that appeared to save soldiers' lives soon passed unofficially from fortress to fortress before being adopted for regular use by colonial forces in state-run military hospitals. Native healers throughout Asia rapidly learned to cater to the particular health concerns of the European troops posted to their shores, and set up apothecaries in garrison

FIGURE 13.1. *Engraving of the Royal Military Hospital of Goa, India*
NOTE: Goa was the administrative hub of the entire eastern Portuguese empire.
This hospital treated some three thousand colonial officials, soldiers,
sailors, merchants, and missionaries each year.
SOURCE: Engraving from António Lopes Mendes, *A Índia Portugueza* (Lisbon, 1886).

towns to vend native medicinal preparations specifically to the *ferenghi* (foreign) colonizers.[6]

Inquiries about indigenous medical knowledge in the colonial missionary context were substantially more complex, intellectually rigorous, and profound. (On the dissemination of knowledge, see Ferreira Furtado in Chapter 9.) While Catholic priests often found themselves, like their martial coreligionists, desperately in need of indigenous cures for tropical maladies, their missionary task relied on the principle that winning native conversions hinged on demonstrating the cultural superiority of Christian Europeans. Throughout the colonial world, therefore, missionaries deliberately touted their knowledge of European medical practices to awe native peoples (typically among the poorest castes who had little access to prestigious professional healers of their own cultures) and attract them to the Church.[7]

Simultaneously, missionaries recognized that native cultures harbored a great store of folk knowledge about efficacious local medicinal plants. The same intellectual proclivities that led missionaries to study indigenous languages and customs (equally strategic knowledge for winning conversions) led them to gather detailed information about native healing arts, remedies, and their ingredients. In time, missionary organizations founded infirmaries and apothecaries in colonial enclaves and developed the foremost European body of expertise about indigenous medicine in the Portuguese colonies.[8] Missionary orders, recognizing the potential for profit from commercializing native drugs, quickly became the principal disseminators of these commodities—and the specialized knowledge of how to prepare and apply them—throughout the Portuguese world. As near monopolists in the global trade of indigenous medicinal substances, missionary orders relied on this revenue to support their evangelical operations in the Portuguese overseas territories.

Medicines originating in India played a particularly significant role in the state-sponsored health care institutions of the Portuguese colonies. Over time, colonial medical authorities in Goa, the administrative capital of the Estado da Índia (the Portuguese empire in Asia), produced numerous official reports about Indian medicines at the request of the Portuguese Conselho Ultramarino (Overseas Council) in Lisbon, the royal body responsible for colonial administration. Such reports (explicated below) became an important conduit of information to crown officials in the metropolis and to medical practitioners in other parts of the empire; they provide a telling gauge of the state of contemporary knowledge about medicinal substances from South Asia and East Africa, and about what techniques were thought to be efficacious.

Further insights about the acquisition and circulation of indigenous Indian medical knowledge within the Portuguese colonial empire can be gleaned by examining records of consignments of medicines shipped from Goa to such destinations as Macau, Timor, Mozambique, Brazil, São Tomé, and Continental Portugal. Colonial health officials generally supplied drug consignments to stock shipboard medical chests or regional colonial hospital facilities within the Estado da Índia and beyond.[9] Merchants and missionaries in Portuguese India, China, and Brazil also supplied consignments of indigenous drugs for trade. Indian healing techniques and other colonial medicinal preparations thus became widely known in Portuguese-controlled enclaves in the Atlantic and Pacific Oceans, far from their indigenous roots, and were fully incorporated into the lexicon of tropical medicine in the Lusophone colonies. Moreover, through trade, professional medical contacts, and the translation of Portuguese medical texts, indigenous healing knowledge from Portuguese-held areas gradually became known across Europe and within contemporary Dutch, English, French, and Spanish maritime enclaves.

EARLY PORTUGUESE TRANSFER OF INFORMATION ABOUT INDIAN DRUGS TO EUROPE

Descriptive works about Asian medicinal plants produced in Portuguese India by Garcia da Orta (1563) and Cristobal da Costa (1578) introduced Europeans to many of the medicinal plants and drugs commonly employed in Eastern healing. Though some Asian medicines had been known in Western Europe since ancient times, Garcia da Orta's work conveyed a new understanding of their multiple uses and characteristics. Cristobal da Costa emulated the template of da Orta's text but improved the detail and accuracy of the earlier work.

As a Spanish-born *converso*,[10] Garcia da Orta moved to Portugal to escape the Inquisition after training in medicine at Salamanca. He practiced in Lisbon before entering Portuguese crown service and sailing to Goa in 1534. Da Orta served as the personal physician to several viceroys and governors of Portuguese India, as well as to the sultan of Ahmadnagar; he enjoyed the friendship and professional collaboration of Hindus, Muslims, and Christians alike.[11]

The culmination of da Orta's labors, *Colóquios dos simples e drogas e cousas mediçinais da India* (Colloquies on the Simples and Drugs of India) (see Figure 13.2), saw rapid and wide distribution in Europe. Published in Goa in 1563, this treatise remained the definitive work on Asian

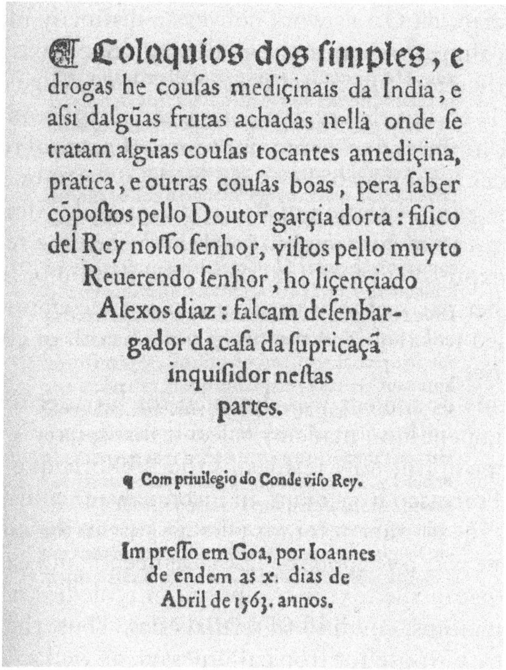

¶ **Coloquios dos fimples, e**
drogas he coufas mediçinais da India, e
afsi dalgūas frutas achadas nella onde fe
tratam algūas coufas tocantes amediçina,
pratica, e outras coufas boas, pera faber
cópoftos pello Doutor garçia dorta : fifico
del Rey noffo fenhor, viftos pello muyto
Reuerendo fenhor, ho liçençiado
Alexos diaz : falcam defenbar-
gador da cafa da fupricaçã
inquifidor neftas
partes.

¶ Com priuilegio do Conde vifo Rey.

Im preffo em Goa, por Ioannes
de endem as x. dias de
Abril de 1563. annos.

FIGURE 13.2. *Frontispiece to* Coloquios dos simples e drogas, *1563*

NOTE: This was the first comprehensive guide to Asian
materia medica produced by a European author.
SOURCE: Diego Garcia da Orta, *Coloquios dos simples e drogas e
cousas mediçinais da Índia* (Rachol Seminary; Goa, India, 1563).

medicine in the Portuguese empire until the nineteenth century. The full text was published only in Portuguese, which limited its circulation. However, much of the original material was translated into Latin by the prominent botanist Charles Lécluse, who in 1567 printed an unauthorized edition of da Orta's text in Antwerp. Incomplete editions in English, French, and Italian followed, pirated from Lécluse's abridged Latin text.[12] Lécluse also appropriated and reprinted work from the Spanish Jesuit physician and botanist Cristobal da Costa, whose *Tractado de las drogas y medicinas de las Indias orientales* (Treatise on the Drugs and Medicines of the East Indies), published in Burgos, Spain, in 1578 followed da Orta's work closely, but expanded upon and corrected some of da Orta's information.[13] Thus, through translations and appropriation, Garcia da Orta's original treatise achieved a very broad circulation.

Colloquies on the Simples and Drugs of India describes fifty-nine different drugs and medicinal preparations, all of them either native to India or observed in use there during the author's perambulations. "India" of course is broadly defined; the geographical area of "the Indies" comprised most of Asia. Despite having been collected and recorded by a Western physician, da Orta's work conveys a distinctly indigenous outlook toward healing; it remains essentially an Indian text.[14]

Another early effort to expand Portuguese knowledge of Indian medicinal plants is an untitled report compiled in 1596 and remanded to the Conselho Ultramarino. This compendium of medical recipes and the Asian substances from which they were made, written by royal order in Goa during the administration of Viceroy Matias de Albuquerque, is a meticulous account of contemporary Malabar healing techniques.[15] It was intended expressly to facilitate the dissemination of Indian healing methods to other parts of the Portuguese maritime empire, where they could be applied to safeguard the precarious health of colonial troops and functionaries.

However, this document was destined for broader distribution. A copy of the compendium, made by a Jesuit missionary a century after its initial composition, now resides in the Bibliothèque Nationale de France. Padre Francisco Rogemunt, in Lisbon awaiting his departure for Macau, copied the document for his personal use in the mission fields of China, where he was posted in the late seventeenth and early eighteenth centuries.[16] Most of the text is composed of remedies and treatments for specific symptoms, conditions, or maladies. Thus, the compendium has a diagnostic purpose for tropical illnesses, as well as a didactic objective regarding medical plants. The latter half of Padre Rogemunt's manuscript includes supplementary remedies and medical information, including Chinese and French language passages, apparently added in the mission fields.[17] The tome shows signs of frequent use in China before being deposited in the French national library. In this remarkable text we see a late-seventeenth-century French Jesuit copying a late-sixteenth-century Portuguese text about Indian medicine for use in China during the early eighteenth century, moving medical information from South Asia to East Asia (where additional medical information was added) and back to Europe.

Clearly, in 1596 the Portuguese viceroy of the Estado da Índia had an additional commercial motivation for creating a comprehensive list of Eastern medicinal recipes. The Portuguese Overseas Council, as well as secular and ecclesiastical trade interests, hoped to market such remedies in their own metropolis and colonial regions in Africa and South America, but also to rival colonial powers in Europe. At that time, Portuguese colonial agents held a virtual monopoly on both the sources for medically efficacious Asian plants and much of the knowledge about how to apply them. Indeed, the earliest trade treaties negotiated in the sixteenth century with Malabar Coast potentates like the Zamorin of Calicut and the kings of Cochin and Quilon (Kollam), which obliged local merchants "to export no drugs or spices without [Portuguese] consent," make clear that this had been their intent from the outset.[18] The Portuguese thus positioned themselves consciously to operate a global conduit for Eastern medical information, expressed in European medical terms and in a European language.

Still another early Portuguese attempt to create a compendium of medicinal plants from South Asia is that of Manuel Godinho de Erédia, a Malay-Portuguese cartographer and adventurer who developed an interest in botany while traveling in the Estado da Índia. His masterful work, an herbal titled *Suma de árvores e plantas da Índia intra Ganges* (Summary of Trees and Plants of India Proper), was compiled in Goa in 1612 but never published until modern times.[19] The manuscript, held in the Belgian Abby de Tongerlo since the mid-eighteenth century, contains seventy-four folio-sized illustrations of Indian plants, the majority having medicinal uses described in accompanying notations (see Figure 13.3). The images, richly colored and finely drawn, are far superior to Cristobal da Costa's earlier woodcuts; Erédia intended to provide Europeans with an accurate idea of each plant's appearance and application. As a mixed-race agent of the Portuguese empire, Erédia was perhaps more sensitive to the cultural context of the materia medica he compiled; the quality of his botanical images was unmatched until the publication of the Dutch botanist Hendrik van Reede's twelve-volume *Hortus Malabaricus* (Garden of Malabar) between 1678 and 1693.[20] (On Erédia's cosmology, see Sheehan in Chapter 12.)

FIGURE 13.3. Butua or Pareira Brava, *colored print, ca. 1612*

NOTE: Native to West Africa and South Asia, this plant was
applied to wounds, abrasions, and inflammations.
SOURCE: Manuel Godinho de Erédia, *Suma de árvores e plantas da Índia intra Ganges*
(compiled in Goa, ca. 1612). Manuscript held in the Abby de Tongerlo, Belgium.

Early travelers from other European colonial powers also circulated their observations about drug markets and medicines in the Estado da Índia. One of the most vivid descriptions of indigenous drug use in Portuguese India comes from the voyage account published in Amsterdam in 1596 by John Huyghen van Linschoten, who lived in Goa from September 1584 to January 1589, gathering information about Portuguese colonial society and trade around the Indian Ocean rim.[21] Describing the market at Cambay (near Diu Island in present-day Gujarat) at the end of the sixteenth century, Linschoten wrote that Portuguese, Persians, Arabs, and Armenians "go there to lade many kinds of drugs, as Amfiom [opium], Camfora, Bangue [*bhang*, similar to hashish], and Sandalwood."[22] Opium, he said, came mostly from Cambay and the Deccan, further inland, but was also shipped from Ormuz, at the mouth of the Persian Gulf, controlled by the Portuguese from 1515 until lost to the English in 1622. Linschoten asserted that the inhabitants of the Malabar Coast consumed opium and other native medicinal substances in great abundance.[23] Linschoten noted that, when prepared for consumption by the wealthy classes, opium was sometimes rolled into balls, often mixed with other *drogas*: mace, cloves, camphor, amber, musk, and *bhang*.[24]

During his time in Portuguese India, the early-seventeenth-century French voyager Francois Pyrard de Laval recorded that some wives of prominent men were known to drug their husbands by mixing opium into food or beverages, thus allowing these women to leave their houses at night in search of other paramours.[25] Philipus Baldaeus, too, noted opium use in Goa during the mid-seventeenth century. In his description of Goa's "Traffick and Manners, and Way of Living of the Portuguese there . . . ," this Dutch Protestant clergyman observed that the main street had abundant rich shops "well-stor'd with . . . drugstery wares."[26]

Portuguese colonial enclaves in India included, almost from their inception, state and missionary health care institutions. By the first quarter of the seventeenth century, Goa could boast of two hospitals operated by the Santa Casa da Misericordia (Holy House of Mercy): Nossa Senhora da Piedade and Todos-os-Santos. The latter, older by a century, had been founded in about 1524 and was open to persons of any race. These were amalgamated in 1681. After 1755, the facilities were open only to white adults or foundlings of European descent. Finally, the Royal Military Hospital was widely praised for its quality, but its facilities were reserved for colonial soldiers (who could be of many races) and officials of the India garrison.[27] Thus, Portuguese India had a lengthy tradition of supporting medical facilities that blended Western and Eastern influences.

During the early stages of colonization at Goa, the Portuguese garrison, missionaries, and traders made use of local medical practitioners in part because of the scarcity of Portuguese *médicos* (physicians). However, Indian physicians, called *vaidyas*, also enjoyed the patronage of their new rulers because they better understood the treatment of tropical disease. Several *vaidyas* held important posts in Goa through the sixteenth and seventeenth centuries, serving as personal physicians to governors, viceroys, and other members of the aristocracy. Indian doctors also could be found treating patients at the Jesuit Colégio de São Paulo and the Convento de Madre de Deus; one even served as the *físico-môr* (chief physician) of the region in the 1640s.[28]

In Macau as well, Portuguese missionaries founded medical installations. For example, by 1563 the Santa Casa de Misericordia, with its celebrated hospital, had been established. Shortly thereafter, the Jesuits founded the Colégio de São Paulo and its adjacent pharmacy. This installation specialized in creating medical compositions with *drogas* from local folk culture. The pharmaceutical expertise of the Jesuit brothers in Macau soon gained a broad reputation. Until the middle of the eighteenth century, European crews and captains doing business in Macau sought the Jesuits' advice and purchased their medicinal preparations; indeed, mariners from all nations considered this small *botica* (apothecary) indispensable to their success in these waters. The Jesuits, for their part, quite openly acknowledged that they owed much of their knowledge to local Chinese healers.[29]

In the seventeenth century, the Portuguese in India began to regulate indigenous physicians, attempting to subordinate local methods to orthodox European practice. *Vaidyas* were required to pass examinations administered by the Portuguese *físico-môr*. A decree of 1618 limited the total number of indigenous healers permitted to practice in Goa to thirty.[30] *Físico-môrs* during the sixteenth and seventeenth centuries, when "Golden" Goa was at its peak of affluence and influence, were almost invariably Portuguese-born physicians trained at the University of Coimbra. Thus, the standard medical practice in Goa under Portuguese administration remained, officially, European.

In 1691, King Pedro II sent two Coimbra-trained doctors to Goa with the express purpose of training locally born students, either of aboriginal or European descent, as medical professionals.[31] These men created a modest training facility in Goa, attached to the Hospital Real Militar, to which even the Leal Senado ("Loyal Senate") in Macau was known to send selected youths to study European medical techniques.[32] However, because one of the European physicians succumbed rapidly to a tropical disease, this initiative was to be short lived.

King Pedro II's policy reveals a conceit for European medical training, regardless of its limitations and the demonstrated efficacy of Asian medicine. Still, crown efforts during the early modern period to train colonial doctors in Goa according to a Western medical curriculum ultimately failed. These circumstances encouraged a reliance on local medicinal practitioners and drugs.[33] Moreover, the preference for European medical training never discouraged the use of indigenous remedies, which experience had shown to be effective. Many local medicines became firmly embedded in the medical culture of Portuguese India and could be found in use at colonial hospitals and missionary infirmaries. This situation continued until the founding of the Goan School of Medicine and Surgery in 1842.

As the focus of Portuguese overseas endeavors shifted to the Atlantic after 1670, the Estado da Índia ceased to be an attractive destination for the majority of European-born and -trained medical professionals. Those Portuguese physicians and surgeons who did emigrate to colonial outposts during this period generally opted to settle in economically vibrant areas like Brazil or Madeira, where their prospects for a comfortable living were brighter than in contemporary Goa. Even some of the more dynamic areas of the Portuguese Atlantic colonies had difficulty attracting skilled physicians and surgeons. In Cachoeira, a colonial river port in the Bahian interior that was a major center for sugar and tobacco production, the Brazilian viceroy complained in 1757 that "there are currently only three or four surgeons and many other apothecaries, and of all of them there is not one that will do" because none of them had been professionally trained at a medical school in Europe.[34] As the number of European-born residents in the eastern colonial enclaves diminished, colonial officials acquiesced to the pragmatic expedience of filling vacant medical posts with native physicians and surgeons (though they had to be of Christian families). Such circumstances contributed to a more experimental openness toward native medicine in the colonies. Simple demographics and the economic eclipse of "Goa Dourada" (Gilded Goa) would ensure a strong native presence among health care professionals in Portuguese India.[35]

For example, during the closing decades of the eighteenth century when Ignácio Caetano Afonso, the chief physician of the Estado da Índia, undertook the treatment of Portuguese soldiers afflicted with the seasonal fevers associated with the onset of monsoon at the Hospital Militar de Goa, he resorted to familiar medicines and techniques at his disposal. While some medicines to combat fevers had been sent from Portugal and were available in the hospital *botica*, the ethnically diverse Indo-Portuguese colonial staff also commonly made use of local plant-based remedies.[36]

Ignácio Caetano Afonso was a native Goan, born into a Portuguese-speaking Indo-European family and raised as one of an elite class in Portuguese colonial society. Though he had little formal medical training, he gained a favorable reputation as a healer. In March 1798, in a letter to the Portuguese secretary of state in Lisbon, the governor of the Estado da Índia described Afonso as "a Brahmin . . . favored with natural talents" for healing. The governor continued to praise Afonso, saying that "notable cures" had been attributed to him, even though he "had not opened any [medical] book for many years."[37] The following year, after Afonso's death, the governor would write that Afonso had "the sense of a *médico*, and practiced for many years, which compensated for the defects of his [medical] education."[38]

While Afonso studied informally at the Hospital Militar de Goa under his predecessor, the Portuguese-born, Coimbra University–trained physician Luís da Costa Portugal (who had made a practice of training promising Goan healers in Western medical techniques[39]), his medical knowledge consisted primarily of native Indian plants and their medicinal applications. In the main, Afonso treated Portuguese soldiers, officials, and colonists with local remedies derived from indigenous Indian drugs—medicines that Afonso had learned from older Goan healers, who had employed them since time out of memory.

The native-born *físico-môr* demonstrated his worth in a variety of ways. His chief responsibilities were to supervise Goa's largest health care institution (including the care and treatment of its three thousand annual patients[40]), its pharmacy, and its garden of indigenous medicinal plants. In 1794, the Conselho Ultramarino, seeking medicines to treat tropical diseases throughout the Portuguese colonies, commissioned Afonso to write a description of all the useful medicinal roots found in Portuguese India. The *físico-môr* produced a twenty-four-page booklet, which he titled *Discripçoens e virtudes das raizes medicinaes* (Descriptions and Virtues of Medicinal Roots).[41] This manuscript included a discussion of five fundamental medicinal roots of the Indian Ocean basin, along with notes about the use and efficacy of their respective plants and seeds. Afonso described various maladies for which these roots could be prescribed, providing details about the preparation of each plant remedy.[42]

One of the medicinal roots to which Afonso referred was the celebrated *pau cobra*, or cobra wood, a name applied to several varieties of plant root known across south India and thought to be effective against snakebite (see Figure 13.4). According to Ignácio Afonso's manuscript, *pau cobra* was known in Ceylon, the plant's native home, as *hampaddu tanah*. In Goa, Afonso wrote that the plant was well known among indigenous herbalists, but it was generally referred to by the Portuguese name.[43]

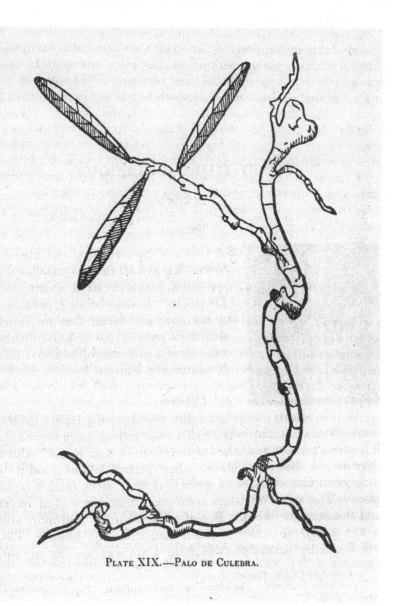

PLATE XIX.—PALO DE CULEBRA.

FIGURE 13.4. Palo de Culebra *or* Pau de Cobra *(Snake Root), woodcut, 1572*

NOTE: A pulp made from the root of this South Asian
plant is a treatment for animal and insect bites.
SOURCE: Woodcut from Cristobal da Costa, *Tractado de las drogas
y medicinas de las Indias orientales* (Burgos, Spain, 1572).

Pau cobra had long held a place in Indo-Portuguese medicine. A discussion of this root appears in Garcia da Orta's text of 1563, in which three varieties are named with origins in Ceylon and south India. Da Orta asserts that at least one of these varieties, called *mordexi*, grew on the "island of Goa" and that south Indian healers also used the root to treat rheumatism, smallpox, measles, and cholera.[44] Sebastião Dalgado's *Glossário luso-asiático* (Luso-Asian Glossary) includes an extensive entry on *pau de cobra* and cites references to this medicine in half a dozen Indo-Portuguese publications of the sixteenth and seventeenth centuries.[45] Dalgado identifies the root with *Aristolochia indica*, *Rauwolfia serpentaria*, and *Strichnos colubrina* (named under Linnaeus's system), all of which have accepted analogous applications in classical Indian Ayurvedic treatment.[46]

CONTROL OF THE PHARMACEUTICAL TRADE
WITHIN THE PORTUGUESE MARITIME EMPIRE

In continental Portugal at the end of the seventeenth century, the great majority of pharmacies were in the hands of that nation's numerous monasteries and operated by the often highly trained brothers of these institutions, be they Jesuit, Dominican, Franciscan, Augustinian, Benedictine, or Carmelite.[47] Only in the larger cities such as Lisbon, Oporto, Coimbra, or Évora were secular pharmacies to be found. These, however, were generally modest concerns; secular pharmacists complained frequently to crown authorities that they could not compete with the monopolistic practices of the great missionary orders, whose purchasing power, established trade and procurement networks throughout the overseas empire, and superior professional reputations combined to impoverish lay pharmacies.[48]

Hence, a virtual monopoly of the lucrative trade in medicinal substances in continental Portugal during the early modern period was controlled by missionary orders or monastic institutions and the colleges associated with them. In the case of medicines arriving from Brazil and the Estado da Índia, Jesuit druggists or apothecaries (*boticários*) in particular enjoyed a clear advantage, as they could rely on their co-religionist associates in Goa, Macau, and Salvador to procure and ship consignments of medicinal plants or prepared medications—such as the prized opium- and mercury-containing *pedras cordeais* (cordial stones)—to their brethren in Portugal.[49] All types of remedies, from cheap tamarind paste to expensive bezoar stones, were common substances in which the Jesuit brothers trafficked on a truly global scale, sending consignments of drugs from India and China to Africa, South America, and Europe. Moreover, the missionary orders relied on this trade for revenue that supported their proselytizing work. The market for colonial medicines in continental Portugal was largely their bailiwick for over two hundred years.

In Lisbon, two of the city's most important pharmacies operated under Jesuit control. These were attached to the Casa Professa de São Roque and the Colégio de Santo Antão. Together, these two pharmacies functioned as the hub of a network of Jesuit *boticas* that extended throughout the Portuguese seaborne empire. Without exaggeration, until the reign of king Dom José (1750–1777), the Jesuits and, to a lesser extent, the Dominicans, helped to drive, direct, and sustain the global market in many of the exotic medicinal plants or animal-based *drogas* arriving in Europe from abroad, partly because of their purchasing might in this profitable trade, but more important because of their presence, influence, and pharmacological expertise at those points in the empire where these rare and desirable substances could be procured.[50]

Another Lisbon monastic pharmacy of great importance and repute was located in the Augustinian monastery of São Vicente de Fora. One of their preparations was a mercury-based "panacea" in pill form, which the monks produced on the premises and shipped in great quantities (with printed dosage instructions) to destinations throughout Portugal and the colonies—its particular use was to combat syphilis.[51] They were best known, however, for their texts on pharmacology and chemistry, which the monks published on an in-house press. In 1704 one of their number, Dom Caetano de Santo António, produced the first continental pharmacopoeia written in vernacular Portuguese.[52] The 1711 edition of this publication was the first medical handbook produced in Portugal to describe the therapeutic application of opium and other medicinal substances from India.[53]

Two extant pharmacy stock lists from the middle of the eighteenth century give us an idea of the relative value of colonial drugs in the Lisbon market. A surviving inventory (taken in 1749–1750) of the pharmacy of the Colégio de Santo Antão provides a fair indication of the volume, price, and importance of imperial drugs available in Portugal at this time.[54] The other inventory, compiled in 1758, is from São Vicente da Fora. This important *botica* kept an assortment of traditional Indian, Chinese, and Brazilian medicinal substances in stock at any given time in the mid-eighteenth century.[55]

The 276 substances listed in the inventory of the *botica* of the Colégio de Santo Antão are divided into categories based on their composition— animal, vegetable, or mineral—and provenance. One hundred and four drugs are classified as vegetable compositions, but only 38 of those came to Lisbon from the Estado da Índia and 10 to 12 came originally from East Africa. Thus, perhaps 26, or about one quarter of all vegetable-based medicinal compounds listed in the Colégio de Santo Antão's inventory came from India, China, or the Spice Islands of the South Pacific. By contrast, 19 vegetable drugs had come from the Americas—almost exclusively from Brazil—while 47, or nearly half, had been gathered in Europe or the Mediterranean basin.[56]

Comparing prices, Asian drugs were in general only marginally more dear than medicinal substances procured from more convenient locations closer to Lisbon, and Asian drugs were actually cheaper on average than medicinal *drogas* originating in South America. Prices were determined not so much by distance traveled as by availability, demand, or difficulty of manufacture.[57] *Drogas* from Asia of animal or mineral origin represented only a tiny fraction—just six or seven items—of the substances found in this pharmacy's stock. Except for a few exotic animal substances (stag horn, bezoar stones), most medicinal drugs imported from Asia on the annual India fleet were plant derivatives.[58]

We may compare the 1750 Colégio de Santo Antão inventory with another completed in 1759 from the *botica* of São Roque in Lisbon. Of the nearly five hundred items listed on the survey, about 10 percent are identifiable as having come from the Asian colonies.[59] Stocks of Asian items tended to be held in modest quantities, ranging usually from a single ounce to a few pounds. By far, the majority of medicinal substances listed were derived from plants of European or Mediterranean origin, reflecting the relative ease of supply for these products, as well as the popular and professional demand for them in Portugal.[60]

Asian medicinal exports to Europe during the early phase of the empire seem to have been limited by relatively low demand, but this picked up in the eighteenth century. Knowledge about Asian curatives was not widespread at the popular or professional level in Portugal until after the publication of Dom Caetano de Santo António's *Pharmacopea lusitana reformada* (Revised Lusitanian Pharmacopeia) in 1711; such knowledge spread only gradually thereafter. In Portugal, folk healers and licensed physicians alike preferred to use locally grown plants or medicines from the European medical tradition almost exclusively, the effects with which they were most familiar.[61] Despite the exotic allure and rumored efficacy of Asian drugs among elites, Portuguese physicians in the home country seem to have resorted to them only rarely, while popular *curandeiros* (folk healers) used them practically not at all.[62]

There were, of course, exceptions: opium, rhubarb, benzoin, *pedras cordeais*, and bezoar stones from Goa and Macau enjoyed a certain popularity (primarily among elites or veterans of colonial service) but, salable as these substances were, they constituted only a minor piece of the total metropolitan pharmaceutical market (see Figure 13.5). Opium use, in the form of the anesthetic laudanum, increased during the Peninsular Campaigns of the Napoleonic Wars, due largely to the drug's popularity among the regimental surgeons of Wellington's Anglo-Portuguese army.

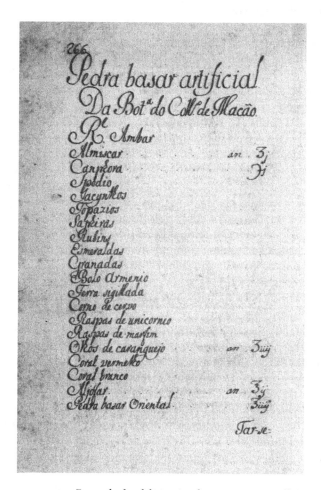

FIGURE 13.5. *Formula for fabricating bezoar stones, mid-1700s*

NOTE: True bezoar stones, once a highly valued medical panacea, form naturally in the stomachs of some animals. To meet demand in Europe, Jesuit apothecaries in Asia created an artificial version using opium and semiprecious stones.

SOURCE: Manuscript of the Jesuit pharmacy of the Colégio de São Paulo (Macau, China, mid-1700s). Archivum Romanum Societatis Iesu, Rome.

Of course, commerce in opium existed outside the framework of religious missionary orders, particularly after the suppression of the Jesuits in 1759 and the massive growth of the opium trade between Gujarat and China. Portuguese army surgeons procured opium for their medicinal requirements; naval ships carried chests of medicines stocked with opium and other Indian medicines; and labor overseers or slave drivers in the Atlantic and Indian Ocean spheres bought opium in quantity to give to their hard-suffering workers.[63]

In the Estado da Índia, government military facilities (as well as private plantations or labor contractors) turned to sources other than the religious orders' institutional pharmacies for bulk opium purchases. Wholesale merchants or port markets in Gujarat could easily have supplied large consignments of opium at an attractive price, but this commerce is largely undocumented. Fiscal documents in Portuguese India reveal that the colonial government typically expended large sums annually to outfit state vessels with appropriate medicines against tropical disease. A chart of state revenues and expenditures for 1762, for example, records that the apothecary of the Hospital Militar de Goa distributed drugs that year worth 5,287 *xerafins* (a substantial government expenditure) to the medicine chests of various vessels of the Portuguese fleet.[64] Most of these medicines, of course, originated in India.

During the seventeenth and eighteenth centuries, some Indian medicinal stocks for Portuguese colonial apothecaries and pharmacies were gathered at Diu or Damão and shipped on to Goa, where the main stores of *drogas* were collected.[65] Often, consignments of drugs were supplied directly by Hindu merchants, wholesalers of medicinal plants who procured bulk quantities of native remedies for Portuguese medical facilities in the Indian colonies.[66] Evidence of Hindu merchants providing regular deliveries of indigenous drugs to Portuguese medical institutions in Goa can be found in the financial records of the Convento do Nossa Senhora de Graça. In 1798 and 1799, two Hindu pharmacists, Rama Xandra Camotim and Segunam Camotim, received large cash sums for medicines and services rendered to the Graça convent hospital.[67]

From Goa, Indian medicinal substances were transshipped and widely distributed, and coasting vessels carried consignments of the popular medicines to Portuguese-held ports along the Malabar Coast as well as to Sri Lanka and destinations in eastern India. The Europe-bound ships of the annual Portuguese India fleet carried cargoes of opium and other Indian drugs to Mozambique, Brazil, and Lisbon. Meanwhile, Indian medicines traveled eastward in Portuguese-licensed vessels to imperial colonies at Malacca, Timor, and Macau, among other destinations.

For example, in 1682, after the Portuguese decided to establish a hospital in Mozambique, a very large consignment of medicines originating from Europe, Brazil, and India was forwarded from Goa to stock the facility's pharmacy.[68] Colonial medical authorities ensured that the new Mozambique hospital would have a supply of the popular *pedras cordeais* and other opiates. Besides painkillers, other Indian medicines in this shipment included rhubarb leaves and pills, medical pastes and unguents prepared with tamarind or aloe, althea ointment, asafetida root, and the curiously named *pirullas hindoos* (Hindu pearls), apparently manufactured with ingredients of Indian origin, copied from a traditional Malabar Coast bolus.[69]

Over a century and a half later, in 1838, various Indian medicines—including opiates, healing balms, pastes, and lotions—were being shipped to East Timor as part of a consignment of drugs requested for the Military and Public Hospital in Dili. Several of the medical preparations had equivalents in traditional Indian methods: examples of these are an unguent made from althea, pain balms including camphor, and the roots and leaves of asafetida.[70] Hospitals, pharmacies, and infirmaries situated throughout the Estado da Índia, as well as medical facilities located in the Atlantic colonies, ordered and consumed a steady supply of traditional Indian medicinal substances from at least the 1550s into the mid-nineteenth century.[71]

The Portuguese also played a role in distributing Malabar Coast remedies to the Muslim Mughal court in the Indian interior. Whenever the Portuguese sent an envoy to negotiate with the Raja in Agra, custom dictated that gifts be exchanged. Among the perfumed herbs, rich fabrics, silver inlaid blades, and potent distilled beverages (*aguardente* and cashew *feni*) sent as tribute, the Portuguese commonly sent typical Hindu-influenced medicines, such as *balsamo apopletico* (apoplectic balsam) for headaches and sandalwood paste for fevers. One typical example of this practice can be found in a record of a shipment from the Portuguese embassy to the court of Raja Sauac Bacinga in December 1737; the itemized medicinal presents filled dozens of jars, bottles, and ornate chests.[72]

Indian princely courts offered a useful vantage point from which to observe the movement of medicinal goods in the Asian trade system. In 1770, a Capuchin priest named Friar Leandro de Madre de Deus was sent to the court at Pune as a missionary, but he also gathered political and commercial information.[73] Two years later, Leandro produced a description of trade routes, commodities, and prices in ports from the Indian Ocean to Macau. Friere Leandro's *Notícias particular do comércio da Índia* (Notices Specific to the India Trade) is a comprehensive overview of this complex trade network, one of the first guides of its kind.[74]

Leandro included medicinal plants, drugs, and curative spices in his description of Indian trade goods; Portuguese colonial merchants sold these Malabar Coast remedies in various ports in China, Annam (Vietnam), and the "East Indies." Most are traditional Indian medicinal substances that had long-accepted uses in classical Ayurvedic, Unani, or Siddha healing, as well as in the local cultures along the southwest coast of India.[75] *Notícias* describes Indian sandalwood, stag horn, and clove oil from Ceylon bringing high profits in Macau when sold as remedies.[76] The gum resin myrrh, purchased in Calicut or Cochin, could be sold for substantial gains as a medicinal ingredient in Malacca or Macau.[77] The balsam or salve made from benzoin, purchased in eastern India, had a profitable market "in every part of the world," according to Leandro's report.[78] Tamarind and pepper, also sold throughout Asia as medicinal substances, left India in the holds of Portuguese merchant vessels.

Notícias, then, provides a contemporary description of Portuguese distribution methods of Indian healing substances in the late eighteenth century, but it is also an information conduit about the drugs themselves. The report was copied and disseminated among merchants and officials of the Estado da Índia as well as informing the Conselho Ultramarino in Lisbon.

The dissemination of information about medicinal substances within the Portuguese empire was not limited to drugs of India. Increasingly during the eighteenth century, as the medicinal plants of the Amazon rain forest and other regions of Brazil became better known, Portuguese colonists in South America took an interest in collecting data about the application of these substances. Naturally the Jesuits, with their preexisting medicinal distribution network, were at the forefront of this endeavor.

After Pombal ordered the expulsion of the Jesuit order in 1759, colonial authorities confiscated and catalogued the goods of this banned brotherhood. In Brazil, the chief customs official wrote a memorandum to the colonial governor on July 30, 1760, referring to goods sequestered at the Colégio dos Jesuítas in Bahia, and to the recipes of certain medications that the padres had kept secret but sold for substantial profits in the colony.[79] The memorandum makes clear that some of the Jesuits' medicines were prepared exclusively with local healing plants. One remedy particular to the region was an antidote for venomous animal bites or poison. Called *Triaga brazílica*, this antidote was so popular among the city inhabitants that revenues from its sale became the financial cornerstone of the entire *colégio*: "With this recipe I am told within [Bahia] were earned three or four thousand *cruzados* and it is certain that this remedy was the primary foundation of this Apothecary, by the great sales that it had . . . and because of its effect."[80]

Other Jesuit remedies in Brazil, however, came from more distant sources. Among these were recipes for medical preparations composed of South Asian indigenous medical plants, which replicated efficacious local remedies from Malabar, Maduri province, and Ceylon, or mixed Indian and Brazilian components into their own innovative concoctions.[81]

The Portuguese monarchy and colonial administrators in Lisbon took an interest in medicinal plants from South America, too. In 1788, in response to a royal order, Brazilian physician and natural scientist Bento Bandeira de Mello prepared a lengthy memorandum on local indigenous medicines. De Mello had been charged with creating an alphabetical list of medicinal plants, fruits, and roots from the captaincies of Pernambuco and Paraíba, with commentary concerning their remedial effects.[82] His annotated list, containing fifty-nine different South American medicinal plants, runs to twenty-four manuscript folios. De Mello sent specimens of many of these plants to the Palácio da Ajuda royal botanical garden in Lisbon.[83]

At São Tomé, a tiny equatorial island off the West African coast, the Portuguese transplanted cinchona trees from the Brazilian Amazon. Quinine, made from the bark of the cinchona tree, is a highly effective natural treatment for malaria. The fertile plantations of São Tomé were soon producing hundreds of tons of cinchona bark for export; quinine in the form of powdered bark or an extract safeguarded European lives throughout the Portuguese tropical colonies.[84] In particular, quinine from São Tomé facilitated Portuguese expansion into the interior of the African colonies during the nineteenth and early twentieth centuries.[85]

Because of continued high mortality among soldiers and colonial officials in the tropics (to say nothing of valuable African slaves shipped as merchandise across the Atlantic and Indian Oceans), imperial authorities in Lisbon maintained their interest in discovering new indigenous remedies from India. In a royal directive dated April 2, 1798, Queen Maria I commissioned the chief surgeon and other *médicos* of the Hospital Militar de Goa to codify their knowledge of indigenous medicine from Portuguese India.[86] The following year, *cirurgião-môr* Dr. José Abriz and his colleagues produced a report, extending to nearly forty manuscript pages, in which they provided detailed descriptions of eleven important roots and plants then in use in the medical facilities of Goa, Damão, and Diu as well as the Portuguese East African colonial holdings.[87]

The Abriz report (dated 1799) constitutes a follow-up to the document submitted in 1794 by the native-born chief physician, Ignácio Caetano Afonso.[88] Afonso's report had clearly attracted some interest among colonial officials in Lisbon, but the members of the Conselho Ultramarino apparently wanted the benefit of a second opinion from a source trained wholly in scientific Western medicine. Abriz and his colleagues, however, were convinced of their Indian-born predecessor's competence, and in fact all of the medicines included in the 1794 Afonso report found an enthusiastic endorsement from the later European-led medical commission. The plants and roots chosen for the 1799 report are notable in that, although they had long been in use as ingredients for remedies along the western coast of India, most of them originated in East Africa (at least according to contemporary Portuguese understanding of their botanical provenance). The Abriz report thus consciously disseminated information about remedies of the greater Indian Ocean basin, rather than just the Indian subcontinent.[89]

Portuguese colonial administrators sought potentially useful and commercially exploitable indigenous remedies well into the nineteenth century. As late as 1830, the new governing charter of the Hospital Militar de Goa stipulated that it was the responsibility of the physicians and surgeons employed there to seek new medical preparations from indigenous sources, investigate their qualities, and report any promising findings to government authorities in the metropolis. The medical staff in Portuguese India was charged with remitting detailed information about "new attempts and discoveries for any remedy or curative whatsoever," and instructed to report specifically on "the preparation, dose, application, and in what cases and circumstances [the medicine] is found useful."[90] The hospital regulations further stipulated that the chief physician and surgeon were to report to Lisbon about the commercial prospects for any new indigenous drugs they discovered, as well as on the "advantages that may be hoped [from the drug] for the alleviation of humanity."[91]

༜

Indigenous Asian, African, and South American medicine thus played a fundamental role in the state-sponsored health care institutions of the far-flung Portuguese colonies. Portuguese imperial agents adopted local medicinal substances and remedies consciously for scientific and commercial ends, as well as for practical reasons, to facilitate and further Portuguese colonial ambitions. During the early modern era, native medicinal preparations and healing techniques therefore became widely known in Portuguese-controlled enclaves in the Atlantic, Indian, and Pacific Oceans, far from their indigenous roots, and were deeply inculcated into the milieu of tropical medicine in the Lusophone colonies, as well as the metropolis. However, early modern Portuguese medical practitioners in the colonies usually were not deeply interested in, and consequently did not absorb, replicate, or disseminate, the broader philosophical context of indigenous healing systems. (Such an interest eventually did awaken, of course, late in the colonial era, but these developments fall outside the scope of this project.)

♈

As a research venture, this chapter is a pioneering effort—unique in its scope and chronological parameters in English-language historiography (and nearly so in the Portuguese). No prior work has made such extensive use of primary sources in an attempt to derive a comprehensive understanding of global indigenous influences on premodern Portuguese colonial medical culture. Historians of colonial medicine have not heretofore appreciated that the Portuguese—whose exposure to African, south Asian, and Chinese ideas concerning healing lasted far longer than any other European nation—tended to be far more receptive during the early modern period to the adoption and dissemination of indigenous medical practices.

While abundant scholarly attention has been directed at the (later) British colonial experience with indigenous medicine in south Asia and other regions,[92] and some recent attempts have been made to gauge the (again, later) impressive Dutch colonial efforts in pharmacological botany,[93] very little has been written about medical exchanges in other European imperial enclaves. Moreover, there is a marked tendency among researchers of this subject to look closely at medical exchanges in one colonial region, but not to consider that region as part of a larger global system of commercial and intellectual exchange. A final impediment to scholarly dissemination of ideas on this matter is that much of the pertinent extant literature is exceedingly rare and not available in English translation.

A copious body of material on colonial medicine exists in the historiography of the Portuguese empire; in 1996, pharmaceutical historian José Pedro Sousa Dias published a forty-page bibliography on the subject containing hundreds of works, mostly in Portuguese.[94] The majority of these, however, are very narrowly focused, concentrating on one geographic area, medical practitioner, or type of disease.[95] Scholarly efforts to be more comprehensive have inevitably suffered from brevity and superficiality, due primarily to the logistical challenges inherent to exploring archives dispersed around the globe.[96]

Nevertheless, a few contemporary researchers have begun to look at global systems of early modern medical exchange by considering healing plants as commodities within a truly imperial framework. Londa Schiebinger's recent book, for example, examines how Dutch, French, and

British scientists either adopted or rejected indigenous medical knowledge employed by Caribbean peoples during the eighteenth century.[97] And Harold J. Cook, director of the Wellcome Trust Centre for the History of Medicine at University College London (UCL) and coeditor of the journal *Medical History*, studies the global movement of early modern medical information by focusing on the Dutch sphere. His inaugural address at UCL, "Medicine, Materialism, Globalism: The Example of the Dutch Golden Age,"[98] and his recent book both consider medicine and natural history through the lens of the seventeenth-century Dutch imperial ascendancy, in an attempt to reassess the relationships between the beginnings of a worldwide trading system and a worldwide exchange of information about nature.[99]

Accordingly, the material contained herein will add to a growing understanding of how healing sciences were transferred from one culture to various others in a globalized European colonial context. As part of a larger historiographical trend, this work aims to contribute to the fields of medical and colonial history by filling gaps in the current literature, presenting new insights and perspectives, and providing much-needed comparative material regarding the healing experiences of the Portuguese, a pioneering colonial power woefully underappreciated in English-language historiography.

Moreover, this chapter is intended to serve as a model for future projects of medical history focused on the other European maritime nations' colonial medical systems. One purpose of these studies will be to examine how Europeans, attempting to advance their colonial enterprise, learned medical techniques through cross-cultural interactions with indigenous peoples. Another goal is to learn exactly what healing methods Europeans ultimately took away from such collaborations. With this approach, the differences between European and non-European concepts of medicine, health, and illness are not as important as assessing what medicines and methods contemporary European colonizers deemed important to absorb from indigenous healing systems, understanding why Europeans found these remedies attractive, and determining how they put native medicine to work toward their own ends, both within the colonial context and at home in the metropolis. Medicine was a tool of empire from the earliest days of European expansion; understanding how Europeans appropriated and employed healing wisdom—from any place and in whatever form—is a key to understanding the imperial process.

CHAPTER FOURTEEN

The Rare, the Singular, and the Extraordinary

Natural History and the Collection of Curiosities in the Spanish Empire

PAULA DE VOS

On the ninth of October 1789, officials at the court of Charles IV in Madrid acknowledged the receipt of twenty-four boxes of "curiosities of nature and of art" remitted to the crown by the viceroy of Peru. The items had been collected by the bishop of Trujillo, Baltasar Jaime Martínez Compañón, during a tour of inspection of his diocese and, according to the officials, represented a very impressive collection of materials that demonstrated his "zealous and discrete curiosity."[1] Royal officials had every right to be impressed. Indeed, the collection of natural and "artificial" curiosities included in the bishop's gatherings reached massive proportions. The twenty-four boxes contained thousands of items gathered from the region. Many of these items reflected the Spanish crown's interest in gathering materials that would lead to economic development in the context of the Enlightenment: two boxes contained almost three hundred different kinds of local medicinal herbs, while other boxes contained samples of minerals, wood, and soil from the region, with information as to their uses as dyes, pigments for paints, furniture, or in cultivating crops. (For a Portuguese perspective on the benefits of collecting, see Walker's essay in Chapter 13.) The bishop also included seeds, roots, and plants of both staple and cash crops: several local varieties of wheat, corn, potato, squash, and beans along with examples of local cacao, almonds, coffee, sugar, cotton, and cascarilla (quinine).

Despite the fact that these "useful" specimens represent the majority of goods contained within the boxes, however, the bishop's collection also included a significant portion of "curiosities," or items whose immediate practical purpose is much harder to detect, and whose presence in various natural history collections constitutes the focus of this essay. Such items included over three hundred clay figurines depicting local gods, local fruits, and local animals. It also contained ethnographic items: textile samples and clothing—including everyday and ceremonial garments, tablecloths, napkins, sheets, and cushions. Along with agricultural instruments were included musical instruments; along with the various types of squash sent was "a plate made out of pumpkins." Three boxes also contained over ninety different types of animals—fish, reptiles, birds, and quadrupeds that had been dried or stuffed. And in box #6, tucked between a cotton blanket and a heavy black stone, lay the "half petrified" bones of a giant—its femur, parts of its sacrum, and one of its molars—from the Province of Huamachuco.

This collection was arguably part of a larger initiative organized by the Spanish crown in the eighteenth century to have local administrators in the Americas and the Philippines collect natural history specimens and send them to Spain. While scholars are well aware of the imperial scientific expeditions that the Spanish crown organized and funded throughout the century, this initiative is less well known and follows in the tradition of the *Relaciones geográficas*.[2] For in addition to funding the formalized expeditions, the crown sent a series of orders to bureaucrats throughout the Spanish empire that led to the collection of over 335 shipments of thousands of natural history specimens that were sent to Madrid over the course of more than seven decades, from 1745 to 1819. The majority of specimens collected, like those sent by Martínez Compañón, were meant for utilitarian purposes. Indeed, over 90 percent of the items collected by colonial bureaucrats were utilitarian—and botanical—in nature: plants that could serve as potential medicines or foodstuffs; cash crops providing new varieties of coffee, tea, or dyestuffs; and trees whose wood might be particularly efficacious in furniture making. As I have argued elsewhere, these botanical specimens fit well within the political economy of the Spanish Enlightenment, which sought to revitalize Spain's domestic fortunes through the cultivation of *fomento*, or development.[3]

The question remains, however, as to what to make of the "curiosities" sent by Compañón, and which constituted the remaining 10 percent of the items collected—about 25 of the 335 shipments. These items, in fact, are much more difficult to categorize: labeled "curiosities," they consisted of *naturalia*, or zoological and mineral specimens in addition to some botanicals; *artificialia*, or items made by human hands—fine art, exotica, and everyday items such as cups, dishes, and utensils; and *preternaturalia*, or marvels of nature, namely monsters (see Table 14.1).[4] The fact that they were even collected, first of all, is significant to a

TABLE 14.1

Date and origin of shipments of curiosities[1]

Origin of Shipment	Year	Type of Materials
Lima	1754	Minerals (3lb. 13oz. bezoar stone)
Cartagena/ Santa Fe	1768	Live animals, shells, animal parts, minerals, ethnographic items
Peru	1772	Live animals, shells, mineral stones, ethnographic items
Quayaquil	1777	Live animals, preserved animals
Manila	1777	Live animals, shells, ethnographic items
Maracaibo	1777	Live animals, shells, animal parts, ethnographic items, petrified items
"Archipelago"	1777	Preserved animals and animal parts, shells
Santo Domingo	1778	Shells, ethnographic items
Havana	1778	Shells
Manila	1778	Live animals, preserved animals, shells/coral, minerals, ethnographic items, petrified items
Manila	1780	Live animals, preserved animals and animal parts, coral, ethnographic items
Nicaragua	1782	Monster (jawbone and teeth of "terrifying monster")
Havana	1785	Live animals
Buenos Aires	1788	Preserved animals
Guatemala	1788	Preserved animals
Lima	1788	Ethnographic items
Cartagena	1789	Live animals, preserved animals and animal parts, monster (hermaphrodite horse)
Mexico	1793	Live animals
Copiapo (Chile)	1796	Shells, minerals, petrified items
Paraguay	1797	Live animals, ethnographic items
Californias	1797	Live animals
Santa Fe	1801	Minerals
Buenos Aires	1801	Minerals
Veracruz	1804	Monsters (hairless cow)
Bataan	No Date	Monsters ("monstrous pig")

1 All the information included in the six tables in this chapter comes from the following sources in the Archivo General de Indias (AGI/S): Lima, 798, "Expediente sobre la remission a Madrid de Aves, Animals, y Esqueletos que ha enviado para S.M. el Virrey de Santa Fe," Estado 52, no. 117, Estado 81, no. 28, Estado 85, no. 34, Estado 85, no. 39, Estado, 81, no. 7, Estado 49, no. 31; Indiferente 1549, "Noticia de las cosas particulares de Isttoria Natural"; Indiferente 1549, "Rason de los Animales Quadrupedos, Bolatiles, y Reptiles"; Indiferente 1549, "Lista General de lo que llevan la Fratgata Astea y Urca Santa Ynes"; Indiferente 1549 "Relación de lo que Conduce Don Pedro Vares" Indiferente 1549, "Lista de los Cajones que remite al Rey Nuestro Senior para su Real Gabinete el Gobernador de Philipinas; Indiferente 1544, "Carta del Intendente Francisco de Paula Sanz, a Antonio Valdés" cited in Index no. 56, p. 21; Indiferente 1545, "Minuta del oficio al Sr. Munoz" cited in Index no. 143, p. 37; Indiferente 1546, "Carta del botánico Martin de Sessé a Pedro de Acuna" cited in Index no. 334, p. 65; Indiferente 1549, "Nota General de lo que contienen los 11 cajones Toscos remitido a Julian de Arriaga"; Indiferente 1549, "Carta de Manuel González Guiral a Antonio Valdés" cited in Index no. 57, p. 22; Indiferente 1544, "Carta del Virrey Caballery de Croix a Antonio Valdés" cited in Index no. 57, p. 22; Indiferente 1549, 17 April 1779, Havana; Indiferente 1549, "Copia de Carta," Mapas y Planos, Estampas 206, Mexico 350, Estampas 232 (1–5), 1804; Indiferente 1550, 1 May 1782, and from Index and "Curiosidades para el Rey."

historiographical tradition that has generally recognized natural history collecting as becoming increasingly specialized, "naturalized," and "scientific" during the eighteenth century. Rather, the fact that the Spanish crown sought to augment rather than diminish its collection of curiosities during this period fits with a growing recognition that curiosities continued to hold an important place in European collections at the time. Second, an examination of the orders specifically targeting the collection of curiosities, and the items sent in response to these orders, sheds light on the contemporary meaning of "curiosity" in eighteenth-century Spain. And finally, I suggest that the reasons for collecting these curiosities are the same reasons applicable from the Renaissance on: for prestige and power, but at this time within the specific context of imperial rivalries during the Enlightenment and an emerging sense of imperial patriotism that played out in the field of natural history.

THE SEARCH FOR CURIOSITIES:
ETYMOLOGY AND HISTORIOGRAPHY

What did it mean to be "curious" in the eighteenth century? As a subject for etymological study, the concept of curiosity seems to have attracted a lot of attention which has revealed that curiosity has a rather complicated past.[5] In ancient Greek and Latin usage, the term was used to connote an inappropriate interest in something, such as that of a busybody who desired to know things not of one's concern.[6] This negative connotation for curiosity continued through late antiquity and into the Middle Ages, when it was increasingly associated with lust and pride due to the writings of St. Augustine.[7]

The fortunes of curiosity began to change, however, in the early modern period when it underwent "two important changes: a shift from the dynamic of lust to that of greed, and . . . an alliance with wonder."[8] The alliance between curiosity and wonder, which had long been associated with a scientific spirit and appreciation of the workings of nature, brought curiosity into the territory of natural philosophy and natural history.[9] The embodiment of curiosity in the sixteenth and seventeenth centuries were the European curiosity cabinets that were meant to represent a "totality" or "microcosm" of the universe, and as such curiosity came to signify a voracious and insatiable quest for knowledge.[10] The cabinets, also referred to as *Kunst- und Wunderkammern*, contained a wide assortment of items and reached their apogee in the baroque culture of the seventeenth century, referred to by Krystof Pomian alternatively as the "Age of Curiosity" and an "exuberant world of curiosity."[11]

For the eighteenth century, however, historians of science have demonstrated a decisive shift away from professional interest in the marvels that the curiosity cabinets contained, and an increasing association of curiosity with vulgarity.[12] This shift goes along with a larger historiography of natural history collecting and cabinets of curiosities, which indicates that collections became more "naturalized," "scientific," and "utilitarian" over the course of the eighteenth century.[13] Scholars from a variety of disciplines, including history of art, history of science, and history of empire as well as anthropology, museology, and literary criticism, generally concur that natural history and the collecting of natural history specimens experienced two important steps toward specialization during this time: first, collections became more "naturalized," meaning that they were increasingly comprised of *naturalia* rather than *artificialia*; and second, a widening distinction between the concepts of utility and curiosity meant that collections increasingly focused on "useful" specimens, while curiosities became the stuff of popular culture, the focus of the amateur.[14]

In response to this narrative, however, scholars have recently raised some challenges to what they term the "macro-perspectival interpretations [that] present the history of collections as a linear development."[15] In an article on "Collections curieuses" of eighteenth-century Paris, Bettina Dietz and Thomas Nutz point to the continued tradition of curiosity collecting among Parisian collectors that problematizes this assumed development and leads to a "historical configuration that does not fit into the usual dualisms of art/science, scientific/nonscientific, or professional/amateur."[16] The collections of Sir Hans Sloane and Joseph Banks in England, André Thouin's collections for the King's garden of Paris, Linneaus's own private collection, and Thomas Jefferson's Indian Hall are just a few examples of the continued importance of curiosities within the official, "scientific" discourse of natural history.[17] These works indicate that curiosity did not in all instances fall away from the interest of the scientific collector, and that an understanding of natural history collecting in the eighteenth century requires a new way of thinking that accounts for the dialectics, rather than the dichotomies, within the practice of collecting.

COLLECTING CURIOSITIES IN THE SPANISH EMPIRE

The collection of natural history specimens in the Spanish empire fits within this emerging historiographical critique, for the Spanish crown placed increasing, rather than decreasing, emphasis on the gathering of "curiosities" in the eighteenth century. As we shall see, its call for curiosities involved the search for the rare and the singular, and included *artificialia*, or items made by human hands, in addition to *naturalia*. Therefore, natural history was not increasingly "naturalized" in eighteenth-century Spain, nor did the collection of curiosities necessarily become more vulgar. Indeed, Spain began instituting the collection of curiosities only in the eighteenth century, largely in response to the establishment of royal institutions designed to display the riches and rarities of its empire. The focus until then had been almost exclusively on the collection of economically useful items and in this way was part of a long tradition of utilitarianism within the empire.

The Spanish empire in fact had had a long history of collecting natural specimens with utilitarian goals in mind, and it appears that the collection of curiosities was only initiated on a significant scale at the beginning of the eighteenth century. Such a realization serves to disrupt the usual attribution of increasing utilitarianism to Francis Bacon, who is often credited with the shift from curiosity to utility in natural history collecting.[18] (On Bacon, see the chapters by Almeida and Sheehan in this volume.) Antonio Barrera-Osorio has clearly demonstrated the empirical and utilitarian research program promoted by the Spanish crown beginning as early as the late fifteenth century, and both he and Jorge Cañizares-Esguerra have argued that Bacon may have in fact derived his inspiration from Spain.[19] (See the contributions of Cañizares-Esguerra and Barrera-Osorio to this volume.) In Cañizares-Esguerra's words, "It is therefore not preposterous to think that Bacon might have had the Spanish empire in mind when he wrote his *New Atlantis*."[20]

Indeed, Spain had a history of collecting natural specimens from the Americas that dated back to the first years of conquest and focused on utilitarian goals, or items that could be put to obvious commercial, medical, or cosmographical and navigational use. Barrera-Osorio has detailed the programs for information-gathering and research that were put into place by the crown in the early sixteenth century, highlighting the practical investigations into the nature of the New World and arguing that "the crown was not as interested in collecting marvels as in gathering information about the new lands and their commodities."[21] (On marvels, see Pimentel in Chapter 5.) To be sure, there were collections of art and of curiosities early on. From the earliest days of conquest, American treasures and fine art streamed into Europe, perhaps the most famous of which were the gifts that Cortes received from Montezuma and then remitted to Charles V. Both Charles V and Philip II were noted for having collected indigenous feather work and precious gems, and collectors also brought (and smuggled) Americana to other parts of Europe, into collections both great and small, including those of the Medici, the

Austrian Habsburgs, Ulisse Aldrovandi (Italian naturalist and creator of Bologna's botanical garden) and even Italian pharmacists.[22] In addition, sixteenth-century naturalists, merchants, and crown officials did traffic in some "curiosities," with parrots, turkeys, tigers, iguanas, and "blue and amber stones" circulated among court insiders.[23] Yet the crown's chief interest early on was in gathering cosmographic, demographic, and geographic information that would serve to solidify its knowledge and thus its hold over the Indies, and when information about New World natural history was sought, it was usually with the intent of exploring its possible commodification.[24] (See Barrera-Osorio in Chapter 11.)

Curious Orders:
The Rare, the Singular, and the Extraordinary

In the eighteenth century, however, a change is apparent in the types of items and information that the crown sought. Although the emphasis still remained on the useful, there are signs that curiosities were gaining in importance and priority for the crown, as evident in a series of orders issued to colonial administrators in the Americas throughout the century. These orders, which requested the gathering and shipment of curiosities, were issued largely in response to the establishment of three different institutions devoted to collecting under three different Bourbon monarchs over the course of the century: the Royal and Public Library established in 1715 under Philip V, the House of Geography and Natural History established in 1752 under Ferdinand VI, and the full-fledged Natural History Cabinet established in 1771 under Charles III.[25]

In issuing these orders, the crown clearly wished to gather together examples of the rare items not to be found in Europe that could be used to serve both patriotic and didactic purposes. All orders specifying the collection of curiosities referred in some way to the desire for things that were atypical or exceptional in some way, whether it be a natural specimen or an item made by human hands. Not just any item of interest was considered curious, however; rather, it had to meet proper specifications of rarity and singularity in order to be deemed worthy of collection. When the crown received a crystal prism from the Indies in 1796, for example, its inspectors found it entirely lacking in the necessary standards for a curiosity. Not only was the crystal not particularly rare, but it had "notable defects" in color and shape, so many that it was not considered worthy of placement in the Natural History Museum of Madrid as it was "not a curiosity at all."[26]

The search for rarity is especially clear in the first explicit and systematic calls for curiosities. These calls began in the early eighteenth century with the orders of Philip V in his efforts to equip the Royal and Public Library of Madrid, which would house collections of coins and other "curiosities" in addition to books.[27] In 1712, he issued a decree requesting books and grammars from the Americas as well as from China, Japan, and the East Indies, and the shipment of "singular and curious things," or "singular and rare things, including a sheet that explains the names of these things, with a note as to each one's properties, their uses, [and] the place they come from."[28]

Orders requesting the collection of curiosities were issued with more regularity in the second half of the century, particularly after the establishment of the Casa de Geografía y Abinete de Historia Natural (Office of Geography and Cabinet of Natural History) in 1752, which was later declared defunct and reconsecrated as the Gabinete de Historia Natural in 1771. These orders largely focused on the collection of *naturalia*. An order of June 6, 1752, for example, requested New World officials to send "all types of minerals, natural productions, and curiosities" to Spain, seeking out "rare, singular, and extraordinary things found in the Indies and remote parts, whether they be minerals, animals, or animal parts, plants, fruits, or any other type which are not very common or are extraordinary due to their type, their size, or other properties."[29] Another circular of 1788 instructed officials as to how to send items of natural history and other "curiosities," and a 1796 order requested specimens gathered from the three kingdoms of nature, especially "rare pieces" for the Natural History Cabinet. A crown order of 1800, finally, requested that American bureaucrats "collect the diverse products of Natural History that are found in [their] districts which, due to their rarity and utility, merit intellectual investigation."[30] Thus these calls for curiosities focused largely on the collection of uncommon or unusual items that would be displayed in the royal institutions.

Curiosities were also to be found in collections of *artificialia*, or items made by human industry.[31] When the new Royal Natural History Cabinet was established in 1771, for example, it was based around a collection bought from Pedro Dávila that included both *naturalia* and "curiosities of art."[32] (On the role of artistic depiction in colonial science, see Bleichmar in Chapter 15.) A listing of these curiosities reveals an amalgamation of things, many of which were items of common material culture, but which would have been unusual to Europeans: clothing, utensils, and weapons of various ancient and modern villages; adornments, shellworks and Chinese porcelain, models and instruments of physics and mathematics, inscribed stones, ancient bronzes and medals, and numerous paintings and miniatures.[33] Orders of 1776 issued by José de Gálvez requesting material specifically for the Natural History Cabinet asked for "curiosities of art, such as idols, paintings, drawings, machines, weapons, clothing, furniture, and other things in the villages that merit particular attention."[34] These items, while common to the villages from which they came, would be singled out due to the fact that "there is no known method or nomenclature [for them] in Europe."[35] Further requests for the everyday items of indigenous material culture are evident in Antonio de Ulloa's request for "antiquities," "ruins of Ancient edifices from pre-Columbian times," common household items (*vasijas*) of all types, agricultural instruments, weapons, idols, adornments, and clothing and textiles, both contemporary and "those of antiquity."[36]

Far from losing its place alongside the collection of useful items, therefore, curiosity not only held its own but also enjoyed a role of increasing importance in natural history collections of the Spanish crown in the eighteenth century. Whereas curiosity had played little part in crown efforts to gather natural specimens and information from the Americas in the sixteenth and seventeenth centuries, it was clearly a factor in collections of the eighteenth century. In addition, the types of objects identified as "curiosities" would include both *naturalia* and *artificialia*, which were specifically identified in the orders. The main defining characteristic for these items would be the extraordinary qualities they possessed—whether they were exceptional, one-of-a-kind specimens, or unusual for being so foreign to European culture, as in the case of *ethnographica*. Whatever qualities were sought, however, the call for curiosities and rarities by Spanish statesmen—those same statesmen who were consciously pursuing Enlightenment goals of social and economic improvement—requires that we recognize that "natural history" at this time included both the useful and the curious, and that curiosities had an important purpose to serve in the context of imperial competition of the period. It can also go a long way toward explaining the eclectic and sometimes very unusual amalgamation of items that were gathered for shipment, and the lens through which colonial officials would have viewed nature.

Curious Collectibles:
The Rare, the Singular, and the Extraordinary

Thus it appears that "curiosity" signified both the rare and the unusual from a European point of view. Yet we cannot know the limits of Spanish curiosity unless we know the material objects that it purported to denote. These objects began to arrive in Spain with increasing regularity the second half of the eighteenth century. Of 335 shipments of natural history specimens, approximately twenty-five contained or were entirely comprised of curiosities, or items without an obvious or specific commercial, industrial, or medicinal application. Sent between 1754 and 1804, they came from all over the Americas and the Philippines as well—from Chile, Paraguay, Argentina, Peru, and Santa Fe (Colombia) to Nicaragua, Mexico, California, and the Caribbean. In order to describe what these shipments contained, I have divided the categories up into those comprising live animals, preserved animals, and animal parts (including shells and corals), minerals (including petrified wood), *ethnographica*, and *preternaturalia*, or monsters.[37]

The search for rare items, as stipulated in the orders, is evident in the types of specimens sent, and the animals and animal parts collected appear similar to those found in other European curiosity cabinets and live menageries of the seventeenth and early eighteenth centuries. In the twenty-five shipments, approximately 132 live animals were sent (see Table 14.2); some were new types of familiar animals, such as monkeys, pigs, turtles, fish, and birds, while others—such as the *tunata*; the *picure*, known for its "extraordinary shape"; and the *cuchicuchi*, which "sleeps all day and at night eats all the lizards it can find"—were completely unknown.[38] Some animals were notable for their brilliant colors, such as the red, blue, and green parrots and the red-breasted doves from Manila, or the Paraguayan parrot with a green body and a head the "color of rouge."[39] And some were noted for their reproductive and nesting habits and for the care of their young, such as the iguana who traveled with its *hijuelo* (baby) in its mouth or the alligator that, after the hatching of her young, became so protective that her "ferocity increased and reached such a level that those crossing the river have to take care, as they say that she will try to upset canoes in order to eject people from them."[40] Others were also noted for their ferocity, like the monkey who traveled from Manila with an iron shackle around his neck or the aforementioned alligator that required careful extraction from the box in which she traveled: officials recommended that those opening the box "take the precaution of putting a lasso around her snout to contain her ferocity."[41]

Most animals were also accompanied with instructions as to their care and feeding, an indication of the significant effort and expense that collecting live specimens required. King Zopilote birds from Cartagena, for example, ate "rotten meat"; various birds from Manila lived alternatively on meat, fish, cooked rice, and cockroaches; various ducks from Guayaquil normally fed on fish, worms, and meat, but these were so domesticated that they would eat whatever was given them.[42] Other animals were not so accommodating: caregivers had to be aware that the tiger cub that traveled in the same shipment from Guayaquil would only dine on fresh, raw meat, for after two days "when the meat begins to have a bad odor," the cub would refuse to eat it.[43] Many traveled

TABLE 14.2

Types of live animals included in shipments of curiosities

Category of Animal	Types	Number Sent
Birds/Flying Creatures (*Volatiles*)	Birds, local varieties Parrots Birds of Paradise Owl Ducks Guacamayos Parakeets Stork Doves Bats	91
Quadrupeds	Pigs Monkeys "Tiger cub" (leopard?) Deer Civets (Argalia cats) Mountain lion	20
Reptiles	Alligators Crocodiles Tortoises Turtles Iguanas	12
Fish and Marine Life	Fish, local varieties Lobsters Crabs	6?
Unknown	Cuchicuchi Picure Tunata	3
	Total live animals:	132

in cages, as was the case for the doves that traveled the long distance from Manila to Madrid "in a cage, each one in its niche in the style of a henhouse."[44] Indeed, many animals did not survive the journey: the official who remitted a shipment of animals from Cartagena in 1789 complained of a sixty-two-day journey that was "very uncomfortable due to calms and contrary winds," during which many animals had died.[45] Those animals that "had escaped the intemperance of the sea" would be cared for on their journey from La Coruña (their port of entry) to Madrid by a servant "who has been charged specifically with the care and feeding of all of [the animals]." Others were not so lucky. The interim governor of Guayaquil, Don Domingo Guerrero, regretted that the male deer of a male-female pair died three days before its scheduled trip to Spain, probably due to the fact that "it was untamable." In addition, the Joyoyo bird he was sending now had only one wing, having "unfortunately broken the other [flying] in the Patio where it was being held."[46]

The animal curiosities contained in the twenty-five shipments also consisted of preserved animals, some of them dissected, and animal parts as well (see Tables 14.3 and 14.4). Preserved whole animals included a bird of paradise and several reptiles and amphibians—snakes stuffed with wool or preserved in alcohol, alligators, iguanas, scorpions, and a "fish-frog" (presumably a tadpole) found in the Orinoco that "is a transmutation from fish to frog . . . this one being in the middle of that stage."[47] There were also varieties of local insects and butterflies as well as fish and marine life, including crabs, shrimp, and sponges. Shells and coral were particularly important collectibles, including "branches" of coral of many different shades—white, black, grey, red, and yellow—and one shipment alone contained almost five hundred shells of all different types, many of which had been set in gold.[48] Perhaps most notable, however, were the "quadrupeds" sent: an anteater, several monkeys, and a "human from pre-Columbian times [*de la Gentilidad*] in the same

TABLE 14.3

Types of preserved animals included
in shipments of curiosities

Category of Animal	Types
Birds	Bird of Paradise (peacock?) Picaflores birds
Quadrupeds	Anteater Human (mummy) Monkeys
Reptiles and Amphibians	Tadpole ("fish frog") Snakes—general; water snake Alligators Iguanas Scorpions
Fish and Marine Life	Crab Fish (general) Bacalao Sponges Shrimp
Insects	Butterflies Tulaneayo insect Brazil beetles

TABLE 14.4

Categories of animal parts included
in shipments of curiosities

Category of Animal Part	Types
Skeletons/bones	Sloth—whole skeleton Manatee—ear and rib bones Dove of Ternate—whole skeleton Bird of Paradise (peacock?)—whole skeleton
Skin/hides/feathers	Condor—feathers Snake—skin Deer of Coromandel—hide Birds—skins with feathers Armadillo—hide
Teeth/tongue/beak	Alligator—teeth Pez Mulier—teeth Fish—tongue Snake—tongue Bird—beak
Limbs	*Guache* (quadruped)—hand, foot, and penis Pig—hoof (*mano*)

way that they buried him, and naturally preserved despite [the passing of] many centuries."[49] Seven types of animals also arrived dissected: a snake, alligator, iguana, worms, various birds, a clam, and an armadillo. And finally, colonial officials gathered and sent many different parts of animals: the skeleton of a sloth, the ear and rib bones of a manatee, the feathers of a condor, snakeskins, and armadillo hides as well as alligator and fish teeth, a bird beak, the tongues of a fish and a snake, a pig's hoof, and the "hand, foot, and penis of a *guache*."[50]

In addition to animals and animal parts, curiosities were also drawn from another of the three kingdoms of nature—the mineral. Minerals included stones, metals, "earth" (various types of soil or sand), and rock salt (see Table 14.5). A number of stones were included, ranging from ordinary stones noted for their color, density, or size—such as the three-pound, thirteen-ounce bezoar stone sent from Lima—to fabulous gems that were chosen for their extraordinary qualities. Both the viceroy of Santa Fe and the governor of Buenos Aires responded to a royal order requesting the "collection and shipment of natural productions, pearls, and precious stones" by sending "fine pearls, both large and small, emeralds, and others . . . found in this region."[51] The viceroy even designated "the most rare and precious Pearls and Emeralds produced in this Kingdom" specifically for "Our Lady the Queen."[52] Among this category were also

TABLE 14.5

Types of minerals included in shipments of curiosities

Origin and Date	Mineral
Cartagena, 1768	Diamonds
Peru, 1772	Mineral stones
Manila, 1778	Cornalina stone
Manila, 1778	White earth
Manila, 1778	Colored/brown earth (*tierra colorada*)
Manila, 1778	Salt stone
Manila, 1778	Black sand (*arenilla negra*)
Manila, 1778	Mineral stones
Manila, 1778	Volcanic ostones of sulfur (*piedras de bolcan de azufre*)
Manila, 1778	Amber
Manila, 1778	Lengua de San Pablo stone
Copiapo, 1796	Gold
Copiapo, 1796	Silver
Copiapo, 1796	Silver
Copiapo, 1796	Silver with arsenic (*plata arsenical*)
Copiapo, 1796	Silver
Copiapo, 1796	Salt stone
Santa Fe, 1801	Pearls
Santa Fe, 1801	Emeralds
Buenos Aires, 1801	Pearls, emeralds, and "other precious stones"

examples of petrifications, including petrified woods, a petrified leaf, and a petrified mushroom. Metals included rare examples of silver and gold. In 1790, for example, Don Pedro de Fraga sent to the Natural History Cabinet "a particular and interesting collection" of virgin gold covered with green copper, which was important not for its monetary value but for "the remarkable nature and rarity of its type."[53]

Thus curiosities were to be found within the kingdoms of nature, but they also fell under the category of *artificialia*, which, as I argued above, remained a significant focus for natural history collection in eighteenth-century Spain. These included a varied assortment of ethnographic items and exotica that, as with the animal and mineral specimens, represented both the rare and the everyday (see Table 14.6). Items of everyday use that would shed light on indigenous customs and material culture included clothing, textiles and thread, and housewares. Textile collections ranged from rare bits of pre-Columbian cloth to a ball of *pita* (agave) thread "that the Motilones Indians spin with great speed and with which they made beautiful stockings" and—as with the bishop of Trujillo's gatherings—entire wardrobes of both everyday and special clothing.[54] The shipments also contained accessories, such as purses decorated with

TABLE 14.6

Types of ethnographic items included in shipments of curiosities

Origin and Year	Specimen	Number
Santa Fe, 1768	Cross	1
Peru, 1772	Feather rug	1
Peru, 1772	"Light machine"	1
Maracaibo, 1777	Fragments of cloth	n/a
Maracaibo, 1777	Crosses	6
Maracaibo, 1777	Tobacco case	1
Maracaibo, 1777	Cups (*pocillos*)	24
Maracaibo, 1777	Cups (*tazas*)	24
Maracaibo, 1777	Bottle	1
Maracaibo, 1777	Sets of glasses	4
Maracaibo, 1777	Inkstands	2
Maracaibo, 1777	Engraved coconuts	7
Maracaibo, 1777	Ball of pita (agave) thread	1
Maracaibo, 1777	*Canuto* (tube, container) *de plata*	1
Santo Domingo, 1777	Tamarind seed	1
Manila, 1777	Small purse decorated with regional coins	1
Manila, 1778	Wooden hats	3
Manila, 1778	Coins	n/a
Manila, 1778	Lanterns of talc (?)	4
Manila, 1778	Walking sticks	4
Manila, 1780	Lanterns	6
Manila, 1780	Lanterns	10
Manila, 1780	Gongs with drumsticks	10
Manila, 1780	Straw hat	1
Manila, 1788	Sweets/candy (*dulces*)	1 box
Lima, 1788	"Manufactures"	1 box
Paraguay, 1797	Toothpicks	n/a

coins from the Malabar Coast and hats "worn by Indians of the province," some made of straw and others of wood.[55] Housewares included a "feather rug," sets of cups, bottles and glasses, and toothpicks made from a type of wood "that does not damage the teeth."[56] Various everyday wares included two inkstands and a tobacco case, four walking sticks, and a mysterious *maquina de luz* (light machine) brought from Peru in 1772.[57] From China came six lanterns "used by the lamas or great priests" and ten others "used by the Mandarins" that were adored with gold, silver, and multicolored silks.[58] The governor of the Philippines also shipped gongs, collectively weighing 226 pounds, that were typically used by the Chinese on board their sampans.[59] Decorations and exotica, finally, were represented by such curiosities as a set of "engraved coconuts," a cross made from the eyetooth of an alligator, and a tamarind seed "mounted in gold and silver, engraved with the face of a *negro*."[60]

The final category of curiosities went "beyond nature," as *preternaturalia*, or monsters, and their inclusion is a testament to continued interest in the marvelous in the eighteenth century.[61] This interest, on the one hand, seems to contradict Lorraine Daston's argument that the eighteenth century experienced a "waning" of the marvelous.[62] On the other hand, however, the tenor of the descriptions included for two of the monsters seems to correspond with what Daston and Park have labeled the era of "repugnance," in which monsters were studied for what they could reveal about the regularities of nature.[63] This is particularly true of the second description, which focuses on the genitalia of a hermaphrodite horse.

Among the shipments of natural and artificial curiosities, then, were four "monstrous" animals of which officials sent notice. Two of these, "a monstrous animal born from a pig" in the province of Bataan, and a "cow born without any hair" in Veracruz, give little information, but two others, the bones of a "terrifying monster" discovered in Nicaragua and a hermaphrodite horse shipped from Cartagena, include more detailed explanations as to the discovery and significance of the specimen. Notice of the "petrified jawbone and teeth" of "a strange monstrous animal" came from a colonial official in Nicaragua in 1782. According to the official, who hoped that the remains would merit "some small place in the Gabinete de Istoria (sic) Natural," this monster had lived in a nearby lagoon and would periodically emerge to devour local livestock.[64] Finally, a horse was sent to Madrid from Cartagena that, according to a report sent along with it, was "a rare Monster," for "its viril member is the opposite of what is usual [*de los demas*], that is, facing backward, and behind it [is] the female [part] which closely resembles rational nature, and formed teats."[65]

CURIOSITIES AND IMPERIAL PATRIOTISM

We now have some idea of what the Spanish crown meant when it re-quested natural history specimens, and particularly the "curiosities of nature and of art," from administrators throughout the empire. The question remains, however, as to why these products were desired. These specimens, whether one-of-a-kind finds or a representative example of a well-known but "exotic" animal or houseware, would not be shipped in bulk as in the case of cash crops, nor would they be further investigated for industrial or agricultural purposes; nor, finally, would they be acclimatized and cultivated on Spanish terrain. They were meant for the most part to be cabinet pieces or gifts of diplomacy, to serve a very different kind of use than their utilitarian partners. According to Krystof Pomian, then, they were "semiophores," or items are taken out of economic circulation, whose value lay not in their practical use or exchange value, but in their meaning.[66]

Thus it could be argued that natural history specimens had a more symbolic than a practical use within Spain's empire. In many ways, they had very important political and diplomatic symbolic meaning and as such were part of a centuries-long European tradition of collecting done in order to confer prestige. Scholars of the eclectic collections found in the curiosity cabinets generally agree that they were used in the service of "cultural politics."[67] Although they originated out of a medical tradition, the various specimens were soon taken up by princely collectors who used them to shore up political power. As a microcosm of the natural world, a "theater of nature," or a "world of wonders in one closet shut," they served as a testament to one's power, status, and authority.[68]

In this way, the political dimension of collecting is clear among the *Wunderkammern* of the sixteenth and seventeenth centuries, and, I would argue, we see it here still very much in place in the eighteenth century. Indeed, the eighteenth- and nineteenth-century establishment of botanical gardens and zoos has come under recent scrutiny by several historians who have pointed to the message of power inherent in these establishments. Scholars of European empires in the nineteenth century have made similar arguments as well. Maya Jasanoff, for example, has argued that imperial collectors were part of the "larger mechanisms of imperial expansion—war, trade, power" in British India and that "by bringing foreign objects to Britain, collectors played an important role in shaping images of the empire at home."[69] Such efforts were important factors, then, in shaping and justifying Britain's role as an imperial power—in its "imperial self-fashioning."[70]

Whether labeled "cultural politics" or "imperial self-fashioning," however, the purpose of collecting curiosities as a mark of prestige, status, and power has remained remarkably similar over at least four centuries of European collecting. What changes, I would argue, is the specific context within which the curiosities were collected, and the specific ways in which they were thought to confer prestige or underwrite imperial power. For the Spanish empire in the eighteenth century, I would suggest that these items constituted a significant way for the crown to fashion itself as both enlightened and progressive. This was particularly important at the time for a number of reasons. First, it was an era that saw continual naval warfare and heightened competition for colonial territory. Within this context of rivalry, Spain was all too aware of its declining fortunes, an awareness that was heightened by the disdain of its imperial neighbors.[71] Despite the inclusion of Spanish intellectuals and statesmen within the European Republic of Letters, French philosophers criticized Spain for its backwardness, superstition, and orthodoxy; and Englishmen complained that "the progress of knowledge in this country must be very slow."[72]

Many of these criticisms, and of Spanish and Spanish American refutations to them, found their support in ideas about natural history. Indeed, theories of natural history and the collection of natural history specimens set the stage upon which arguments about decline played out. Natural history found its place within these arguments largely due to the writings of the Count of Buffon (1707–1788), a French naturalist and director of the Jardin du Roi in Paris, whose thirty-six–volume *Histoire naturelle* (published between 1749 and 1788) argued that the American continents, being new lands and thus both humid and cold overall, could only produce degenerate flora and fauna—of which its human beings constituted a significant example.[73] This well-known episode, referred to by historians as "the dispute of the New World," produced impassioned rebuttals from patriots in the Americas.

This dispute was, I would argue, a major factor in explaining the increasing attention paid to the collection of curiosities. For within this context, the collection of curiosities became particularly vital—a proof of the value of one's colonial territories. In this way, arguments over natural history reflected the broader political, military, and economic rivalries taking place between European imperial powers at the time. According to Susan Deans-Smith, "Increasing rivalries among museum and *gabinete* (cabinet) directors to acquire the most unique and rare specimens and artifacts reflected a cultural variation of European political and imperial rivalries of the day."[74] In this way, arguments over natural history reflected the broader political, military, and economic rivalries taking place between European imperial powers at the time. Thus Spain had added incentive to carry out natural history investigations with an added focus on curiosities, and it had the colonial territory, the personnel, and the long-standing tradition of information gathering to support it.

CHAPTER FIFTEEN

A Visible and Useful Empire

Visual Culture and Colonial Natural History in the Eighteenth-Century Spanish World

DANIELA BLEICHMAR

An unsigned portrait painted at the turn of the nineteenth century in the city of Santa Fe de Bogota, in present-day Colombia, depicts José Celestino Mutis (1732–1809), one of the foremost botanists working in the Spanish Americas in the last decades of the eighteenth century (see Figure 15.1). Trained as a physician and surgeon in Spain, in 1761 a twenty-nine-year-old Mutis traveled to America as personal physician to the newly appointed viceroy to the New Kingdom of Granada (a territory corresponding to present-day Colombia and parts of Venezuela, Ecuador, and Panama). Mutis never returned to Europe, remaining in New Granada until his death in 1809, aged seventy-seven. During his first twenty years in America, Mutis earned a living as a physician and a professor of mathematics and astronomy; he also conducted botanical and zoological investigations. In 1783, with the crown involved in a wide-ranging and large-scale program of investing in scientific pursuits, Mutis received permission and funds to direct the Royal Botanical Expedition to the New Kingdom of Granada (1783–1810). Mutis and his patrons envisioned that the expedition would promptly yield useful and valuable information in the form of natural commodities that Spain could use to break the trade monopolies held by European competitors; the naturalist and his team diligently attempted to locate American varieties of cinnamon, tea, pepper, and nutmeg, as well as new types of cinchona, the valuable antimalarial.[1]

There is no question of Mutis's genuine interest in economic botany, as his journals and correspondence attest. Nevertheless, he devoted enormous efforts to another end, one with less obvious economic or utilitarian applications: the production of visual representations of American plants. Over the years, the expedition employed over forty artists, thirty of them working simultaneously at one point. While Mutis and a handful of botanical collaborators penned only about 500 plant descriptions, the artistic team created a staggering total of almost 6,700 finished folio illustrations of plants and over 700 detailed floral anatomies, far more than any other European expedition and an astonishing number of hand-produced objects for a workshop anywhere at the time. Given the existence of this extensive visual archive, the enormous efforts that Mutis made to recruit, train, and supervise his painters (going as far as creating a drawing school for young boys through which the expedition could satisfy its own demand for painters), and the frequent discussion of the production and uses of natural history images in his journals and correspondence, it is clear that images were of central importance to the expedition's exploration of American nature.[2]

FIGURE 15.1. *Portrait of José Celestino Mutis, ca. 1800*
SOURCE: Oil on canvas, 124 × 92 cm, anonymous. Real Jardín Botánico de Madrid.

The portrait of Mutis suggests that his high regard for representation extended to his attitude toward observation. The painting depicts a naturalist deeply engaged in the pursuit of his craft. Mutis sits before a work table, fixing his focused gaze on the viewer with weary patience, as if we had just burst into the study of muted grays and browns and interrupted his silent labor. He has lifted his head, but his body remains hunched over in concentration, eager to resume the examination of the flower he holds up toward him. A branch of the same plant lies ready to be pressed between the pages of a notebook and become an herbarium specimen; books scattered around the table serve as sources of corroboration in describing and classifying the plant. The instrument that Mutis holds in his right hand—a magnifying lens—connects the naturalist's own instrument—his eyes—to his subject of study; it also functions as an attribute, a symbol of the acute observational capacities that characterize him as a botanist. This is not mere looking but rather expert, disciplined, methodical observing. The flower Mutis so carefully considers is also an attribute: it is a *Mutisia clematis*, the New Granadan species that Carl Linnaeus the younger named in the botanist's honor.[3] Thus, the portrait celebrates Mutis's talents as botanical discoverer and connects them to his capacities as an observer.[4] The painting elucidates that for Mutis—as for other naturalists at the time—representation and observation were intimately connected, part of a comprehensive understanding of visuality both as an epistemological method appropriate for studying natural history and as a tool useful for the various tasks and demands of scientific work, including training, funding, patronage, exploration, discovery, transportation, and persuasive communication.

This essay will discuss the purposes and activities of late-eighteenth-century Spanish natural history as it operated on a global stage, focusing on botany in particular. I will examine two aspects whose relationship is often less than obvious to modern-day readers, accustomed as we may be to differentiate between "pure" and "applied" science and to consider that art and science constitute rather different, perhaps even disconnected, pursuits—two distinctions that did not exist in the early modern period. First, I will describe the discourse of utility that surrounded natural history in the second half of the eighteenth century. Second, I will address the role of visuality—a term I use to indicate the practices and results of both observation and representation. I argue that both utility and visuality account for the push given to the natural sciences in the Spanish empire in this period as well as for their practices.

Moreover, both of them served symbolic and concrete purposes. Thus, I do not suggest that utility constituted a "hard" reason of economic and political gains while visuality provided a "soft" decorative or symbolic background of empire. Rather, they had similar and related functions. On the one hand, naturalists and their patrons had high expectations for the economic and scientific benefits that natural history could bring: a renewed and more powerful empire, and one that could hold its own on a fiercely competitive international stage. On the other hand, they both understood the project as fundamentally one of seeing: seeing new and wonderful products scattered throughout the empire (cinnamon! pepper! tea!). Seeing anew the same plants that had for centuries grown in the colonies—that is, seeing them for the useful and profitable natural commodities that they truly were—as well as seeing anew those long-held dominions, rediscovering them as bountiful paradises of botanical productivity and profit rather than as territories littered with depleted mines. Seeing across time, to connect the reign of Charles III (r. 1759–1788) to the glorious days of Philip II (r. 1556–1598) and the work of botanists like Mutis and his colleagues to that of early traveler Francisco Hernández (discussed in Chapter 11). And seeing across space, using a few travelers as surrogate eyes for the many naturalists and administrators who remained in Madrid, as well as using images as paper avatars that allowed men scattered throughout the empire to examine for themselves distant, fragile, or impermanent floras.

UTILITY

Because the examination and methodical investigation of the natural productions of my American dominions are advisable for my service and for the good of my vassals, not only to promote the progress of the physical sciences but also to banish doubts and adulterations in matters of medicines, dyes, and other important arts; and to increase commerce; and in order that herbaria and collections of natural products be formed, describing and delineating the plants that are to be found in those fertile dominions of mine; [and] to enrich my natural history cabinet and court botanical garden.[5]

Thus Charles III, king of Spain, Naples, and Sicily, commanded Spanish naturalists Hipólito Ruiz and José Pavón in 1777 to lead a botanical expedition through Peru and Chile. He reiterated those reasons in a similar document of 1787, in which he agreed to fund Spanish naturalist Martín de Sessé's proposal to conduct a botanical expedition in New Spain. Investigating nature would serve practical purposes of great interest to the empire. At the time, plants, animals, and minerals provided valuable commodities for use in medicine and industry, pitting

European powers against one another as they competed to profit from trade in foodstuffs and luxury products like sugar, coffee, tea, tobacco, ginger, nutmeg, or pepper; medicinal substances like cinchona or guaiacum (to treat malaria and syphilis, respectively); and dyestuffs like indigo or cochineal.[6] Thus, the royal orders emphasize the importance of resolving controversies about the identity of products and of fostering commerce.

Throughout the second half of the eighteenth century, Spanish political theorists, imperial administrators, and naturalists repeatedly expressed the utilitarian potential of economic botany—as did their counterparts across Europe and throughout outposts of the empire. (See also De Vos's discussion of this theme in Chapter 14.) Carl Linnaeus, the Swedish reformer of botanical classification and taxonomy, hoped his nation could cease to buy the products from the British, French, Dutch, and Spanish, dreaming of sugarcane and coffee plantations in Scandinavia.[7] European scientific voyagers crisscrossed the oceans with cargoes of seeds, live plants (whose lives often turned out to be frustratingly short), collections of naturalia and ethnographic materials, and page upon page of written descriptions and drawings.[8] Botanical gardens and natural history cabinets served as training centers, sources of funding, and sites for experimentation; they also acted as nexuses in the widespread networks of correspondence and exchange that characterize the social economy of natural history at the time.[9] The lure of profitable nature attracted thriving imperial powers as well as rising and deteriorating ones. It proved particularly alluring in the Spanish empire, where it offered the promise of revisiting past days of imperial glory: if the discovery of the New World, and New World silver, had turned Spain into a world power, the rediscovery of American nature might improve its standing in the international stage in the late eighteenth century.

The potential profits of economic botany in a climate of fierce international competition are repeatedly evoked in Spanish writings of the 1760s, '70s, and '80s.[10] Influential economist and statesman Pedro Rodríguez de Campomanes (1723–1802) suggested a revision of Spanish Atlantic trade policies, redirecting the crown's earlier interest in American precious metals toward the exploitation of natural resources.[11] He supported his argument by examining the ways in which other European nations ran their colonies, with England given as the most illustrative and worrisome example, and analyzing trade in tobacco, sugar, and cocoa.[12] Beyond this examination of well-known commodities, Campomanes felt particularly eager to identify new materials that could be exploited for profit. He was not alone in his thinking. Prominent botanist Casimiro Gómez Ortega, director of Madrid's Royal Botanical Garden from 1771 until 1801, claimed in 1770 that, "examining its true interests, [Spain] prefers to the laborious American gold and silver mines other fruits and natural products that are easier to acquire and no less useful in increasing prosperity and wealth."[13] Eager to enlist the widespread network of imperial administrators into a vast army of natural history collectors, in 1779 Gómez Ortega distributed throughout Spain and the colonies printed instructions for transporting live plants. "The vegetable riches of Spanish America," he explained, "have over the mineral ones the advantage that they can be propagated and multiplied ad infinitum once they are possessed and naturalized" in the peninsula (see Figure 15.2).[14] Both the metropolis and the colonies would benefit from economic botany, as Gómez Ortega explained the following year in a memoir promoting American pepper: "It is useless to possess the most benign and fertile territories in the world, if we do not attempt to profit from the natural productions that they grant us, extending knowledge and consumption of them within the country, and fostering its extraction through free trade [*libre comercio*]. Without these measures, the most expansive territories become sterile deserts, as useless to their colonists as to the metropolis."[15]

Gómez Ortega concluded a 1777 report to José de Gálvez (1720–1787), his close ally and at the time minister general of the Indies, by suggesting that sending twelve naturalists and as many chemists or mineralogists to investigate American nature would yield "a greater utility than a hundred thousand men fighting to add a province to the Spanish empire."[16] Along similar lines, in 1800 Spanish botanist Antonio José Cavanilles, author of a six-volume *Flora of Spain* and director of Madrid's Royal Botanical Garden (1801–1804) after Gómez Ortega, sketched the recent history of botany in the language of conquest: "Whether from the joy of Botany, or the conviction of the utility that it brings to States, each day new supporters enlisted under her flags. Many of them, as conquerors of vegetal riches, went forth to reconnoiter new countries braving risks and hardship. The celebrated [Michel] Adanson took Senegal on his own; [Pehr] Loeffling, Spain and America; [Fredrik] Hasselquist, Palestine; and [Peter] Forsskål, Arabia."[17] Cavanilles described many of these valiant conquerors as martyrs who perished in the service of botany, with Hasselquist dying in Smyrna, Loeffling in South America, and Forsskål succumbing to pest in Yemen. Cavanilles's language of martyrdom and Gómez Ortega's choice of twelve as the appropriate number of naturalists and mineralogists to send to the Indies underscore that natural investigation was symbolically related not only to military but also to spiritual conquest. While the association of botanical exploration and missionary activity was not unique to Spanish naturalists—Linnaeus referred to the

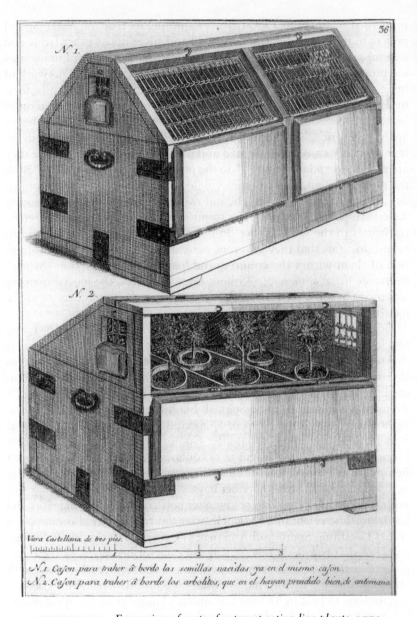

FIGURE 15.2. *Engraving of crates for transporting live plants*, 1779

SOURCE: Casimiro Gómez Ortega, *Instrucción sobre el modo más seguro y económico de transportar plantas vivas por mar y tierra a los países más distantes ilustrada con láminas. Añádese el método de desecar las plantas para formar herbarios* (Madrid, 1779).

disciples he scattered around the globe as his "apostles"[18]—it did have a particularly significant valence in the Spanish imperial context: the spiritual conquest of the New World began with the arrival of twelve Franciscan missionaries to New Spain in 1524.

Naturalists shared this enthusiasm for the commercial potential of botany and eagerly pursued the patronage opportunities it presented. José Celestino Mutis articulated this potential very explicitly in the two related proposals he sent to Charles III in May 1763 and June 1764. Recently arrived in New Granada, Mutis argued for the necessity of funding a natural history expedition that he would lead through the kingdom. He listed in great detail the flora that could be newly or more efficiently exploited in America, mentioning dyes for the textile industry, glues for producing paints, woods for making furniture and instruments, and materia medica (products with medicinal applications). He went into great detail about the exploitation of wild cinnamon and cinchona, expressing horror at the wasteful way in which the latter was exploited in the region of Loja and suggesting he would search for it in other regions of South America, design a strategy to maximize production, and direct a successfully administered *estanco* (monopoly). Throughout the proposal, Mutis often repeated the words "glory," "utility," and "profit."[19] He described the expedition not as a completely novel idea but as the continuation of an interrupted investigation begun with Philip II and Francisco Hernández, in this way cleverly relating the current king to his illustrious predecessor. Natural history exploration was connected to empire at both concrete and symbolic levels.

The economic and political promise of natural history provoked a flurry of scientific activity: between 1760 and 1808, Spain sponsored fifty-seven expeditions to its colonies—an impressive number, particularly considering their sustained neglect by an Anglophone scholarly tradition that has devoted great attention to their British and French counterparts (see Table 15.1).[20] While natural history was a focus of interest, travelers also conducted astronomical observations, drew maps and coastal charts, carried out geographical explorations, and participated in military and commercial maneuvers. These expeditions functioned within a complex system of institutions and personnel, a "colonial machine"—to use James McClellan and François Regourd's evocative phrase—comprising botanical gardens, natural history cabinets, pharmacies, observatories, naval schools, and a widespread network of contributors adhering to shared rules of correspondence and exchange that determined the types of materials to be sent and the ways to do so.[21] From the late 1770s onward, Madrid dispatched to its colonial possessions not only voyagers but also multiple questionnaires and requests for

TABLE 15.1

Natural history expeditions in the Spanish Empire, 1750–1810

Expedition	Dates	Areas	Naturalists	Artists	Images
Limits expedition to Orinoco	1754–1756	Orinoco (Venezuela)	Pehr Löfling	Juan de Dios Castel	~ 200
Botanical expedition to Chile and Peru	1777–1788	Peru and Chile	Hipólito Ruiz, José Pavón, Joseph Dombey	José Brunete, Isidro Gálvez	~ 2,400
Botanical expedition to New Kingdom of Granada	1783–1810	New Granada (Colombia, Venezuela, Ecuador)	José Celestino Mutis and associates	Salvador Rizo, Francisco Matis, and over 40 others	~ 7,300
Expedition to the Philippines	1786–1794	The Philippines	Juan de Cuéllar	Anonymous local artist(s)	80
Circumnavigation (Alejandro Malaspina)	1789–1794	South, Central, North America; Australia; the Philippines	Tadeus Haenke, Luis Née, Antonio Pineda	José Guío, José del Pozo, Francisco Pulgar, José Cardero, Tomás de Suria, Fernando Brambila, Francisco Lindo, Juan Ravenet, José Gutiérrez	~ 1,000
Botanical expedition to New Spain	1786–1803	Mexico and Guatemala	José Mociño, Martín de Sessé, Vicente Cervantes	Atanasio Echeverría, Vicente de la Cerda	~ 2,000
Expedition to Cuba	1796–1802	Cuba	Baltasar Manuel Boldó, José Estévez	José Guío (from previous Malaspina expedition)	~ 70
Expedition to Ecuador	1799–1808	Ecuador	Juan Tafalla	Francisco Pulgar, Francisco Xavier Cortés	~ 220

reports from the Council of Indies, the Royal Botanical Garden (f. 1755), and the Royal Cabinet of Natural History (f. 1776).[22]

Government officials, colonial administrators, and naturalists alike tirelessly invoked utility, repeating the word as a mantra. It appears again and again in official correspondence between José de Gálvez (head of the Council of Indies 1776–87) and Casimiro Gómez Ortega, in Gómez Ortega's own exchanges with the many naturalists he supervised, in Gálvez's letters and mandates to colonial officials, and in the responses generated by such requests. It lies at the core of publications such as Gómez Ortega's memoir promoting an American pepper, *Historia natural de la malagueta o pimienta de Tavasco* (Natural History of Malagueta, or Tabasco Pepper; Madrid, 1780), which suggests that patriotic Spaniards should satisfy their appetite for the spice with this variety, rather than divert to Dutch coffers profits that should rightfully remain in the homeland. And it set the pace for naturalists' research: Hipólito Ruiz published a detailed study suggesting four American substitutes for natural commodities exploited by the Dutch, French, and British, and another about cinchona varieties in Peru.[23] José Celestino Mutis claimed to find American versions of nutmeg, tea, and cinnamon; he labored for decades to identify and distinguish regional varieties of cinchona, and became embroiled in fierce (and ultimately bitterly disappointing) priority disputes.[24]

Utility is without question an important aspect in the study of eighteenth-century colonial natural history. It provides a central explanatory factor for understanding the extraordinary efforts and funds devoted to its pursuit, as well as the specific practices in which naturalists were involved. Given the constant preoccupation with utility and profit in eighteenth-century European natural history, it is not surprising that historians have devoted great attention to examining this aspect of colonial science, even using it as a standard by which to judge an expedition's "success" or lack thereof. Yet, there are many aspects of eighteenth-century colonial and imperial natural history that cannot be understood—indeed, that make very little sense—when thinking narrowly in utilitarian terms. The most significant one is the exorbitant visual productivity. The scientific "colonial machine" at work in the Spanish world in the second half of the eighteenth century aimed to produce useful and profitable information and commodities. Thus it is hard to know what to make of the thousands and thousands of images that it churned out in the form of finished paintings and drawings, working sketches, and printed engravings, and of the constant references to observation made by naturalists, administrators, and artists, in both textual and pictorial form. Were visual materials—both observations and representations—considered useful? If not, did they serve another purpose? Or were they simply a by-product of this machine, perhaps a bit cumbersome but ultimately of little importance, no more than a puff of smoke?

VISUALITY

Every single eighteenth-century Spanish scientific expedition, without exception, employed artists and produced astonishing numbers of images. Indeed, illustrations represent the bulk of their work: they drew many more images than wrote descriptions or collected objects. The expeditions yielded very few publications—Ruiz and Pavón were the only travelers to print a thorough *Flora* of the region they explored, and the number of memoirs and shorter publications is limited.[25] And although naturalists generated much manuscript material that they did not publish (as demonstrated by the many memoirs, reports, and descriptions that did not reach print), the expeditions dedicated the vast majority of their funds, time, and personnel to the production of visual material. In some cases, draughtsmen outnumbered naturalists, most notably in Mutis's expedition to New Granada, which privileged images to the point that its painting workshop became the nucleus of the project. Mutis's correspondence, both personal and official, is dominated by his concern with the operation of the workshop, the quality of its images in relation to printed European botanical illustrations, and their value as material evidence. The colonial machine was a visual apparatus.

The central importance of images to colonial science is clear from the attention paid to them at every point of an expedition, from the very first planning stages through the hiring of personnel, the work of naturalists and artists on the field, the shipments sent from the colonies, and their enthusiastic reception in Madrid. When Ruiz and Pavón set out for Chile and Peru in 1777, the royal orders they received not only addressed the expedition's utilitarian goals but also required them to "describe and delineate [draw]" the plants they encountered. The royal orders given to Martín de Sessé a few years later reiterated the importance of producing visual representations, instructing him to "methodically examine, draw, and describe the natural productions of my fertile dominions in New Spain."[26] Likewise, the July 1789 royal orders requesting that colonial administrators produce informative reports about the natural commodities of their regions, mentioned previously, noted that reports should include drawings of the natural specimens described. The Spanish crown hoped that expeditions would produce not only useful knowledge of valuable commodities but also a visual archive of observations and illustrations.

The inventories of multiple shipments sent to Madrid from throughout the empire attest to the efforts made to satisfy this visual appetite. Ruiz and Pavón sent almost 250 colored drawings in a first shipment dispatched in April 1779, and many more followed: roughly 200 images in 1780, 1,000 in 1784, 600 in 1787, and almost 600 more that they took with them on their return voyage to Spain in 1788.[27] The other expeditions operated in a similar fashion, periodically shipping images to Madrid's Botanical Garden and Natural History Cabinet. The high status of images in relation to the many other materials the expeditions produced is apparent from Casimiro Gómez Ortega's response to Ruiz and Pavón's first shipment. Surveying the six crates packed with dried plants, seeds, bulbs, roots, and natural curiosities, and the single crate containing almost two hundred drawings of plants and about sixty floral anatomies—by then the live plants had withered—Gómez Ortega reported to José de Gálvez: "the most valuable [items] are the two hundred and sixty drawings, and the herbals."[28]

Given the centrality of visual material to the perceived success of an expedition, the process of organizing one involved the careful selection of competent artists. In 1776, recommending that the Spanish crown sponsor the botanical expedition to Chile and Peru, Gómez Ortega pronounced it "absolutely necessary that the botanists be accompanied by two draughtsmen." Ignacio de Hermosilla, director of the San Fernando Royal Academy of Fine Arts (f. 1752), personally supervised a competition to locate the two most skilled draughtsmen among the academy's students. The artists selected through this procedure, Fernando José Brunete and Isidro de Gálvez, thus went from copying classical casts and painting mythological or historical scenes in Madrid classrooms to making botanical drawings as they traveled in South America. They did not suffer from lack of resources, setting out well-provisioned with drawing materials and receiving the same yearly salary as Ruiz and Pavón, a clear indication of their high status within the project. Nor did they lack direction: Gómez Ortega prepared a set of written instructions outlining their tasks in great detail, exactly as he did for the expedition's naturalists.[29]

The instructions consist of eight articles covering the content and style of the illustrations the draughtsmen would produce, the day-to-day organization of work, and the relationship between artists and naturalists. Like Gómez Ortega's initial memo, the instructions indicate that although artists were "absolutely necessary" members of the expedition, to the point of being compensated at the same rate as Ruiz and Pavón, the botanists clearly had greater authority and status within the expedition. Gómez Ortega began by repeating his prescription that the artists should copy nature faithfully, not adorning it with anything coming from their own imagination, and limiting themselves to drawing whichever parts the botanists indicated as important for "knowing and distinguishing" each plant. Only fresh plants should be depicted, with each image including both the plant's portrait and the detailed anatomy of its fruit and flower. The standard size to be used for all drawings would be decided by the botanists, who would provide a model to demonstrate the appropriate dimension to the artists. This would ensure an overall coherence of the images as a series, and facilitate the quick and economical production of engraved plates for publication.

The instructions also imposed a working procedure that subjugated artistic process to the specific goals and operation of the expedition. In order to maximize productivity within the fast pace of travel, the artists should limit themselves to drawing in pen and ink, adding color only when a plant had remarkably beautiful, eye-catching, or peculiar hues, and even then coloring a single flower, fruit, and leaf, which would serve as a model to complete the image's illumination back in Spain.[30] Any spare time should be dedicated to botany, not art, by helping the naturalists to form herbaria, organize their manuscripts, or other such tasks. The botanists' control over the artists' production and use of time also extended to regulating their bodies by mandating where and when they should travel, as well as to controlling their productivity by allocating work supplies. The desired images were botanical objects, not artistic creations, hence the insistence on depicting those characters considered important by the botanists and on avoiding a decorative painterly style—a key concern of natural history illustration at the time. Botanical considerations mandated the content, style, even the size of the image. While the draughtsman acted as the expedition's hand, hired to produce the images that naturalists desired, the naturalist served as its eye, selecting the object to be depicted, indicating which traits to focus on and which to disregard, and imposing the particular vision with which to approach and represent nature. In this and other expeditions, naturalists daily supervised and directed artists, evaluating whether a drawing qualified as finished and satisfactory or whether it needed correction. Figure 15.3, for example, shows naturalist Luis Née's correction to José Guío's watercolor drawing of a plant, written directly across the image: "The fruit should be green: the painter made an error." Botanists considered that without their guidance artists were useless, since "making a perfect drawing does not have to do with representing the visible parts of a plant

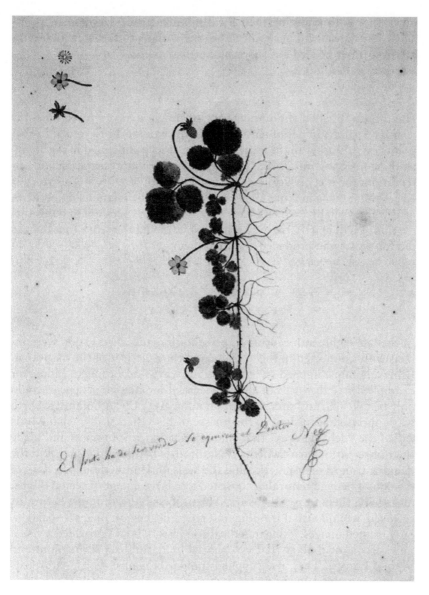

FIGURE 15.3. Rubus radicans Cav. (Rosaceae), *watercolor, with annotation*

NOTE: The annotation by naturalist Luis Nee to José Guío's drawing states: "The painter made an error: the fruit should be green" (*El fruto ha de ser verde se equivocó el pintor*).
SOURCE: Archivo del Real Jardín Botánico de Madrid, VI, 40.

but rather with knowing their location, direction, size, and shape."[31] This view considered naturalists the true authors of the drawings, with artists as their needed but subordinate amanuensis.

&

Having established the importance of image making to eighteenth-century Spanish imperial natural history, I would like to return to two points raised earlier in the essay. The first, briefly addressed in my analysis of Mutis's portrait, is the connection between representation and observation, for it is my argument that the role of visuality in this story is larger than image making alone. The second is the relationship between utility and visuality, which I have suggested as the axes guiding this discussion. I will clarify these two issues by tracing the work that images did for those who produced and used them.

THE WORK THAT IMAGES DO:
USEFUL VISUALITY

In the late eighteenth century, representation and observation were intrinsically connected in the persona and the practices of the naturalist. Being a botanist implied seeing in specific ways—according to Mutis, for example, three distinct but almost identical-looking plant species could only be told apart if "surveyed with botanical eyes."[32] A botanist's worth was appraised according to the finesse of his visual skills, what Mutis called the "delicate eyes" that characterized great botanists and made their observations trustworthy.[33] Naturalists considered visual skill the defining trait of their practice and the basis of their method. The day-to-day job of being a naturalist consisted to a large extent of seeing things, and seeing them in specific ways. The process of becoming a naturalist revolved around visual training and relied heavily on images—particularly printed images.[34] Illustrated natural history books provided a visual and verbal vocabulary that was shared by naturalists throughout and beyond Europe. They also supplied standards against which naturalists could gauge the value of their own work, as well as models for them to emulate or react against. Thus, books helped to define and arbitrate a community of competent and relevant reader-practitioners.[35]

Printed images also defined the traveling naturalist's job by demarcating what they needed to accomplish in their voyages, namely, to describe any local productions not included within the European printed inventory of global nature, to rectify any discrepancies, and to resolve incomplete or erroneous descriptions. Books provided naturalists with the illustrations they needed to approach nature, with parameters for producing new images, and with a medium for presenting their own contributions to natural history. Printed illustrations allowed travelers to ascertain whether the plants and animals they encountered were truly unknown, in which case they could be introduced into the literature and linked to their discoverer's name—ideally, accompanied by an image. The traveling naturalist's way of seeing involved a constant back-and-forth between the book and the field, using books to interpret what they saw on a field and producing their own texts and images to respond to what they read as well as to contribute new information. In the eighteenth century, images provided an entry point into the exploration of nature, functioned as a key instrument for producing knowledge, and constituted the foremost result of natural investigations. Images operated at every point of a trajectory that moved from the collection of natural data to its incorporation into a global inventory of nature through textual description and visual representation.[36]

The process of making a representation, of course, tends to involve an act of observation. Naturalists drew often; often, they drew well. Their letters, journals, notes, and memoirs are peppered with sketches. Their eyes and hands worked in tandem, aiding each other: drawing allowed them to see, to think, and also to communicate. When working out a taxonomical point for themselves, naturalists produced notes that tend to include diagrams to clarify the problem at hand as well as its solution. When writing letters, especially to correspondents who, as was often the case, had never been in the region they were describing, naturalists appended drawings to get their point across. For example, when characterizing four varieties of the same plant by differentiating among leaf shapes, using words to describe their forms—spear-shaped, oblong, heart-shaped, oval—did not suffice. An accompanying drawing proved necessary to make the point clear beyond error (see Figure 15.4). Taxonomy was a problem that concerned both utility and visuality: naturalists used their own and others' observations and illustrations in their attempts to determine whether these four different plants, growing in four different regions of South America, could all be legitimately traded as cinchona, or whether a plant commonly deemed "pepper" in South America could really be considered the same species as the "true" pepper that the Dutch imported from East Asia.

The fact that naturalists were happy with casual sketches in their own notes but preferred to send colleagues more skilled images produced by artists provides yet another reason for their great care in selecting and overseeing draughtsmen. It is telling that although the naturalists from

FIGURE 15.4. *Illustration of four types of cinchona plants*
SOURCE: Real Jardín Botánico de Madrid, III, 2, 3, 110.

the various Spanish expeditions only rarely corresponded with one another, the very few letters that they did exchange invariably mentioned
their painters. In late 1788, naturalist Vicente Cervantes replied to a
letter in which Mutis asked about the talents of Atanasio Echeverría
and Vicente de la Cerda, two Mexican artists who had recently joined
the expedition led by Sessé. Mutis, then eagerly trying to locate more
artists for his expedition, wondered about the Mexican painters' talents, considering whether it would be possible for him to hire artists
from New Spain or other regions of Spanish America rather than from
Spain. Cervantes answered the query not only with words but also by

sending images. "I satisfy Y.M.'s discreet curiosity about [our] draughts-men," he wrote to Mutis, "with two [drawings of] plants produced by the pair [of new painters] who we have obtained from this [Fine Arts] Academy of San Carlos [in Mexico City] . . . They are very tender and docile youths, and very sharp in their work. Given these qualities, and the foundations they demonstrate, we trust they will not pale in compari-son with those that might have come from Peru, who I would not judge more advanced."[37] Similarly, the lone letter that naturalist Luis Née, a member of the Malaspina circumnavigation expedition, sent to Mutis addressed the incorporation of a botanical draughtsman to the team. Writing from Ecuador in 1790, Née expressed satisfaction at the art-ist and described him as good, patient, possessing the "foundations" of botany, and skilled at defining all the parts of a plant, especially the fruit and flower. "The drawings I have taken care to direct until today," Née enthused, "include nothing beyond what is necessary for any systematist to know the [plant's] class and order. Adding a methodical description seems sufficient to know the plant that is being presented."[38] Thus, Née agreed with Ruiz that artists were crucial members of an expedition but ultimately subservient to the naturalist's authority.

The capacity of images to visually incarnate and mobilize plants that remained in key ways unseen and unknown, even three centuries after Spaniards had first encountered New World nature, accounts for much of their significance. Naturalists found the usefulness of images extended to their social functions: images were central to the gift ex-change economy that regulated almost any type of transaction at the time. When soliciting additional funds, reassuring an anxious patron or courting a new one, requesting that an acquaintance based in Eu-rope send recently published books (illustrated, of course) that would aid them in their work, or attempting to enter into correspondence with a world-renowned botanist, naturalists knew to send pictures along with their letters. Mutis, for instance, sent several shipments of drawings and herbarium specimens to Carl Linnaeus, the ultimate arbiter of botanical worth at the time.[39] Mutis also had repeatedly need to present samples of his artists' work in order to reassure increasingly impatient officials in New Granada and Madrid of the expedition's progress. When Casimiro Gómez Ortega pronounced the 260 images and the herbals—that is, the visual materials—the most valuable items in the shipment that Ruiz and Pavón sent to Madrid, he immediately suggested that they be bound in albums and presented to the king. They constituted tangible evidence that, throughout the empire, naturalists, artists, and administrators strived to carry out his orders to produce information that was not only useful but also visible.[40]

CONCLUSION

The Spanish eighteenth-century natural history expeditions can be understood as projects to make the empire's nature visible. This is not a mere figure of speech: images represent the most important material used and produced by the expeditions. Without exception, every single expedition, whether it was focused on natural history, astronomy, geography, navigation, exploration, or cartography, employed draughtsmen and produced numerous images as part of its stated goals. The natural history expeditions regularly had at least as many and often more artists than naturalists. At home or abroad, European naturalists used images in their daily work, and wrote abundantly about them in their journals and correspondence. Pictures deserved special mention in the inventories of collections shipped back to Europe, and frequently received the most attention as crates were unpacked. When traveling naturalists sought to honor a patron, scientific or administrative, or needed to ask a favor, images constituted the preferred instrument of persuasion. Images were not ornamental secondary products but the central concern of many expeditions, at times the most important: the bulk of the Spanish expeditions' output consisted of illustrations.

This obsessive interest in images was by no means a Spanish peculiarity, but common to European natural history in general and to the domestication of foreign nature in particular. The manufacture and use of images was a central practice through which nature, particularly foreign and exotic nature, was investigated, explained, and possessed by early modern Europeans.[41] Illustrations were not a pleasant but inconsequential by-product of natural investigation. Rather, they constituted both a vital technique and arguably one of the more important results of natural history as a field of study. By the late eighteenth century, natural history was a fundamentally visual discipline that used images in all its practices and spaces. Images were deployed in training, research, and communication, be it published or manuscript. Educated as observers and representers, naturalists and artists constructed a visual culture of natural history based on standardized ways of viewing nature and on pictorial conventions guiding its depiction. Naturalists moved constantly between the world of objects "out there" in the field and the world of objects "in here" in collections. Images bridged the gap between exterior and interior, the field and the collection, by offering a hybrid domesticated space, a paper nature that was always and perfectly available for virtual exploration.

The transatlantic circulation of people, specimens, and images is further elucidated by juxtaposing the portrait of José Celestino Mutis in the beginning of this chapter with another painting produced by an artist from his workshop (see Figure 15.5). This portrait depicts Antonio José Cavanilles, a renowned Spanish botanist and author, the director of the Royal Botanical Garden of Madrid between 1801 and 1804, and a long-time correspondent and supporter of Mutis. Although this portrait, like the one of Mutis, is not signed, it can be confidently attributed to Salvador Rizo, who was the expedition's head artist and Mutis's second-in-command. While Mutis was most probably painted from life, Rizo's portrait of Cavanilles must have been based on an engraving, since neither the botanist nor the painter ever crossed the Atlantic. The painting celebrates both naturalist and artist, and also illustrates Mutis's expectations as to how the expedition's images would be used in botanical practice. Cavanilles is shown in profile, sitting before a work table. His left hand points to a botanical illustration that is recognizable as one of the expedition's own, with the name "Rizoa" visible at the top of the image. Cavanilles carefully follows the drawing's outline with his left hand, while his right hand is occupied with pen and paper on the composition of the plant's

FIGURE 15.5. *Portrait of Antonio José Cavanilles, ca. 1800*
SOURCE: Oil on canvas, 86 × 66 cm, Salvador Rizo
(attributed). Museo Nacional, Bogotá, Colombia.

botanical description based on its physical attributes, as prescribed by the Linnaean system. Thus, the Spanish botanist is able to sit at his desk in Europe and conduct "firsthand" observations of foreign nature, using the image to classify and name American flora. As Mutis explained in a letter, "no plant, from the loftiest tree to the humblest weed, will remain hidden to the investigation of true botanists if represented after nature for the instruction of those who, unable to travel throughout the world, without seeing plants in their native soil will be able to know them through their detailed explanation and living image."[42] The illustration, then, not only stands in for the object it represents but also supplants the very act of travel: the nondescript indoor setting of both portraits makes it impossible to know that Mutis conducted his observations in Bogotá, New Granada, while Cavanilles sat at his desk in Madrid, Spain. The portraits also erase time, collapsing sequential acts of travel, transport, observation, and description into simultaneous events.

Visuality—the term I have used to address the practices and results of both observation and representation—made the empire's nature identifiable, translatable, transferable, and appropriable. By defining nature as a series of transportable objects whose identity and importance were divorced from the environment where they grew or the culture of its inhabitants, illustrated natural histories rejected the local as contingent, subjective, and translatable. Instead, they favored the dislocated global as objective, truthful, and permanent. The natural history illustration, with its flower always in bloom, its fruit permanently ripe, its animal caught in clarity and perpetuity, was at once the instrument, the technique, and the result of natural history as a field of study.

Afterword

NOBLE DAVID COOK AND ALEXANDRA PARMA COOK

The contributors to this volume have convincingly demonstrated that the rise of European science in the early modern era is not solely the result of the figures working north of the Pyrenees: Copernicus, Linnaeus, Bacon, Harvey, Descartes, or the Italians Galileo or Da Vinci and other similar pioneers. Rather, events in the north leading to the so-called scientific revolution were precipitated and nurtured by a series of major probings and discoveries in the fourteenth and fifteenth centuries that were moving most markedly in the Iberian Peninsula. Already in 1963, John H. Parry pointed out the key characteristics of these developments in his stimulating study of the "age of reconnaissance."[1] In Spain and Portugal, the very foundations of mathematics and astronomy that were key elements in the early stages of the scientific revolution were arguably the most advanced of Europe in the fifteenth century. The Arabic system of numbering was the system most practiced in Iberia, rather than the Roman system still widely used in continental Europe at the time, the consequence of seven centuries of much of the peninsula being under the domination of Islam. And it would be to Islamic centers of learning, such as Cordoba, that medieval north European students of science and medicine would travel to study. Exposed to Jewish and Islamic scholarly works as well as preserved translations of some of the great Hellenistic philosophers, geographers, astronomers, and physicians, these centers played an important role in the dissemination of scientific knowledge. Even with the Christian "Reconquista" of the peninsula reaching a critical stage in the mid-thirteenth century, the transfer of knowledge continued apace.

The authors of all the chapters build from the foundation set by the navigators, traders, fishermen, and spreaders of the "word"—the sponsors, from slavers to princes, of the mid-fifteenth century. All of the contributors recognize the connection between commercial interests and discovery, and they stress the link between the "practical" and the more "esoteric" intellectual developments that occurred during the long period collectively here examined, from 1490 to 1790. The authors of this volume share the belief that technological and scientific breakthroughs are closely linked. They do not claim that modern science exists in the Iberian world during this period; rather, they collectively show that those working under Iberian jurisdictions were just as advanced during much of the three centuries as their counterparts to the north, and that their contributions to change, especially to the eighteenth century "Enlightenment," were just as significant. Moreover, several of the authors successfully demonstrate the impact of the Americas on the development of early modern European science and medicine and the interaction, rather than a unilateral flow, between the "core" and the "colonial periphery."

Two distinct periods are covered in this anthology. The first is the period of discovery and settlement, and the second is the period in which transformations in the building of knowledge take place as a result of the foundations, the scientific revolution, and the age of Enlightenment. No distinct chronological barrier is crossed; instead, the shift from one stage to the next is more dependent on fields. For navigation, the first stage comes with the increasing use of navigational charts, of mathematics and tables—such as that of Abraham Zacuto—for calculations of geographical position, and the increasing use of instruments such as the astrolabe, quadrant, and sextant in conjunction with a better understanding of astronomy and the systematic fleshing out of physical geography from the coasts into the interior of new continents. The second stage will come in the eighteenth century with the chronometer for use in longitudinal positioning, better optical instruments, and new developments in ship configuration and fitting, permitting quicker and more sustainable long-distance voyages.

In medicine, the first period is characterized by the collection and cataloguing of potentially valuable products that can be employed in curing. As early as the second Columbus voyage, university-trained Dr. Diego Alvarez Chanca was assigned to the group not only to cure sicknesses of the participants but also to search out indigenous plants of value.[2] Cataloguing of the new animals and plants, in the Aristotelian sense, was a key part of the process, and there are a host of major Spanish and Portuguese figures who made substantial contributions. Early efforts often aimed to be inclusive: Gonzalo Fernández de Oviedo y Valdés intended a complete history, but only Part 1, the *Sumario*, covering predominantly the geography and the flora and fauna of the Indies, finished in 1526 and printed in 1535, was published during his lifetime. Perhaps the earliest compilation of New World medicinal plants comes from the indigenous physician, Martín de la Cruz in Mexico. It was written in Náhuatl in 1552 and translated into Latin by Juan Badiano and accompanied by excellent illustrations.[3] The use of American plants to treat diseases fascinated men such as Dr. Nicolás Monardes of Seville, who collected medicinal plants from returning overseas travelers and attempted to grow some of them in his garden. He spent years studying the medical propensities of the new plants, testing many on his own patients, and finally publishing his results in 1574.[4] Also in the 1570s and sponsored by Philip II, came the systematic and massive compilation of Mexican medicine by Dr. Francisco Hernández. The Portuguese counterparts brought back to Europe knowledge of medicines and medical practices not only from their South American colony but also from Asia, thus laying important foundations that would lead to new developments in later centuries.

One of the aims of participants in the era of Iberian expansion was to find precious metals. A general knowledge existed of the techniques of extraction of metals from ores. Certainly, a long tradition of mining prevailed in the kingdoms of Castile from Roman times. Deposits of gold, silver, copper, zinc, lead, sulfur, coal, iron, and mercury were scattered. Both the Spanish and Portuguese undertook a systematic search for these and other valuable commodities overseas. Their knowledge of mining techniques and metallurgy was solid. The impact of mining on the rise of science was probably most direct in the extraction of silver from its ore. It would be experienced miners who would assist in finding and then developing refined methods to extract metal from ore, as Bartolomé de Medina did with amalgamation for silver extraction in New Spain in the early 1550s. Challenges faced miners in America, especially at the richest lode of all, Potosí, where the lack of combustibles at high altitudes or other sources for heating the ore to extract the metal, plus an inadequate labor supply, led to further experimentation with mercury. Using this process, the amalgam of mercury bound with silver could be extracted from crushed ore with very low heat, and part of the mercury given off as a gas could be captured and recycled as it cooled. The process was introduced into Potosí in the 1570s, and continuous experimentation took place to find the best combinations of mercury and other elements to extract the silver. That experimentation and testing was a hallmark of metallurgy, one of the practical sciences that most early flowed into what might be considered the true sciences. Mathematics here too was important, along with weighing and measuring, as various compounds were tried. These outlines would continue and in the eighteenth century be expanded by a new series of mining expeditions to the Americas sponsored by a new ruling dynasty in Spain.

In the cataloguing of the physical and natural realms of the freshly encountered universe, new materials came into being. Some were useful, studied, and adopted. The search accelerated in the eighteenth century, with the scientific expeditions sponsored by Charles III and Charles IV in Spain, as well as the Marquis de Pombal in Portugal. If not practical, stones might end up in the "cabinet of wonders" to be seen and admired for their unique characteristics. Many of the developments that occurred in the physical sciences in the eighteenth century were again related to the practical, in this case especially to the military needs of a pressured and increasingly indefensible Iberian world. Wars of the period were a constant drain on the peninsula, internally and abroad. The consequence was the need to quickly apply all the techniques of "modern military science" possible to protect Spain's and Portugal's interests. The training of military cadets in mathematics, ballistics, and chemistry as well as military architecture was supported, both on the peninsula and in the colonies, at universities. Major strides in mapmaking took place. One would not say that Iberian breakthroughs occurred as a result of these efforts, but clearly recently published Enlightenment-period texts were studied and the knowledge was applied to bolster the defense of imperial interests and the mechanical and agricultural arts.

David Goodman in the first chapter of this anthology stresses the "surge of publication on science, medicine, and technology in colonial Spanish America" in the last twenty years. In part, he points out that the increase was the consequence of the Consejo Superior de Investigaciones Científicas (CSIC) and its role in support of "scientific" historical inquiry leading up to and including the quincentenary of the "encounter." Goodman set the stage with his critical review of works on science published through the 1990s, and what we have in this volume is the culmination of the subsequent phase, the solid work by scholars focusing directly on the issues of the nature and significance of scientific developments in the Iberian world, both culling archival sources that were not thoroughly used by their predecessors and carefully examining the new works of numerous scholars who have contributed much to our core of knowledge.

Standard texts of the history of science and technology written a half century ago rarely mention the achievements of the Portuguese, as Palmira Fontes da Costa and Henrique Leitão point out in Chapter 2, and they eloquently remind us that even now there has been no real examination of "Portuguese imperial science." The major contribution of the Portuguese is in nautical science, and even here the work of the Lusitanians is often unappreciated. Yet early on the Portuguese had come to challenge the work of Ptolemy and many other Hellenistic sources on geography. The authors argue that João de Castro even took "an 'experimental' and 'modern' approach to the study of nature." As Onésimo T. Almeida notes in Chapter 4, Castro measured, kept notes, double-checked, and "theorize[d] about the role of experience in the process of knowledge acquisition." In the mid-1550s, Castro and mathematician Pedro Nunes held extensive discussions over the value of theory versus practice. Almeida argues further that the Portuguese produced a substantial number of logbooks and manuals not yet well known, and that they quickly discarded the teaching of the ancient authorities, since especially the experience of navigators had proven them wrong. The Spanish would also experiment and modify. As Kevin Sheehan ably shows in Chapter 12, invention frequently was the only road to survival while crossing the large expanse of the Pacific Ocean, for both Portuguese and Spanish travelers. The 1605–1606 voyage of P. Fernández de Quirós is a case in point. For drinking-water supply he had invented an apparatus to extract water from seawater, converting two to three *botijas* daily using heat from a ship brazier. During Quirós's career, he recorded over fifty often well-detailed reports on the Pacific and its navigation, preparing in the process some two hundred maps as well as charts and globes. Earlier, according to Almeida, Castro's *Roteiro* covering the route from Lisbon to Goa and beyond provides a stunning example of the value of this type of work for opening up the way to the rich resources of south and southeast Asia. As Sheehan suggests, in both the Spanish and Portuguese cases, patronage—royal, church, or other—was important in the support and organization of successful major ventures.

Given the greater population and wealth of Castile, as well as Spain's drive to consolidate its administration under Isabel and Ferdinand, it is not a surprise that greater royal supervision was wielded over the process of the creation of empire than in Portugal. Within a decade of the discovery and settlement of an island base in the Caribbean, the House of Trade (Casa de la Contratación), as pointed out by María M. Portuondo in Chapter 3, was created by the crown, and roughly twenty years later, Charles V established the Council of the Indies. Rather than a haphazard and inconsistent policy, the Spanish created a bureaucratic structure fully capable of administering much of two continents and hundreds of islands in the far Pacific. The revenues that flowed into the metropolis, thanks especially to the mineral wealth and the royal *quinto* on mining production (a levy of a fifth of the proceeds), allowed the bureaucracy to support trained cosmographers, chroniclers, chief pilots, a naval school, and much more. This permitted systematic collection and storage of information for state needs. As Portuondo argues, the ability to order and collect the information on the Indies was so great that the "young discipline of cosmography would strain under the avalanche of new knowledge that resulted from the discovery of the New World." The result was a massive corpus of information covering geography, natural history, ethnography, cartography, mathematics, and astronomy that flowed toward the *Casa* and Council, literally filling archives and libraries and providing ample meat for generations of later historians to digest. Volumes are needed to adequately cover the individual contributions of the key figures of the period, men like Alonso de Santa Cruz; Juan López de Velasco; the physician Francisco Hernández, sent to Mexico to collect information on native American medicine; Juan de Herrera, who established the Royal Mathematics Academy in Madrid (1584); and so many more key contributors mentioned elsewhere in this anthology. As Portuondo suggests, Juan de Herrera's significance lies in the stress on mathematics for the practical, including surveying, architecture, mechanics, and cartography. By setting up the academy he evaded the conservatism of the university "authorities." Reliance on observation was required in the new era, not the study of the ancient sources that had virtually nothing to say about the new lands and their resources. Men at the time adapted: "they developed an epistemic criteria that demanded facts based on empiricism and corroborated eyewitness testimony, ways of ascertaining matters of fact that became the hallmark of seventeenth-century scientific practices."

Antonio Barrera-Osorio in Chapter 11 shifts the significance process from the role of Court, *Casa*, and Council even further toward the practical. He argues that the expedition undertaken by Francisco Hernández (1571–77), although sponsored by Philip II and his bureaucracy, really began in the streets of Seville, not at court. It was in this bustling port city, filled with its bankers and merchants in search of profit, as well as the sailors and ship captains and the whole of the infrastructure needed for the provisioning of the expeditions, that we witness the critical evolutionary transformation. Barrera-Osorio shows "how artisans, technologies, and commodities brought changes in sixteenth-century epistemological practices in the context of long-distance empires." Experimentation and invention were needed for these ventures to be successful, and the revenues taken from the products that were encountered had to be ample to support the process. It would be merchants, working with or under the direction of physicians and surgeons—who often were themselves merchants interested in profit as well as healing—who would ultimately lead to discoveries. Dr. Nicolás Monardes established his own botanical garden and filled it with New World medicinal plants in Seville a decade before Dr. Hernández set out for Mexico. Experiments were constantly being made in the colonies, and the role of the practical—the empirical—was critical. Even the House of Trade fostered the practical, supporting efforts to build better bilge pumps, to dredge silt from estuaries blocking shipping, to endorse and direct new searches for products of value, and to set up a school for the training of ship pilots. Barrera-Osorio goes so far as to say that the Casa de la Contratación was "a pioneering model for the scientific academies of the seventeenth century."

As we have seen, the search for new medicines is one of the principal aims that consumed the imagination of investors in commerce, and no wonder, given the high rate of mortality in the period. Timothy Walker in Chapter 13 points out how the Portuguese in the East were driven to experiment with indigenous medicine by the heavy mortality from tropical diseases in their overseas settlements. In the beginning, the Portuguese Jesuits tried to prove to locals that European medicine was better, using proof of the superiority of all things "Christian" to help in their efforts to convert, but it quickly became evident that European remedies were ineffective in the new environment. The Portuguese began a systematic search not only for new medicines but also for medical practices that might better protect their settlers. Regardless of the reasons, both Iberian countries saw from the beginning the potential benefits to be derived from new medicines. In spite of their relatively smaller size, the Portuguese were very effective in publicizing their new "discoveries." According to Walker, the Conselho Ultramarino in Lisbon (similar to the Spanish Casa de la Contratación) sent numerous requests for information on knowledge of "medicinal substances from south Asia and east Africa, and about what techniques were thought to be efficacious." Thanks to shipping and administrative as well as other records on two continents, Walker was able study not only the medicines but also the doctors and hospitals that were set up in the Portuguese trading outposts within the empire. Two major sixteenth-century contributions resulted from the efforts of the Portuguese: the treatises of Garcia da Orta (1563) and of Cristobal da Costa (1578). In a widely disseminated text, Da Orta described fifty-nine drugs made from Asian medicinal plants. The Portuguese version was translated into Latin in 1567 and printed in Antwerp, The Netherlands, followed soon by English, French, and Italian translations. Da Costa's work was first published in Burgos, Spain, and was also translated and widely consulted. As Walker concludes, "medicine was a tool of empire from the earliest days of European expansion."

Yet the flow in the direction of a more secular approach was not continuous. Juan Pimentel in Chapter 5, for example, challenges the reader to question, "Is it possible to talk of 'baroque science'?" Much of the seventeenth century, especially in Spain, has been depicted by cultural historians, taking the cue from José Antonio Maravall, as the era of the baroque. In architecture, music, literature, and painting, it is the excessive, the florid, the convoluted which characterize the age. It is a mix

of zealous religiosity and at the same time the increasing secular, anticlerical outlook, that will intertwine and titillate the spirit. To explore
the issue, Pimentel examines Juan E. Nieremberg (1595–1658), born in
Madrid to German parents who came to Spain as part of the court entourage of Maria of Austria. A polyglot, conversant in Hebrew, Greek,
and Latin, he assumed administration of the academy of mathematics
that had been set up by Philip II, and after taking holy orders in 1623,
began to teach natural history. Although much of his encyclopedic work
dealt with America, he never traveled there. His two most important
treatises provide a good overview of seventeenth-century knowledge, especially his *Historia Natural, Maxime Peregrinae* published in Ambers
in 1635. As Pimentel points out, although Nieremberg filled his works
with information on the flora and fauna, and the peoples encountered by
the Europeans, his was a search for God in the natural world. The "lost"
peoples had to be incorporated into the Biblical account; the natural
and religious were parts of the same universe. According to Pimentel,
Nieremberg represents "preterimperial science."

Anna More in Chapter 6 also focuses on seventeenth-century figures:
two scientists engaged in a debate over the nature of the 1680 comet as
seen in Mexico. The setting is the colonial sphere, not the metropolis,
and the debate is fierce and public, between the noted Mexican Carlos
de Sigüenza y Góngora and the equally famous Austrian Jesuit Eusebio
Kino, known for his missionary work in the northern borderlands.
The debate between the two men took the form of treatises, rather
than face-to-face confrontation, although they did meet and converse,
and parted as enemies. The "duel" was one over the significance of the
natural event, with the American providing a largely scientific analysis
of the comet, while the Jesuit defended the religious view; for him,
it foretold disaster. The Mexican's approach was mathematical and,
as More argues, more empirical than Kino's. Sigüenza accepts reason as
an avenue toward understanding reality, as established by mathematics,
whereas Kino's approach represented the divine. For Anna More, the
significance of the debate lies in that it "shows the extent to which scientific reason in colonial Mexico was related to the geopolitics of core
and periphery in early modern science."

Several authors in this anthology deal predominantly with the eighteenth century. Fiona Clark in Chapter 8 also focuses on an individual example, in this case, the Mexican José Antonio Alzate y Ramírez
(1737–1799), a priest with an interest in the new ideas, especially science.
From a position of what Clark calls "enlightened Catholicism," he was a
major figure in the colony's Republic of Letters. As editor of the *Gazeta
de Literatura de México* (1788–95), Alzate y Ramírez saw the transla-

tion and dissemination in Mexico of many European works, especially those of the French. In 1771 he became a corresponding member of the Paris Académie des Sciences (Academy of Sciences), the only Mexican to be so honored. His focus was on the publication of works in the fields of public health and medicine, weather, mining, and technology—largely coming from French sources but also from other European countries as well as Peruvian and Cuban periodicals.

The cataloguing of the flora and fauna of the Iberian worlds reached its apogee in the eighteenth century. Stimulated by the exemplary work of Linnaeus, scientific expeditions under state sponsorship were undertaken in various regions of the Americas and the Pacific. The results were spectacular. The contribution of José Celestino Mutis, as Daniela Bleichmar demonstrates in Chapter 15, is one of the best examples. Still a young man and trained in medicine, Mutis traveled to New Granada as part of the entourage of a new viceroy; he never returned to Spain. Between 1783 and 1810 he directed the Royal Botanical Expedition. The team he trained and led combined scientific description with careful artwork; the result was about 6,700 illustrations of plants and some 700 floral details. As the author points out, art and applied science were closely linked in the period. And here, too, a link exists between the utility that is the potential economic value for empire, and the eighteenth-century expeditions. There were many others, including the expedition to Chile and Peru under Hipólito Ruiz and José Pavón (1777–88), one of the few whose results were published in the period. Illustrating the flora and fauna of empire, as well as other human and natural resources, and the systematic collection of examples to be transported for study and display in the metropolis were elements in the process, as Paula De Vos documents in Chapter 14. It is difficult to say what motivated one of the more prolific figures—the bishop of the Peruvian city of Trujillo, Baltasar Jaime Martínez Compañón—but during his long tenure in the diocese, he amassed a dozen volumes of illustrations of the material, biological, and cultural artifacts of the district that were sent back to Spain. De Vos suggests that during both periods the reasons for collecting were power and prestige, "but at this time within the specific context of imperial rivalries during the Enlightenment and an emerging sense of imperial patriotism that played out in the field of natural history."

Notably absent in eighteenth-century Iberian America are works on political economy, for obvious reasons. Antonio de Nariño in Colombia would suffer the consequences for his dissemination of French political writings of the era. Discussion of nature, medicines, and precious resources was safe in all regions. And the various "scientific" societies established in various parts of the Americas, modeled on Iberia's examples, did similar work—notably in Cuba, Guatemala, Peru, and Nueva Granada. In Brazil, as Júnia Ferreira Furtado in Chapter 9 shows us, the physician José Rodrigues Abreu (1682–1755) lived in Minas Gerais and wrote four volumes on medicine, stressing empiricism and experimentation. Rodrigues Abreu also provided an imaginary geography of gold discoveries in Brazil, in which he discarded theories of the "old authorities" as he underscored the value of "seeing" over "hearsay."

Martha Few in Chapter 7 reminds us that, in the case of medicine, the church could still at times take umbrage. How would the institution react in eighteenth-century Guatemala, for example, if certain rituals were practiced? Could deformations of infants be caused by incantations or spells? This was an issue the Inquisition would be keen to investigate. What was the impact of the mixture of Amerindian, African, and European practices? Did it lead to medical *mestizaje*? In medicine, as elsewhere, there was a shift toward a more "modern" approach, but the direction was not consistently forward. In all areas in the century, the state was an interested party in developments. This is certainly true in cartography (as argued by Nuria Valverde and Antonio Lafuente in Chapter 10), where we see a flood of new maps being prepared, based on better measurements and close observation. Improved technology made the maps especially significant, and by the late eighteenth century they suddenly "acquired an enormous diplomatic value." The maps served to set the boundaries of empire, boundaries that were increasingly challenged in the eighteenth century.

At the beginning of this volume, the editors pose the question of why the contributions of the Iberian experience have been largely ignored by historians of science. The answer is not simple. For a variety of reasons, historians of science have focused on the discoveries of mostly "northern" Europeans. An obvious reason is nationalism; one of the traditional functions of history, many have argued, has been nation building. Contributing also is a shift of power away from the Iberian countries, very successful in the sixteenth century, yet falling far behind the countries north of the Pyrenees by the late seventeenth century. The actions of English, French, and Dutch explorers as they entered jurisdictions controlled by the Iberians, and came to carve out their own colonies, opened opportunities for them to play a similar role in the accumulation of knowledge. The activities and successes of the northern Europeans accelerated in the eighteenth century, in spite of efforts of the Spanish Bourbon monarchy to block their encroachments and reestablish authority. But neither Spanish nor Portuguese "enlightened" administrators were able to stem the tide. Even with important contributions to natural and medical science and cartography in the eighteenth century, Spain and Portugal in the nineteenth century were seriously challenged. The Napoleonic invasion of the Iberian Peninsula, the subsequent colonial rebellions, and the resulting internal instability left little energy for state sponsorship of major efforts. The rapid changes in science and technology associated with the industrial revolution were delayed in the Iberian world. The rise of modern historical scholarship in the late nineteenth century began in northern Europe, and part of these historians' interest was explication of the rise of science. In an age of nationalism, it should not be surprising that their own figures will be the first to be studied. More important, the widespread image of Spain and Portugal as backward and intolerant societies dominated by the Inquisition led to the belief that no significant scientific endeavors took place in the Iberian realms. Although much work, mostly written in Spanish and Portuguese, has been done in recent decades refuting this stereotype, it has not yet reached the wider academic audience. One of the many merits of this volume, in addition to the original work done by the contributors, is precisely the diffusion among English speakers of the most recent scholarship on science and medicine in Spain, Portugal, and Latin America.

Reference Matter

Notes

Preface

1. A recent publication by one of the volume's contributors addresses this problem at length: Antonio Barrera-Osorio, *Experiencing Nature: The Spanish American Empire and the Early Scientific Revolution* (Austin: University of Texas Press, 2006).

2. It is telling that many exceptions to this historiographical blind spot come from the contributors to this volume. Chapters 1 and 2 of the collection provide a thorough historiographical review of work on science in the Spanish and Portuguese empires.

3. See Roy MacLeod, "Introduction," in *Nature and Empire: Science and the Colonial Enterprise, Osiris* 15 (2000), 6; Shula Marks, "What Is Colonial about Colonial Medicine? And What Has Happened to Imperialism and Health?" Presidential Address, *Social History of Medicine* 10, no. 2 (1997), 205–19; and Michael Worboys, "Tropical Diseases," in *Companion Encyclopedia of the History of Medicine*, ed. W. F. Bynum and Roy Porter, 2 vols. (London and New York: Routledge, 1993), 1:22–44, among others decrying this scholarly omission.

4. See for instance Londa Schiebinger and Claudia Swan (eds.), *Colonial Botany: Science, Commerce, and Politics in the Early Modern World* (Philadelphia: University of Pennsylvania Press, 2005); Pamela Smith and Paula Findlen (eds.), *Merchants and Marvels: Commerce, Science, and Art in Early Modern Europe* (New York: Routledge, 2001); John Brewer and Roy Porter (eds.), *Consumption and the World of Goods* (London and New York: Routledge, 1993); and David Philip Miller and Peter Hans Reill (eds.), *Visions of Empire: Voyages, Botany and Representations of Nature* (Cambridge: Cambridge University Press, 1996).

Introduction

1. See also Jorge Cañizares-Esguerra, *Nature, Empire, and Nation* (Stanford, CA: Stanford University Press, 2006).

2. See, for example, Richard Westfall, *The Construction of Modern Science: Mechanisms and Mechanics* (New York: Cambridge University Press, 1977).

3. On these new practices and economies, see Amy Butler Greenfield, *A Perfect Red* (New York: HarperCollins, 2005); Londa Schiebinger and Claudia Swan (eds.), *Colonial Botany: Science, Commerce, and Politics in the Early Modern World* (Philadelphia: University of Pennsylvania Press, 2005); Richard Drayton, *Nature's Government* (Yale University Press, 2000); Susan Scott Parrish, *American Curiosity* (Chapel Hill: University of North Carolina Press, 2006).

4. Unfortunately, the only study on the impact of alchemy in the Atlantic mining economy is limited to New England; see William Newman, *Gehennical Fire* (Cambridge, MA: Harvard University Press, 1994).

5. On the Black Legend, see William Maltby, *The Black Legend in England* (Durham, NC: Duke University Press, 1971); Julián Juderías, *La leyenda negra estudios acerca del concepto de España en el extranjero* (1943; Barcelona: Araluce, 2003); Ricardo García Cárcel, *La leyenda negra* (Madrid: Alianza Editorial, 1992).

6. On Muñoz's archival research, see Jorge Cañizares-Esguerra, *How to Write the History of the New World* (Stanford, CA: Stanford University Press, 2001), chap. 3.

7. María M. Portuondo, "Secret Science: Spanish Cosmography and the New World, 1570–1611" (PhD diss., Johns Hopkins University, 2005).

8. Fernando Bouza, *Corre manuscrito* (Madrid: Marcial Pons, 2001).

9. Adrian Johns, *The Nature of the Book: Print and Knowledge in the Making* (Chicago: University of Chicago Press, 1998).

10. David Shields, *Civil Tongues and Polite Letters in British America* (Chapel Hill: University of North Carolina Press, 1997).

11. Various authors in this collection give widely different figures for the number of Spanish expeditions to the New World in the eighteenth century (De Vos, 27; Blechmar, 57; Goodman, 54). José de la Sota Ríus has identified 64 expeditions until 1807; see his "Spanish Science and Enlightenment Expeditions," in *Spain in the Age of Exploration, 1492–1819*, ed. Chiyo Ishikawa (Seattle, WA: Seattle Art Museum, 2004).

12. On the ways religious images worked in the Catholic world, see Mary Carruthers, *The Book of Memory* (Cambridge: Cambridge University Press, 1990); Jeffrey Chipps Smith, *Sensuous Worship* (Princeton, NJ: Princeton University Press, 2002); Jeffrey F. Hamburger, *The Visual and the Visionary: Art and Female Spirituality in Late Medieval Germany* (New York: Zone Books, 1998).

13. See, for example, Ramón Mujica Pinilla, "'Reading without a Book': On Sermons, Figurative Art, and Visual Culture in the Viceroyalty of Peru," in *The Virgin, Saints, and Angels: South American Paintings 1600–1825*, ed. Suzanne Stratton-Pruitt (Stanford, CA: Iris and Gerald Canto Center, 2006), 41–66.

14. For a lucid interpretation of the millenarian, artisan roots of these traditions and their impact among Atlantic Huguenots, see Neil Kamil, *Fortress of the Soul: Violence, Metaphysics, and Material Life in the Huguenots' New World, 1517–1751* (Baltimore: Johns Hopkins University Press, 2005). For the impact of these ideas among British colonists, see Newman, *Gehennical Fire*.

15. Adraina Romeiro, *Um visionário na corte de d. João V: Revolta e milenarismo nas Minas Gerais* (Belo Horizonte: Universidade Federal de Minas Gerais,

2001); Plínio Freire Gomes, *Um herege vai ao paraíso: Cosmologia de um ex-colono condenado pela Inquisicão (1680–1744)* (São Paulo: Companhia das Letras, 1997).

16. On this patristic and medieval tradition of prefigurative, typological readings, see Friedrich Ohly, *Sensus Spiritualis: Studies in Medieval Significs and the Philology of Culture* (Chicago: Chicago University Press, 2005).

17. For an interpretation of the Escorial as the fulfillment of the Temple of Solomon, see Rene Taylor, *Arquitectura y Magia* (Madrid: Ediciones Siruela, 1999).

Chapter One

1. Roger B. Merriman, *The Rise of the Spanish Empire in the Old World and in the New*, 4 vols. (New York: MacMillan, 1918–34), 1:vii.

2. Henry Kamen, *Spain's Road to Empire: The Making of a World Power, 1492–1763* (London: Allen Lane, 2002).

3. Failure to register this increase is an omission in the otherwise comprehensive source by Paula H. Covington et al. (eds.), *Latin America and the Caribbean: A Critical Guide to Research Sources* (Westport, CT: Greenwood Press, 1992). In its nine hundred pages, "science" is hardly ever mentioned. Another reference work does much better: Barbara Tenenbaum et al. (eds.), *Encyclopaedia of Latin American History and Culture*, 5 vols. (New York: Simon & Schuster, MacMillan, 1996). But for comprehensive coverage the indispensable guide is the outstanding *Handbook of Latin American Studies*, scanning publications in its annual volumes: vols. 1–13 (Cambridge, MA: Harvard University Press, 1935–51); vols. 14–40 (Gainesville: University of Florida Press, 1951–78); vols. 41– (Austin and London: University of Texas Press, 1979–). Currently the annual volumes are alternately devoted to Social Sciences and Humanities. For history of science, medicine, and technology, the biennial Humanities volume is the one to consult, using the index entries for "science," "scientists," "medicine," "medicinal plants," and "expeditions" and the invaluable introductory, editorial survey-essay titled "General History," as well as the brief introductory comments for each of the bibliographical sections devoted to the various regions of colonial Spanish America.

4. Elizabeth Hill Boone, "Introduction," *Native Traditions in the Postconquest World: A Symposium at Dumbarton Oaks, 2nd through 4th October 1992*, ed. Elizabeth Hill Boone and Tom Cummins (Washington, DC: Dumbarton Oaks Research Library, 1998), 1.

5. José Luis Peset and Tomás Gómez, introduction to papers presented at the Franco-Spanish conference "Ciencias y Técnicas en la América Española del Siglo XVIII," held in Madrid, November 1987, and published in *Asclepio* 39, no. 2 (1987). Gómez's sentiments and initiative, fulfilled in the conference, are recorded on pp. 5–7 and 9–11.

6. José Luis Peset (ed.), *Ciencia, vida y espacio en Iberoamérica*, 3 vols. (Madrid: CSIC, 1989), 1:ix–xii.

7. Ibid.

8. All this intellectual excitement in Spain, a full seven years before the advent of the quincentenary, is recorded in Peset's preface of January 1989 in José Luis

Peset (ed.), *Ciencia, vida y espacio en Iberoamérica*, 3 vols. (Madrid: CSIC, 1989), 1: ix–xii; this preface summarizes the fruit of the Consejo's "Mobilization Programme" of research.

9. Mariano Peset, "Prologo," *Doctores y escolares: II Congreso Internacional de Historia de las Universidades Hispánicas, Valencia, 1995*, 2 vols. (Valencia, Spain: Universitat de Valencia, 1998), 1:31.

10. María Luisa Rodríguez Sala (ed.), *El eclipse de luna: Misión científica de Felipe II en Nueva España* (Huelva, Spain: Universidad de Huelva, 1998).

11. María Isabel Vicente Maroto and Mariano Esteban Piñeiro, *Aspectos de la ciencia aplicada en la España del siglo de oro* (Salamanca, Spain: Junta de Castilla y León, 1991).

12. Jesús Bustamante, "El conocimiento como necesidad de estado: Las encuestas oficiales sobre Nueva España durante el reinado de Carlos V," *Revista de Indias* 60, no. 208 (2000), 33–55.

13. René Acuña (ed.), *Relaciones geográficas del siglo XVI*, 10 completed vols. (Mexico City: UNAM, 1982–).

14. Raquel Álvarez Peláez, *La conquista de la naturaleza americana* (Madrid: CSIC, 1993).

15. Francisco de Solano (ed.), *Cuestionarios para la formación de las relaciones geográficas de Indias: Siglos XVI/XIX* (Madrid: CSIC, 1988).

16. Ana Olivera, "Riesgo y salud en los cuestionarios americanos (Siglos XVI–XIX)," in ibid., lxv–lxxviii.

17. Pedro de Valencia, *Obras completas*, vol. 5, *Relaciones de Indias, 1: Nueva Granada y Virreinato de Perú*, ed. Francisco Javier Fernández and Jesús Fuente Fernández (León: University of León, 1993). For the linguistic analysis, see p. 53f; for the quotation of Valencia's regrets on Spaniards' neglect of Indian medicinal remedies, see p. 350.

18. José Pardo Tomás and María Luz López Terrada, *Las primeras noticias sobre plantas americanas en las relaciones de viajes y crónicas de Indias (1493–1553)* (Valencia: Universitat de València/CSIC, 1993).

19. José María López Piñero and José Pardo Tomás, *Nuevos materiales y noticias sobre la historia de las plantas de Nueva España de Francisco Hernández* (Valencia, Spain: Universitat de València/CSIC, 1994).

20. Carmen Benito-Vessels, "Hernández in Mexico: Exile and Censorship?" in *Searching for the Secrets of Nature: The Life and Works of Dr. Francisco Hernández*, ed. Simon Varey, Rafael Chabran, and Dora Weiner (Stanford, CA: Stanford University Press, 2001), 41–52.

21. Henry Kamen, *Spain in the Later Seventeenth Century, 1665–1700* (London: Longman, 1980).

22. John Lynch, *Bourbon Spain, 1700–1808* (Oxford: Blackwell, 1989), 247.

23. Ibid., 253.

24. Ibid., 336.

25. Ibid., 367.

26. Manuel Selles and Antonio Lafuente, "La formación de los pilotos en la España del siglo XVIII," in *La ciencia moderna y el nuevo mundo*, ed. José Luis Peset (Madrid: CSIC, 1985), 149–91; Selles and Lafuente, "Sabios para la

armada: El curso de estudios mayores de marina en la España del siglo XVIII," in Peset, *Ciencia, vida y espacio en Iberoamérica*, 3:485–504; Selles, "La astronomia náutica en la España ilustrada: El tratado de Mendoza y Ríos," *Asclepio* 39, no. 2 (1987), 33–47.

27. Dolores González-Ripoll Navarro, *A las órdenes de las estrellas: La vida del marino Cosme de Churruca y sus expediciones a América* (Madrid: CSIC, 1995); and Navarro, "La formación académica y práctica de los marinos del siglo XVIII: Cosme de Churruca (1761–1805), un oficial científico," in *La ciencia española en ultramar: Actas de las I Jornadas sobre "España y las expediciones científicas en América y Filipinas,"* ed. Alejandro Díez Torres et al. (Aranjuez, Spain: Doce Calles, 1991), 313–23.

28. Antonio Lafuente and José Luis Peset, "Las academias militares y la inversión en la España ilustrada (1750–1760)," *Dynamis* 2 (1982), 193–209.

29. Horacio Capel et al., *De Palas a Minerva: La formación científica y la estructura institucional de los ingenieros militares en el siglo XVIII* (Barcelona and Madrid: Serbal and CSIC, 1988).

30. Ibid., 315f.

31. J. Omar Moncada Maya, "Ciencia en acción: Ingeniero y ordenación del territorio en Nueva España en el siglo XVIII," in *Mundialización de la ciencia y la cultura nacional: Actas del Congreso Internacional "Ciencia, Descubrimiento y Mundo Colonial,"* ed. Antonio Lafuente, Arturo Elena, and María L. Ortega (Aranjuez, Spain: Doce Calles, 1993), 219–33.

32. Ángel Guirao de Vierna, "Análisis cuantitativo de las expediciones españolas con destino al Nuevo Mundo," in Peset, *Ciencia, vida y espacio en Iberoamérica*, 3:65–93.

33. María Luisa Martínez de Salinas Alonso, "Ciencia y real hacienda: Notas para un acercamiento a la financiación de las expediciones científicas," in Díez Torres et al., *La ciencia española en ultramar*, 197–207.

34. R. Rodríguez Nadal and A. González Bueno, "Las colonias al servicio de la ciencia metropolitana: La financiación de las 'Floras Americanas' (1791–1809)," *Revista de Indias* 55 (1995), 597–634.

35. Jean-Pierre Clément, "Réflexions sur la politique scientifique française vis-à-vis de l'Amérique espagnole au siècle des Lumières," in *Nouveau Monde et renouveau de l'histoire naturelle*, 3 vols., ed. Marie-Cécile Bénassy et al. (Paris: Université de la Sorbonne Nouvelle, 1986), 1:131–59.

36. Antonio Lafuente and Antonio Mazuecos, *Los caballeros del punto fijo: Ciencia, política y aventura en la expedición geodésica hispanofrancesca al virreinato del Perú en el siglo XVIII* (Barcelona and Madrid: Serbal and CSIC, 1987).

37. Antonio Lafuente, "Una ciencia para el estado: La expedición geodésica hispano-francesca al virreinato del Perú (1734–1743)," *Revista de Indias* 43 (1983), 542–629.

38. Juan Pimentel, *La física de la monarquía: Ciencia y política en el pensamiento colonial de Alejandro Malaspina (1754–1810)* (Aranjuez, Spain: Doce Calles, 1998).

39. María Higueras Rodríguez, *Catálogo crítico de la documentación de la*

Expedición Malaspina del Museo Naval, 3 vols. (Madrid: Museo Naval, 1985–1994).

40. R. Cerezo Martínez et al. (eds.), *La expedición Malaspina (1789–1794)*, vols. 1– (Madrid, 1987–). Nine volumes of a projected twelve have been published.

41. Andrés Galera Gómez, *La Ilustración española y el conocimiento del Nuevo Mundo: Las ciencias naturales en la Expedición Malaspina (1789–1794); La labor científica de Antonio de Pineda* (Madrid: CSIC, 1988).

42. Marcelo Frías Núñez, *Tras El Dorado vegetal: José Celestino Mutis y la Real Expedición Botánica del Nuevo Reino de Granada (1783–1808)* (Seville: Diputación Provincial de Sevilla, 1994).

43. For discussion of the artists and the operation of the studio created for the expedition, see the assessment by an art historian of the Universidad Complutense, Madrid: Carmen Sotos Serrano, "Aspectos artísticos de la Real Expedición Botánica de Nueva Granada," *Mutis y la Real Expedición Botánica del Nuevo Reyno de Granada*, 2 vols., ed. María Pilar de San Pío Aladrén (Barcelona: Villegas, 1992), 1:121–57.

44. D. A. Brading, "Bourbon Spain and Its American Empire," in *The Cambridge History of Latin America*, vol. 1, *Colonial Latin America*, ed. Leslie Bethell, 389–439 (Cambridge: Cambridge University Press, 1984), 420f.

45. Marie Helmer, "La mission Nordenflycht [sic] en Amérique espagnole (1788): Échec d'une technique nouvelle," *Asclepio* 39, no. 2 (1987), 123–44; John Fisher, "Tentativas de modernizar la tecnología minera en el virreinato del Perú: La mision minera de Nordenflicht (1788–1810)," in *Minería y metalurgia: Intercambio tecnológico y cultural entre América y Europa durante el período colonial español*, ed. M. Castillo Martos (Seville: Muñoz Moya y Montraveta, 1994); Carlos Contreras and Guillermo Mira, "Transferencia de tecnología minera de Europa a los Andes," in Lafuente et al., *Mundialización de la ciencia y la cultura nacional*, 235–49.

46. Serena Fernández Alonso, "Los mecenas de la plata: El respaldo de los virreyes a la actividad minera colonial en las primeras décadas del siglo XVIII; El gobierno del Marqués de Casa Concha en Huancavelica (1723–1726)," *Revista de Indias* 60 (2000), 345–71; Miguel Molina Martínez, *Antonio de Ulloa en Huancavelica* (Granada, Spain: University of Granada, 1995).

47. Carlos Sempat Assadourian, "La bomba de fuego de Newcomen y otros artificios de desagüe: Un intento de transferencia de tecnología inglesa a la minería novohispana, 1726–1731," *Historia Mexicana* 50 (2001), 385–457.

48. Robert C. West, "Early Silver Mining in New Spain, 1531–1555," in *In Quest of Mineral Wealth: Aboriginal and Colonial Mining and Metallurgy in Spanish America*, ed. Alan K. Craig and Robert C. West (Baton Rouge: Louisiana State University, 1995), 119–35; P. Bakewell, "Introduction," in *Mines of Silver and Gold in the Americas*, ed. P. Bakewell (Aldershot, England: Ashgate Variorum, 1997), xv–xvi, where Bakewell also suggests the value of traditional African metallurgical skills possessed by black slaves.

49. Peter Bakewell, *Miners of the Red Mountain: Indian Labor in Potosí, 1545–1650* (Albuquerque: University of New Mexico Press, 1984).

50. Jeffrey A. Cole, *The Potosí Mita, 1573–1700: Compulsory Indian labor in the Andes* (Stanford, CA: Stanford University Press, 1985).

51. Kendall W. Brown, "Workers' Health and Colonial Mercury Mining at Huancavelica, Peru," *The Americas* 57 (2001), 467–96. The stimulus of environmental concern is also evident in a recent volume of essays on the depletion of forest resources in colonial Spanish America: M. Lucena Giraldo (ed.), *El bosque ilustrado: Estudios sobre la política forestal española en América* (Madrid: Instituto Nacional para la Conservación de la Naturaleza, 1991).

52. José Sala Catala, "Vida y muerte en la mina de Huancavelica durante la primera mitad del siglo XVIII," *Asclepio* 39, no. 1 (1987), 193–204.

53. Linda A. Newson, *Life and Death in Early Colonial Ecuador* (Norman: University of Oklahoma Press, 1995).

54. Suzanne Austin Alchon, *Native Society and Disease in Colonial Ecuador* (Cambridge: Cambridge University Press, 1991).

55. Noble David Cook and W. George Lovell (eds.), *Secret Judgements of God: Old World Disease in Colonial Spanish America* (Norman and London: University of Oklahoma Press, 1991).

56. Marcelo Frías Nuñez, *Enfermedad y sociedad en la crisis colonial del antiguo régimen: Nueva Granada en el tránsito del siglo XVIII al XIX; las epidemias de viruelas* (Madrid: CSIC, 1992); Angela T. Thompson, "To Save the Children: Smallpox Innoculation, Vaccination and Public Health in Guanajuato, Mexico, 1797–1840," *The Americas* 49 (1993), 431–55; José G. Rigau-Pérez, "The Introduction of Smallpox Vaccine and the Adoption of Immunization as a Government Function in Puerto Rico," *Hispanic American Historical Review* 69 (1993), 393–423; Andrew L. Knaut, "Yellow Fever and the Late Colonial Public Health Response in the Port of Veracruz," *Hispanic American Historical Review* 77 (1997), 619–44; Martha Eugenia Rodríguez and Francisco Balbuena, "Las inhumaciones y la legislación sanitaria en la ciudad de México, siglo XVIII," in *Estudios de historia de la medicina: Abordajes e interpretaciones*, ed. Ana Rodríguez de Romo and Xóchitl Martínez Barbosa (Mexico City: UNAM, 2001), 89–98.

57. David Cahill, "Financing Health Care in the Viceroyalty of Peru: The Hospitals of Lima in the late colonial period," *The Americas* 52 (1995), 123–54.

58. Juan Marchena Fernández, *Oficiales y soldados en el Ejército de América* (Seville: Escuela de Estudios Hispano-Americanos de Sevilla, 1983); and Carmen Gómez Pérez, "Niveles sanitarios en la ciudad Americana del siglo XVIII: Las series de documentación militar," in Peset, *Ciencia, vida y espacio en Iberoamérica*, 1:31–52.

59. D. Soto Arango et al. (eds.), *Científicos criollos e ilustración* (Aranjuez, Spain: Doce Calles, 1999), 11.

60. Patricia Aceves, "La difusión de la química de Lavoisier en el Real Jardín Botánico de Madrid y en el Real Seminario de Minería (1788–1810)," *Quipu* 7 (1990), 5.

61. Juan José Saldaña (ed.), *Historia social de las ciencias en América Latina* (Mexico City: Miguel A. Porrúa, 1996).

62. Lafuente et al., *Mundialización de la ciencia y la cultura nacional*, 18.

63. Anthony Pagden, "Identity Formation in Spanish America," in *Colonial Identity in the Atlantic World, 1500–1800,* ed. Nicholas Canny and Anthony Pagden (Princeton, NJ: Princeton University Press, 1987), 51–93; D. A. Brading, *The First America: The Spanish Monarchy, Creole Patriots, and the Liberal State, 1492–1867* (Cambridge: Cambridge University Press, 1991).

64. For the case that illiterate Amerindians could not have devised humoral theory thus it must have been brought over by the Spaniards, see George M. Foster, "On the Origin of Humoral Medicine in Latin America," *Medical Anthropology Quarterly: New Series* 1 (1987), 355–93; for an indigenous origin, lost in the mists of time, see Ellen Messer, "The Hot and Cold in Mesoamerican Indigenous and Hispanicized Thought," *Social Science and Medicine* 25 (1987), 339–46. As for the mysterious *puquios,* the case for their construction by indigenous people around 500 CE is argued by Katherine Schreiber and Josué Lancho Rojas, "Los puquios de Nasca: Un sistema de galerías filtrantes," *Boletín de Lima* 59 (1988), 51–62. And for the case against their pre-Hispanic invention, see Monica Barnes and David Fleming, "Filtration-Gallery Irrigation in the Spanish New World," *Latin American Antiquity* 2, no. 1 (1991), 48–68.

65. Carlos González-Sánchez, "Los libros de españoles en el virreinato del Perú, siglos XVI y XVII," *Revista de Indias* 56 (1996), 7–47; Teodoro Hampe Martínez, *Bibliotecas privadas en el mundo colonial: La difusión de libros e idéas en el virreinato del Perú (siglos XVI–XVII)* (Frankfurt am Main and Madrid: Vervuert and Iberoamericana, 1996).

66. Renán Silva, *Los ilustrados de Nueva Granada 1760–1808: Genealogía de una comunidad de interpretación* (Medellín, Spain: Fondo Editorial Universidad EAFIT, 2002), 245f.

67. Margarita Menegus (ed.), *Universidad y sociedad en Hispanoamérica: Grupos de poder; siglos XVIII y XIX* (Mexico City: Universidad Autónoma de Mexico, 2001), publishing papers of a joint conference of the Universitat de València and Universidad Nacional Autónoma de México held in 1995.

68. Celina Lértora Mendoza, "La enseñanza de la física en el Río de la Plata: Tres ejemplos sobre la situación en el siglo XVIII," in *Claustros y estudiantes,* ed. M. Peset and S. Albiñana, 2 vols. (Valencia, Spain: Universidad de Valencia, 1989), 1:379–410.

69. Enrique González González, "El rechazo de la Universidad de México a las reformas ilustrados (1763–1777)," *Estudios de Historia Social y Económica de América* 7 (1991), 94–124.

70. Juan Bosco Amores Carredano, "La sociedad económica de la Havana y los intentos de reforma universitaria en Cuba (1793–1842)," *Estudios de Historia Social y Económica de América* 9 (1992), 369–80.

71. Renán Silva, *Universidad y sociedad en el Nuevo Reino de Granada: Contribución a un análisis histórico de la formación intelectual de la sociedad colombiana* (Santafé de Bogotá, Columbia: Banco de la República, 1992), 335.

72. Jean-Pierre Clément, *El mercurio peruano, 1790–1795,* 2 vols. (Frankfurt am Main and Madrid: Vervuert and Iberoamericana, 1997–98), 1:95.

73. Ibid., 1:227, 243, with subtle analysis of the meaning of "nation" on p. 227f. For Clément's argument that the long article in *El Mercurio Peruano* on

Peru's coca plant is an assertion of national identity, see his "La coca du Perou ou la passion botanique au XVIIIe siècle," in Bénassy et al., *Nouveau Monde et renouveau de l'histoire naturelle*, 1:65–84.

74. José Luis Peset, *Ciencia y libertad: El papel del científico ante la independencia americana* (Madrid: CSIC, 1987), 46f.

75. For example, see the analysis in John Lynch, "The Origins of Spanish American Independence," *The Cambridge History of Latin America*, vol. 2, *From Independence to c. 1870*, ed. Leslie Bethell (Cambridge: Cambridge University Press, 1985), 3–50, esp. 42f.

76. Thomas F. Glick, "Science and Independence in Latin America (with Special Reference to New Granada)," *Hispanic American Historical Review* 71 (1992), 307–34.

77. Rafael Sagrado Baeza, "Ciencia, viajes e independencia," *Memorias de la Academia Mexicana de la Historia* 37 (1994), 29–63.

78. Suzannne Austin Alchon, *Native Society and Disease in Colonial Ecuador* (Cambridge: Cambridge University Press, 1991), 124f.

79. Carlos Viesca Triviño, "El códice de la Cruz-Badiano, primer ejemplo de una medicina mestiza," in *El mestizaje cultural y la medicina novohispana del siglo XVI*, ed. J. L. Fresquet Febrer and J. M. López Piñero (Valencia, Spain: Instituto de Estudios Documentales e Históricos sobre la Ciencia, Universitat de València/CSIC, 1995), 71–90. This volume is dedicated to the elimination of "ethnocentrism," seen as "one of the most persistent obstacles to the study of history of medicine and science," ibid., 9.

80. Julia M. H. Smith, "Introduction. Regarding Medievalists: Contexts and Approaches," in *Companion to Historiography*, ed. Michael Bentley (London and New York: Routledge, 1997), 115, quoting Peter Novick, *That Noble Dream: "The Objectivity Question" and the American Historical Profession* (Cambridge: Cambridge University Press, 1988).

81. A. G. Hopkins, "Development and the Utopian Ideal, 1960–1999," in *The Oxford History of the British Empire*, vol. 5, *Historiography*, ed. Robin W. Winks (New York: Oxford University Press, 1999), 646.

82. Ibid., 649. See also, in the same volume, the penetrating assessment by D. A. Washbrook, "Orients and Occidents: Colonial Discourse Theory and the Historiography of the British Empire," 596–611.

83. Alan Knight, "Latin America," in *Companion to Historiography*, ed. Michael Bentley (London and New York: Routledge, 1997), 748.

84. Renán Silva, *Los ilustrados de Nueva Granada: 1760–1808: Genealogía de una comunidad de interpretación* (Medellín, Spain: Fondo Editorial Universidad EAFIT, 2002), 13.

85. John Lynch, *Latin America between Colony and Nation: Selected Essays* (Basingstoke and New York: Palgrave, 2001), 9–10.

86. Lyman L. Johnson and Susan M. Socolow, "Colonial History: Essay," in *Latin America and the Caribbean: A Critical Guide to Research Sources*, ed. Paula H. Covington et al. (Westport, CT: Greenwood Press, 1992), 322.

87. Elizabeth Hill Boone, "Introduction," in *Native Traditions in the Postconquest World: A Symposium at Dumbarton Oaks, 2nd through 4th October*

1992, ed. Elizabeth Hill Boone and Tom Cummins (Washington, DC: Dumbarton Oaks Research Library, 1998), 1–9.

88. Miguel León-Portilla, "Las comunidades mesoamericanas ante la institución de los hospitales para indios," *Boletín de la Sociedad Mexicana de Historia y Filosofía de la Medicina* 6 (1983), 193–217. But while postmodernist influence may have encouraged the author, his "reverse image" here might be independent of postmodernism because already in the late 1950s he had coined the phrase "vision of the vanquished" for the title of an anthology illustrating the conquest of Mexico from the viewpoint of the conquered.

89. Antonio Lafuente and Antonio Mazuecos, *Los caballeros del punto fijo: Ciencia, política y aventura en la expedición geodésica hispanofrancesca al virreinato del Perú en el siglo XVIII* (Barcelona and Madrid: Serbal and CSIC, 1987), 11; Antonio Mazuecos, "Conciencia de la expedición geodésica al reino de Quito," in Díez Torres et al., *La ciencia española en ultramar*, 124.

90. Juan Pimentel, *La física de la monarquía: Ciencia y política en el pensamiento colonial de Alejandro Malaspina (1754–1810)* (Aranjuez, Spain: Doce Calles, 1998), 34.

91. Fermín del Pino Diaz, "Por una antropología de la ciencia: Las expediciones ilustradas españolas como 'Potlatch' reales," in *Ciencia y contexto histórico nacional en las expediciones ilustradas a América*, ed. Fermín del Pino Diaz (Madrid: CSIC, 1988), 173–86. Del Pino's student applied the interpretation to particular expeditions: Fernando Monge Martínez, "La honra nacional en las expediciones de Cook y Malaspina: Una visión antropológica," in ibid., 187–98.

92. Batia B. Siebzehner, *La universidad americana y la ilustración: Autoridad y conocimiento en Nueva España y el Río de la Plata* (Madrid: Mapfre, 1994), 188–90.

93. Jorge Cañizares-Esguerra, *How to Write the History of the New World: Histories, Epistemologies and Identities in the Eighteenth-Century Atlantic World* (Stanford, CA: Stanford University Press, 2001), 9, 348.

94. Jeremy Black, *Visions of the World: A History of Maps* (London: Mitchell Beazley, 2003), 16, 19.

95. All of these papers are conveniently brought together in J. B. Harley, *The New Nature of Maps: Essays in the History of Cartography* (Baltimore and London: Johns Hopkins University Press, 2001).

96. Ibid., 35, 37.

97. Ibid., 158.

98. Ibid., 106.

99. Ibid., 63.

100. Ibid., 107.

101. Ibid., 53.

102. Ibid., 67.

103. Ibid., 105.

104. J. Brian Harley, "Rereading the Maps of the Columbian Encounter," *Annals of the Association of American Geographers* 82 (1992), 524, 526.

105. Barbara E. Mundy, *The Mapping of New Spain: Indigenous Cartography*

and the Maps of the Relaciones Geográficas (Chicago and London: University of Chicago Press, 1996), xix.
106. Ibid., 72.
107. Ibid., 213.
108. Ibid., 216.
109. Duccio Sacchi, *Mappe dal Nuovo Mondo: Cartografie locali e definizione del territorio in Nuova Spagna (secoli XVI–XVII)* (Milan: Franco Angeli, 1997), 165.
110. Ibid., 289.
111. Thomas F. Glick, "Imperio y dependencia científica en el siglo XVIII español e inglés: La provisión de los instrumentos científicos," in Peset, *Ciencia, vida y espacio en Iberoamérica*, 3:49–63. Glick here announces (p. 49) that his student, Marta Ardila, was preparing a comparative study of science in eighteenth-century Spanish and British American colonies, but I am unaware of any subsequent publication.

Chapter Two

1. The literature on the history of the Portuguese empire is immense. The following English-language indications are useful and easily accessible, although without any intention of their being complete or systematic. By Charles Ralph Boxer, these slightly dated but still very important works should be mentioned: *The Portuguese Seaborne Empire, 1415–1825* (London: Hutchinson, 1969); *The Golden Age of Brazil, 1695–1750* (Berkeley and Los Angeles: University of California Press, 1969); *The Christian Century in Japan, 1549–1650* (Berkeley and Los Angeles: University of California Press, 1967). More recent studies of great importance are the following: Bailey W. Diffie and George D. Winnius, *Foundations of the Portuguese Empire, 1415–1580* (Minneapolis: University of Minnesota Press, 1977); A. J. R. Russell-Wood, *A World on the Move: The Portuguese in Africa, Asia, and America, 1415–1808* (Manchester: Carcanet Press, 1992); Sanjay Subrahmanyam, *The Portuguese Empire in Asia, 1500–1700* (London: Longman, 1993).
2. The subject is too vast to be summarized here. The following bibliographies confirm this assessment and provide extensive lists of relevant works: Alfredo Pinheiro Marques, *Guia de história dos descobrimentos e expansão portuguesa* (Lisbon: Biblioteca Nacional, 1987); *Repertório bibliográfico da historiografia portuguesa, 1974–1994* (Coimbra: Instituto Camões, Faculdade de Letras de Coimbra, 1995); Artur Teodoro de Matos and Luis Filipe F. Reis Thomaz (eds.), *Vinte anos de historiografia ultramarina portuguesa, 1972–1992* (Lisbon: Comissão Nacional para as Comemorações dos Descobrimentos Portugueses, 1993).
3. There is no overall view of the history of science in Portugal, but the following bibliography can be very helpful: Rómulo de Carvalho, "Bibliografia das obras de autores nacionais publicadas durante o século XX que se ocupam das actividades científica e técnica dos portugueses nos séculos anteriores," in *História e desenvolvimento da ciência em Portugal (séc. XX)*, vol. 3 (Lisbon:

Academia das Ciências de Lisboa, 1992), 1781–1922. Reprinted in *Actividades científicas em Portugal no século XVIII* (Évora: Universidade de Évora, 1996), 683–840. The most recent production on the history of science is listed in Conceição Tavares and Henrique Leitão, *Bibliografia de história da ciência em Portugal, 2000–2004* (Lisbon: Centro de História das Ciências da Universidade de Lisboa, 2006).

4. For an overview of the development of this field of study, see the essay by Luís de Albuquerque, "Historiografia sobre a náutica portuguesa dos descobrimentos," in *A historiografia portuguesa de Herculano a 1950: Actas do colóquio* (Lisbon: Academia Portuguesa da História, 1978), 357–69.

5. Henrique Lopes de Mendonça, *Estudos sobre navios portugueses dos séculos XV e XVI*, 2nd ed. (1892; Lisbon: Ministério da Marinha, 1971); and Francisco A. Marques de Sousa Viterbo, *Trabalhos náuticos dos portugueses nos séculos XVI e XVII* (1896, 1900; Lisbon: Impresa Nacional Casa da Moeda, 1988).

6. Such is the case, for example, with the wild exaggerations around the so-called "School of Sagres." For important analysis of the origin and repercussion of this myth, see W. G. L. Randles, "The Alleged Nautical School Founded in the Fifteenth Century at Sagres by Prince Henry of Portugal, Called the 'Navigator,'" *Imago Mundi* 45 (1993), 20–28. See also Francisco Contente Domingues, "Horizontes mentais dos homens do mar no século XVI: A arte náutica portuguesa e a ciência moderna," in *Viagens e viajantes no Atlântico quinhentista: Primeiras jornadas de história ibero-americana*, ed. Maria da Graça M. Ventura (Lisbon: Edições Colibri, 1996), 203–18.

7. Armando Cortesão and Avelino Teixeira da Mota, *Portugaliae Monumenta Cartographica*, ed. Alfredo Pinheiro Marques, 6 vols. (Lisbon: Imprensa Nacional-Casa de Moeda, 1988). See the extensive bibliographies of Albuquerque in Alfredo Pinheiro Marques, *Luís de Albuquerque na historiografia portuguesa: A serenidade e a convicção* (Coimbra and Figueira da Foz: Centro de Estudos do Mar, 1998).

8. See for example, Reijer Hooykaas, "Science in Manueline Style: The Historical Context of D. João de Castro's Works," in *Obras completas de D. João de Castro*, 4 vols. 1968–1982, ed. Armando Cortesão and Luís de Albuquerque (Coimbra: Academia Internacional da Cultura Portuguesa, 1982), 4:231–426; Reijer Hooykaas, "The Portuguese Discoveries and the Rise of Modern Science," *Boletim da Academia Internacional da Cultura Portuguesa* 2 (1966), 87–107 [republished in *Selected Studies in the History of Science* (Coimbra: Acta Universitatis Conimbrigensis, 1983), 579–98]. Hooykaas was not the only one to call attention to these facts. Another expert was David Waters; see "Portuguese Nautical Science and the Origins of the Scientific Revolution," *Boletim da Academia Internacional da Cultura Portuguesa* 2 (1966), 165–91.

9. The *cosmógrafo-mor* and the problem of nautical teaching in general are analyzed in A. Teixeira da Mota, "Os regimentos do cosmógrafo-mor de 1559 e 1592 e as origens do ensino náutico em Portugal," *Memórias da Academia das Ciências de Lisboa (Classe de Ciências)* 13 (1969), 227–91. See also Nuno Valdez dos Santos, *Setecentos anos de estudos navais em Portugal* (Lisbon: Academia de Marinha, 1985).

10. Luís Manuel Ribeiro Saraiva, "A Companhia de Jesus e os historiadores da matemática portuguesa," in Nuno da Silva Gonçalves (ed.), *A Companhia de Jesus e a missionação no Oriente: Actas do Colóquio Internacional, 21–23 Abril 1997* (Lisbon: Brotéria, Fundação Oriente, 2000), 311–30.

11. António Leite, "Pombal e o ensino secundário," in *Como Interpretar Pombal? No bicentenário da sua morte*, ed. Manuel Antunes et al. (Lisbon: Brotéria, 1983), 165–81. The essential reference for any study of the Jesuits in Portugal is still the massive though somewhat dated multivolume work by the Jesuit historian Francisco Rodrigues, *História da Companhia de Jesus na assistência de Portugal*, 4 vols. in 7 bks. (Porto: Apostolado da Imprensa, 1931–1950). A recent, and excellent, contribution is Dauril Alden, *The Making of an Enterprise: The Society of Jesus in Portugal, Its Empire, and Beyond, 1540–1750* (Stanford, CA: Stanford University Press, 1996).

12. See Gian Paolo Brizzi (ed.), *La Ratio Studiorum: Modelli culturali e pratiche educative dei Gesuiti in Italia tra cinque e seicento* (Rome: Bulzoni, 1981); Frederick A. Homann (ed.), *Church, Culture and Curriculum: Theology and Mathematics in the Ratio Studiorum* (Philadelphia: Saint Joseph's University Press, 1999).

13. The pioneer study about this Jesuit mathematical course is Luís de Albuquerque, "A 'Aula da Esfera' do Colégio de Santo Antão no século XVII," *Anais da Academia Portuguesa de História* 21 (1972), 337–91. This is today largely superseded by the works of Ugo Baldini. See, by Baldini: "As assistências ibéricas da Companhia de Jesus e a actividade científica nas missões asiáticas (1578–1640): Alguns aspectos culturais e institucionais," *Revista Portuguesa de Filosofia* 54 (1998), 195–245; "The Portuguese Assistancy of the Society of Jesus and Scientific Activities in Its Asian Missions until 1640," in *História das ciências matemáticas: Portugal e o Oriente* (Lisbon: Fundação Oriente, 2000), 49–104; "L'insegnamento della matematica nel Collegio di S. Antão a Lisbona, 1590–1640," in *A Companhia de Jesus e a missionação do Oriente* (Lisbon: Fundação Oriente/Brotéria, 2000), 275–310; "The Teaching of Mathematics in the Jesuit Colleges of Portugal from 1640 to Pombal," in *The Practice of Mathematics in Portugal: Papers from the International Meeting Organized by the Portuguese Mathematical Society, Óbidos, 16–18 November, 2000*, ed. Luís Saraiva and Henrique Leitão (Coimbra: Imprensa da Universidade de Coimbra; Acta Universitatis Conimbrigensis, 2004), 293–465.

14. See references in previous note.

15. Resina Rodrigues, "Física e filosofia da natureza na obra de Inácio Monteiro," in *História e desenvolvimento da ciência em Portugal (até ao Século XX)* (Lisbon: Academia das Ciências de Lisboa, 1986), 1:191–242; Ana Isabel Rosendo, "Inácio Monteiro e o ensino da matemática em Portugal no século XVIII" (master's thesis, Universidade do Minho, Braga, 1996); Ana Isabel Rosendo, "O compendio dos elementos de mathematica do P. Inácio Monteiro," *Revista Portuguesa de Filosofia* 54 (1998), 319–53.

16. João Manuel S. A. Miranda, "Alguns aspectos do intercâmbio científico e cultural entre a Academia das Ciências de Petersburgo e a comunidade dos

"Jesuítas Matemáticos" em Pequim nas décadas de 30–50 do século XVIII," in Silva Gonçalves, *A Companhia de Jesus e a missionação no Oriente,* 331–64.

17. Henrique Leitão, "Os primeiros telescópios em Portugal," in *Actas do 1º Congresso Luso-Brasileiro de História da Ciência e da Técnica* (Évora, Portugal: Universidade de Évora, 2001), 107–18; Henrique Leitão, "Jesuit Mathematical Practice in Portugal, 1540–1759," in *The New Science and Jesuit Science: Seventeenth Century Perspectives,* ed. Mordechai Feingold (Dordrecht, Netherlands: Kluwer, 2003), 229–47.

18. Luís Saraiva (ed.), *History of Mathematical Sciences: Portugal and East Asia,* vol. 1 (Lisbon: Fundação Oriente, 2000) and vol. 2 (Lisbon: EMAF-UL, 2001); Luís Miguel Carolino and Carlos Ziller Camenietzki (eds.), *Jesuítas, ensino e ciência* (Casal de Cambra, Portugal: Caleidoscópio, 2005).

19. Juan Casanovas and Philip C. Keenan, "The Observations of Comets by Valentine Stansel, a Seventeenth Century Missionary in Brazil," *Archivum Romanum Societatis Iesu* 62 (1993), 319–30 and three works by Carlos Ziller Camenietzki: "O cometa, o pregador e o cientista: António Vieira e Valentim Stansel observam o céu da Bahia no século XVII," *Revista da Sociedade Brasileira de História da Ciência* 14 (1995), 37–52; "Savants du bout du monde: Les Jesuites astronomes de Salvador," in "Symposium: Mission et diffusion des sciences européennes en Amérique et en Asie—Le cas Jésuite (XVIe–XVIIe siècles)" (21st International Congress on the History of Science and Technology, Mexico City, 2001), *Archives Internationales d'Histoire des Sciences* 52, no. 148 (2002), 147–58; and "The Celestial Pilgrimages of Valentin Stansel (1621–1705), Jesuit Astronomer and Missionary in Brazil," in *The New Science and Jesuit Science: Seventeenth Century Perspectives,* ed. Mordechai Feingold (Dordrecht, Netherlands: Kluwer, 2003), 249–70.

20. For exemple, see the classical works: Serafim Leite, "Diogo Soares SI, matemático, astrónomo e geógrafo de Sua Majestade no estado do Brasil (1684–1748)," *Brotéria* 45 (1947), 596–604 and Jaime Cortesão, "A missão dos padres matemáticos no Brasil," *Studia* 1 (1958), 123–50. Or the more recent studies: André Ferrand de Almeida, *A formação do espaço brasileiro e o projecto do novo atlas da América portuguesa (1713–1748)* (Lisbon: Comissão Nacional para as Comemorações dos Descobrimentos Portugueses, 2001); Manuel Fernandes Thomaz and Isabel Malaquias, "Aspectos científicos das expedições de demarcação de limites na América meridional," in *Actas do 1º Congresso Luso-Brasileiro de História da Ciência e da Técnica* (Évora, Portugal: CEHFC-Universidade de Évora, 2001), 201–13.

21. The standard history of the Jesuits in Brazil is still a good source of information: Serafim Leite, *História da Companhia de Jesus no Brasil,* 10 vols. (Lisbon and Rio de Janeiro: Portugália/INL, 1938–1950) [reprinted in 4 vols. (São Paulo: Edições Loyola, 2004)]. More recent studies are the following: Paulo de Assunção, *A terra dos Brasis: A natureza da América portuguesa vista pelos primeiros jesuítas (1549–1596)* (São Paulo: Annablume, 2001); Fernando Santiago dos Santos, "Os Jesuítas, os indígenas e as plantas brasileiras: Considerações preliminares sobre a triaga brasílica" (master's thesis, Pontifícia Universidade Católica de São Paulo, 2003).

22. Bernardino António Gomes, *Elogio histórico do Pe. João de Loureiro* (Lisbon: Typographia da Aacdemia Real das Sciencias de Lisboa, 1865); E. D. Merrill, "A Commentary on Loureiro's 'Flora cochinchinensis,'" *Transactions of the American Philosophical Society* 2, no. 24 (1935), 1–13, 19–23, 28–29, 33–35, 38–49.

23. Carlos França, "Os portugueses do século XVI e a história natural do Brasil," *Revista de História* 15 (1926), 3–119.

24. Dorotheus Schilling, *Os Portugueses e a introdução da medicina no Japão* (Coimbra: Instituto Alemão da Universidade de Coimbra, 1931); Arlindo Camilo Monteiro, *De l'influence portuguaise au Japon* (Lisbon: Seara Nova, 1934); Luís de Pina, *Evangelização e medicina portuguesa no Japão quinhentista* (Coimbra: Tipografia Gráfica de Coimbra, 1950); Charles Boxer, "A Note on the Interaction of Portuguese and Chinese Medicine in Macao and Peking (16th–18th Centuries)," in *Medicine and Society in China*, ed. J. Z. Bowers and E. F. Purcell (New York: Josiah Macy Jr. Foundation, 1974), 22–39.

25. For example, the actions of Portuguese astronomers at the court of Jai Singh. See A. Delduque da Costa, "Os padres matemáticos no observatório de Jaipur," *Oriente Português* 4 (1932), 58–64; Amândio Gracias, "Uma embaixada científica portuguesa à corte dum rei indiano no século XVIII," *Oriente Português* 19–21 (1938), 187–202; G. Moraes, "Astronomical Missions to the Court of Jaipur, 1730–1743," *Journal of Bombay Royal Asiatic Society* 27 (1951), 61–65; Virendra N. Sharma, "Jai Singh, His European Astronomers, and the Copernican Revolution," *Indian Journal of the History of Science* 17 (1982), 333–44; Virendra N. Sharma and Lila Huberty, "Jesuit Astronomers in Eighteenth Century India," *Archives Internationales d'Histoire des Sciences* 34 (1984), 99–107.

26. See for example the work by Ines Zupanov, *Disputed Mission: Jesuit Experiments and Brahmanical Knowledge in Seventeenth-Century South India* (New Delhi and New York: Oxford University Press, 1999); M. N. Pearson, "Hindu Medical Practice in Sixteenth-Century Western India: Evidence from Portuguese Sources," *Portuguese Studies* 17, no. 1 (2001), 100–13; A. Salema (ed.), *Ayurveda at the Crossroads of Care and Cure: Proceedings of the Indo-European Seminar on Ayurveda Held at Arrabida, Portugal, in November 2001* (Lisbon and Pune, India: Centro de História de Além-Mar, Universidade Nova de Lisboa, 2002); I. Zupanov, "Drugs, Health, Bodies and Souls in the Tropics: Medical Experiments in Sixteenth-Century Portuguese India," *Indian Economic and History Review* 39, no. 1 (2002), 1–43.

27. See José Lopes Dias, "O Renascimento em Amato Lusitano e Garcia de Orta," *Estudos Castelo Branco* 4 (1964); Charles Boxer, *Two Pioneers of Tropical Medicine: Garcia d'Orta and Nicolás Monardes* (London: Hispanic and Luso Brazilian Councils, 1963); R. N. Kapil and A. K. Bhatnagar, "Portuguese Contributions to Indian Botany," *Isis* 67 (1976), 449–52; Luís Filipe Barreto, *Garcia de Orta e o diálogo civilizacional* (Lisbon: Instituto de Investigação Científica e Tropical, 1983).

28. Carolus Clusius, *Aromatum et simplicium aliquot medicamentorum apud indos nascentium historia* (Antwerp, Belgium, 1567), 4.

29. A detailed comparative study of de Orta's *Colloquies* and Clusius's *Aromatum et simplicium* is still a desideratum.

30. A second edition was printed in 1872 edited by Adolfo de Varnhagen, followed by another in 1891 (first volume) and 1892 (second volume). This was commented and edited by Conde de Ficalho and become the standard edition of the *Colloquies*. Conde de Ficalho was also the author of the first biography on the eminent Portuguese physician and naturalist: *Garcia de Orta e o seu tempo* (Lisbon: Imprensa Nacional, 1886). A more recent biography is provided in Jaime Walter, *Garcia da Orta: Relance da sua vida* (Lisbon, 1963).

31. Luís de Pina, *Contribuição dos portugueses quinhentistas para a história da medicina do Oriente: Nota preliminar* (Lisbon: Sociedade Nacional de Tipografia, 1938), 295.

32. Carlos França "Os portugueses da renascença, a medicina tropical e a parasitologia," *O Instituto* 73 (1926), 1–18.

33. Luís de Pina, *As ciências na história do império colonial português, séculos XV a XIX* (Porto: Imprensa Portuguesa, 1945), 4.

34. Jorge Canizares-Esguerra, "Iberian Science in the Renaissance: Ignored How Much Longer?" *Perspectives on Science* 12, no. 1 (2004), 86–124.

35. In his *In Discoridis Anazarbei de medica materia libros quinque enarrationes* (1553), Amato Lusitano presents the description of new materia medica from the Orient. Unlike Garcia de Orte, he had not visited or lived in territories from the New World. He obtained plant specimens and information from Portuguese and Venetian navigators. See José Lopes Dias, *Comentários ao "Index Discorides" de Amato Lusitano* (Castelo Branco, Portugal: Gráfica de S. José, 1968).

36. Carlos França, "Os portugueses do século XVI e a História Natural do Brasil," *Revista de História* 15 (1926), 3–119; Luís de Pina, *Contribuição dos portugueses quinhentistas para a história da medicina do Oriente: Nota preliminar* (Lisbon: Sociedade Nacional de Tipografia, 1938); Luís de Pina, *Flora e fauna brasílicas nos antigos livros médicos portugueses* (Coimbra: Coimbra Editora, 1944). A more recent study on the multidirectional flow and acclimatizing of plants is José Mendes Ferrão, *A aventura das plantas e os descobrimentos portugueses* (Macau: Comissão Territorial de Macau para as Comemorações dos Descobrimentos Portugueses, 1996).

37. Alberto C. Correia, *O ensino de medicina e cirurgia em Goa nos séculos XVII, XVIII e XIX: História do ensino médico-cirúrgico no Hospital Real de Goa, antes da fundação da Escola Médico-Cirúrgica de Nova-Goa* (Bastorá: Tipografia Rangel, 1941); Luís de Pina, *Expansão hospitalar portuguesa ultramarina: Séculos XVI e XVII* (Porto: Tipografia Porto Médico, 1943). More recent contributions on charitable institutions include Laurinda Abreu, "O papel das Misericórdias dos 'lugares de além-mar' na formação do império português," *História, Ciências, Saúde—Manguinhos* 8, no. 3 (2001), 591–611 and Isabel Sá, "Shaping Social Space in the Centre and Periphery of the Portuguese Empire: The Example of the Misericordias from the Sixteenth to the Eighteenth Centuries," *Portuguese Studies* 13 (1997), 210–21.

38. William J. Simon, *Scientific Expeditions in the Portuguese Overseas Ter-*

ritories (1783–1808): The Role of Lisbon in the Intellectual Scientific Community of the Late Eighteenth Century (Lisbon: Instituto de Investigação Científica Tropical, 1983). A general survey of the topic is provided in David M. Knight, "Travels and Science in Brazil," *História, Ciências, Saúde—Manguinhos* 8, supplement (2001), 809–22.

39. R. Carvalho, *A história natural em Portugal no século XVIII* (Lisbon: Biblioteca Breve, 1987), 39–62.

40. See José Luís Cardoso, "From Natural History to Political Economy: The Enlightened Mission of Domenico Vandelli in Eighteenth-Century Portugal," *Studies in the History and Philosophy of Science* 34 (2003), 781–803.

41. See João Carlos P. Brigola, *Colecções, gabinetes e museus em Portugal no século XVIII* (Lisbon: Fundação Calouste Gulbenkian, Fundação para a Ciência e a Tecnologia, 2003).

42. José Luís Cardoso, "Introdução," in *Memórias económicas da Academia Real das Ciências de Lisboa, para o adiantamento das artes, e da indústria em Portugal, e suas conquistas (1789–1815)*, ed. J. L. Cardoso (Lisbon: Banco de Portugal, 1990–91), xvii–xxxiii; Oswaldo Munteal Filho, "Todo um mundo a reformar: Intelectuais, cultura ilustrada e estabelecimentos científicos ilustrados em Portugal e no Brasil, 1779–1880," *Anais do Museu Histórico Nacional* 29 (1997), 87–108.

43. On the practice of science within a military context, see João Carlos P. Brigola, *Professores da Academia Real de Marinha (1801–1837): Militares, cientistas e políticos* (Lisbon: Academia de Marinha, 1993); José Luís Assis, "A militarização da ciência nas viagens de exploração científica no séc. XVII," *Revista Militar* 12 (2000), 1107–22; Beatriz Siqueira Bueno, "Desenho e desígnio— o Brasil dos engenheiros militares," *Oceanos* 41 (2000), 40–58; Margarida Tavares da Conceição, "A praça de guerra: Aprendizagem entre a Aula do Paço e a Aula de Fortificação," *Oceanos* 41 (2000), 24–38; Silvino da Cruz Curado, "Contributo dos engenheiros militares para a estruturação do Brasil na segunda metade do século XVIII," in *Actas: IX colóquio dos militares e a sociedade portuguesa* (Lisbon: Comissão Portuguesa de História Militar, 2000), 159–75.

44. The bibliography on this subject includes Carlos França, *Doutor Alexandre Rodrigues Ferreira (1756–1815): História de uma missão científica ao Brasil no século XVIII* (Coimbra: Imprensa da Universidade, 1922); Arthur C. F. Reis, "Um cientista luso-brasileiro na identificação da Amazônia," *Boletim da Sociedade de Geografia de Lisboa* 9 (1972), 175–87; Napoleão Figueiredo, *Alexandre Rodrigues Ferreira naturalista da Amazônia no século XVIII: In memoriam de Cristóvão Santos* (Braga, Portugal: Livraria Cruz, 1982); Osvaldo Rodrigues da Cunha, *O naturalista Alexandre Rodrigues Ferreira: Uma análise comparativa de sua viagem filosófica (1783–1793) pela Amazônia e Mato Grosso com a de outros naturalistas posteriores* (Belém, Brazil: Museu Paraense Emílio Goeldi, 1991); Russell Mittermeier, *Philosophical Journey: A Rediscovery of the Amazon, 1792–1992* (Rio de Janeiro: Index, 1992).

45. Ângela Domingues, *Viagens de exploração geográfica na Amazônia em finais do século XVIII: Política, ciência e aventura* (Funchal, Portugal: Secretaria Regional do Turismo, Cultura e Emigração; Centro de Estudos de História

do Atlântico, 1991). On cartographic expeditions, see also André Ferrand de Almeida, *A formação do espaço brasileiro e o projecto do novo atlas da América portuguesa (1713–1748)* (Lisbon: Comissão Nacional para as Comemorações dos Descobrimentos Portugueses, 2001); André Ferrand de Almeida, "Entre a guerra e a diplomacia: Os conflitos luso-espanhois e a cartografia da América do Sul (1702–1807)," in *A Nova Lusitânia*, ed. João Carlos Garcia (Lisbon: Comissão Nacional para as Comemorações dos Descobrimentos Portugueses, 2001). *Imagens cartográficas do Brasil nas colecções da Biblioteca Nacional (1700–1822)*, Catálogo bibliográfico e Ilustrações (Lisbon: Comissão Nacional para as Comemorações dos Descobrimentos Portugueses, 2001), 37–65.

46. Miguel Ferreira de Faria, *Imagem útil: José Joaquim Freire (1760–1847), desenhador topográfico e de história natural* (Lisbon: Universidade Autónoma de Lisboa, 2001).

47. Contributions to these studies include also Ronald Raminelli, "Do conhecimento físico e moral dos povos: Iconografia e taxonomia na viagem filosófica de Alexandre Rodrigues Ferreira," *História, Ciências, Saúde—Manguinhos* 8, supplement (2001), 969–92; and Ermelinda Moutinho Pataca, "A confecção de desenhos de peixes oceânicos das 'viagens philosophicas' (1783) ao Pará e à Angola," *História, Ciências, Saúde—Manguinhos* 10, no. 3 (2003), 979–91.

48. Silvia Figueiroa and Clarete da Silva, "Enlightened Mineralogists: Mining Knowledge in Colonial Brazil," *Osiris* 15 (2000), 174–89. See also Silvia Figueiroa, Clarete da Silva, and Ermelinda Moutinho, "Aspectos mineralógicos das 'Viagens Filosóficas' pelo território," *História, Ciências, Saúde—Manguinhos* 11, no. 3 (2004), 713–29.

49. Clarete Paranhos da Silva, *O desvendar do grande livro da natureza: Um estudo do mineralogista José Vieira Couto, 1798–1805* (São Paulo: Fapesp/AnnaBlume/Unicamp, 2002).

50. Maria Margaret Lopes, Clarete Paranhos da Silva, Silvia Fernanda de M. Figueirôa, and Rachel Pinheiro, "Scientific Culture and Mineralogical Sciences in the Luso-Brazilian Empire: The Work of João da Silva Feijó (1760–1824) in Ceará," *Science in Context* 18, no. 2 (2005), 201–24.

51. Janet Browne, "Natural History Collecting and the Biogeographical Tradition," *História, Ciências, Saúde—Manguinhos* 8, supplement (2001), 959–67.

52. Lycurgo Santos Filho, *História geral da medicina brasileira* (São Paulo: Editora da Universidade de São Paulo, 1991).

53. Márcia Moisés Ribeiro, *A ciência dos trópicos: A arte médica no Brasil do século XVIII* (São Paulo: Editora HUCITEC, 1997).

54. Vera Regina Beltrão Marques, *Natureza em Boiões: Medicina e boticários no Brasil setecentista* (Campinas, Brazil: Editora da UNICAMP, 2000).

55. On the membership and activities of this Academy, see Augusto da Silva Carvalho, *As academias científicas do Brasil no século XVIII* (Lisbon: Ottosgráfica, 1939).

56. Ângela Domingues, "Para um melhor conhecimento dos domínios coloniais: A constituição de redes de informação no império português em finais de setecentos," *História, Ciências, Saúde—Manguinhos* 8, supplement (2001), 823–38.

57. On the problem of the diffusion of natural historical information in relation to the Portuguese empire, see also Lorelai Kury, "Homens de ciência no Brasil: Impérios coloniais e circulação de informações," *História, Ciências, Saúde—Manguinhos* 11, supplement (2004), 109–29.

Chapter Three

1. Víctor Navarro Brotóns (ed.), *Jerónimo Muñoz: Introducción a la astronomía y la geografía*, Colleción Oberta (Valencia, Spain: Consell Valencià de Cultura, 2004).

2. See also Antonio Barrera-Osorio, *Experiencing Nature: The Spanish American Empire and the Early Scientific Revolution* (Austin: University of Texas Press, 2006).

3. After the first printed Latin edition of 1475, the *Geography* became the model for most subsequent cosmographical works. For an introduction to Renaissance cosmography, see J. Lennart Berggren and Alexander Jones, eds. and trans., *Ptolemy's Geography: An Annotated Translation of the Theoretical Chapters* (Princeton, NJ: Princeton University Press, 2000). The study of Renaissance cosmography has traditionally focused on the study of maps, while the textual component—the descriptive geography—has been somewhat neglected. Anthony Grafton's survey of early modern texts about the discovery of the New World is a valuable introduction to the field. Grafton, *New World, Ancient Texts* (Cambridge, MA: Belknap Press, 1992).

4. William B. Ashworth, "Natural History and the Emblematic World View," in *Reappraisals of the Scientific Revolution*, ed. David Lindberg and Robert Westman (Cambridge: Cambridge University Press, 1990), 301–32.

5. José Pulido Rubio, *El piloto mayor de la Casa de la Contratación de Sevilla* (Seville: Tipografía Zarzuela, Teniente Borges 7, 1950), 979–83. For a comprehensive sampling of current scholarship about the Casa de la Contratación, see Antonio Acosta Rodríguez, Adolfo Luis González Rodríguez, and Enriqueta Vila Vilar (eds.), *La Casa de la Contratación y la navegación entre España y las Indias* (Seville: Universidad de Sevilla; Consejo Superior de Investigaciones Científicas; Fundación El Monte, 2003).

6. For an in-depth study of the political and scientific implications of the Treaty, see Luis Antonio Ribot García (ed.), *El Tratado de Tordesillas y su época*, 2 vols. (Madrid: Junta de Castilla y León, 1995); and Eduardo Trueba and José Llavador, "Geografía conflictiva en la expanción maritima Luso-española, siglo XVI," *Revista de Historia Naval* 15, no. 58 (1997).

7. For a comprehensive survey of navigation techniques used during the voyages of discovery, see Luis de Albuquerque, *Astronomical Navigation* (Lisbon: Comissão Nacional para as Comemorações dos Descobrimentos Portugueses, 1988); and E. G. R. Taylor, *The Haven-Finding Art: A History of Navigation from Odysseus to Captain Cook* (London: Hollis & Carter, 1956).

8. Alison Sandman, "Cosmographers vs. Pilots" (PhD diss., University of Wisconsin, 2001), 283–88. Ursula Lamb explored this topic earlier in "Science

by Litigation: A Cosmographic Feud," *Cosmographers and Pilots of the Spanish Maritime Empire* (Aldershot, England: Ashgate Variorum, 1995), 40–57.

9. "Nombramiento de Jerónimo de Chaves como cosmógrafo y catedrático de cosmografía de la Casa de la Contratación" (5 December 1552), Archivo General de Indias (AGI), Contratación 5784, libro 1, fol. 95–95v. For examples of the kind of examinations given to prospective pilots, see "De la Cosmografía: Examen de pilotos," in *Colección de documentos inéditos relativos al descubrimiento, conquista y organización de las antiguas posesiones españolas de ultramar*, ed. Real Academia de la Historia, 25 vols. (1885–1932; Nendeln, Liechtenstein: Kraus, 1967), vol. 25, chap. 15.

10. When traveling from east to west as in a route similar to the one taken by Spanish galleons sailing to the Indies, the compass needle deviates from North in a manner that suggests the deviation is proportional to longitudinal distance. This apparent correlation prompted generations of Spanish cosmographers to attempt to build an instrument that used the phenomenon to determine longitude at sea. For an early discussion, see Martín Fernández de Navarrete, *Disertación sobre la historia de la náutica*, ed. Carlos Seco Serrano, Biblioteca de Autores Españoles, vol. 77 (Madrid: Ediciones Atlas, 1954–55). Many of the original instrument proposals presented to the Council of Indies are in AGI, Patronato 262.

11. Marie Ange Etayo-Piñol, "Medina y Cortés o el aprendizaje de las técnicas de navegación en Europa en el siglo XV," *Revista de Historia Naval* 16, no. 64 (1998), 43n3.

12. Martín Cortés, *Breve compendio de la sphera y de la arte de navegar* (Seville: Casa Antón Álvarez, 1551), fol. 68.

13. Pablo Emilio Pérez-Mallaína Bueno, *Spain's Men of the Sea: Daily Life on the Indies Fleets in the Sixteenth Century*, trans. Carla Rahn Phillips (Baltimore: Johns Hopkins University Press, 1998), 83–92.

14. For a biographical study, see José Pulido Rubio, *El piloto mayor de la Casa de la Contratación de Sevilla* (Seville: Tipografía Zarzuela, Teniente Borges 7, 1950, 1950), 639–711 and Mariano Esteban Piñeiro, "Los cosmógrafos al servicio de Felipe II," *Mare Liberum* 10 (1995), 525–39.

15. Josep Lluís Barona and Xavier Gómez i Font (eds.), *La correspondencia de Carolus Clusius con los científicos españoles* (Valencia: Seminari d'estudis sobre la ciència, 1998), 76–78, 125–26.

16. In the late eighteenth century, Fernández de Navarrete cited two published navigation maps: Carta de Marear (Seville, 1579 and 1588); and Martín Fernández de Navarrete, *Biblioteca marítima española*, 2 vols. (Madrid: Viuda de Calero, 1851; repr., New York: Burt Franklin, 1968), 686.

17. "Título de catedrático de la cátedra de la ciencia de cosmografia de la ciudad de Sevilla para Rodrigo Zamorano por tiempo de cinco años por muerte de Lic. Ruiz" (20 November 1575), Biblioteca Colombina y Capitular (BCC), vol. 38, fol. 296. The appointment was renewed in 1579 and became permanent in 1584 (ibid., fols. 293, 294).

18. "Rodrigo Zamorano cosmógrafo y catedrático de la Casa de la Contratación de Sevilla sobre se le acreciente el salario que tiene" (14 May 1582), AGI, Patronato 262, ramo 11, fol. 1v.

19. Zamorano's book was later translated into English by Edward Wright in *Certaine Errors in Navigation* (London, 1610).

20. The first was a three-minute error in the declination of the sun during the equinoxes. The tables also corrected the values of the sun's elevation during the solstices and gave new times for the sun's entry into the equinoxes. Rodrigo Zamorano, "Al lector," *Compendio de la arte de navegar* (Seville, 1581).

21. Antonello Gerbi, *Nature in the New World: From Christopher Columbus to Gonzalo Fernández de Oviedo*, trans. Jeremy Moyle (Pittsburgh: University of Pittsburgh Press, 1985), 226–31. For a detailed study of Oviedo's utilitarian interest, see José Pardo Tomás and María Luz López Terrada, *Las primeras noticias sobre plantas americanas en las relaciones de viajes y crónicas de Indias, 1493–1553* (Valencia: Instituto de Estudios Documentales e Históricos sobre la Ciencia, Universitat de València, CSIC, 1993), 86–97.

22. In 1536, Santa Cruz presented to the Casa de la Contratación an instrument that relied on the effect of magnetic variation on the compass needle to determine longitude, as well as a map showing these relative to longitude. Mariano Cuesta Domingo (ed.), *Alonso de Santa Cruz y su obra cosmográfica*, 2 vols. (Madrid: CSIC, 1983), 1:62.

23. Archivo General de Simancas (AGS), Estado 121, fols. 1–22.

24. "The Book of Longitudes," in Cuesta Domingo, *Alonso de Santa Cruz*, 1:139–202. Archived at the Biblioteca Nacional de España, ms. 9441.

25. Alonso de Santa Cruz, *Islario general*, in ibid., 1:330.

26. The extent of Santa Cruz's scientific work would have remained unknown to modern scholars but for the fortuitous inventory of an old leather trunk. The inventory lists over 338 maps, including two atlases: a lost map atlas, and the *Islario general* with 120 maps. The trunk also contained chronicles, geographic descriptions, and several cosmographical and astrological treatises. These documents became the nucleus of a group of reference material passed from cosmographers and chroniclers of the Council of Indies to their successors well into the seventeenth century. "Minuta del inventario de los papeles de la antigua gobernación de Nueva España y Perú, que quedaron por muerte de Alonso de Santa Cruz, cosmógrafo de Su Majestad" (12 October 1572), AGI, Patronato 171, no. 1, ramo 16, fols. 1–10v.

27. As cited in M. I. Vicente Maroto, "Alonso de Santa Cruz e el oficio de cosmógrafo mayor de Consejo de Indias," *Mare Liberum* 10 (1995), 517.

28. Consejo de Indias, *Ordenanzas reales del Consejo de Indias: Gobernación y estado temporal* (Madrid: Casa de Francisco Sanchez, 1585). For some studies on the laws, see Ismael Sánchez Bella, *Dos estudios sobre el código de Ovando* (Pamplona, Spain: Universidad de Navarra, 1987); and Juan Manzano Manzano, *Historia de las recopilaciones de Indias*, 2 vols. (Madrid: Ediciones Cultura Hispánica, 1950).

29. The *Geografía* was first edited in 1894 by Justo Zaragoza. It was published again with a comprehensive introductory study in Juan López de Velasco, *Geografía y descripción universal de las Indias*, Biblioteca de Autores Españoles, vol. 248 (Madrid: Ediciones Atlas, 1971). Recent studies include Walter D. Mignolo, *The Darker Side of the Renaissance: Literacy, Territoriality, and*

Colonization (Ann Arbor: University of Michigan Press, 1995), 243–55; Jean-Pierre Berthe, "Juan López de Velasco (ca. 1530–1598) cronista y cosmógrafo mayor del Consejo de Indias: Su personalidad y su obra geográfica," *Relaciones* 19, no. 75 (1998), 141–72; and Ricardo Padrón, *The Spacious Word: Cartography, Literature and Empire in Early Modern Spain* (Chicago: University of Chicago Press, 2004).

30. López de Velasco was probably imitating a similar questionnaire used in Castile. For more on the history of questionnaires in Spain as tools to collect information from newly discovered territories, see Raquel Álvarez Peláez, "Etnografía e historia natural en los cuestionarios oficiales del siglo XVI," *Asclepio* 41, no. 2 (1989), 103–26 and Francisco de Solano (ed.), *Cuestionarios para la formación de las relaciones geográficas de Indias, siglos XVI–XIX* (Madrid: CSIC, 1988).

31. Some of the more comprehensive collections of *relaciones* include Marcos Jiménez de la Espada, *Relaciones geográficas de Indias: Perú*, Biblioteca de Autores Españoles, vols. 183–185 (Madrid: Real Academia Española, 1965); René Acuña (ed.), *Relaciones geográficas del siglo XVI*, 10 vols. (Mexico City: UNAM, 1982); Pilar Ponce Leiva (ed.), *Relaciones histórico-geográficas de la audencia de Quito: Siglos XVI–XIX*, 2 vols. (Quito: Abya-Yala, 1991–92); Mercedes de la Garza (ed.), *Relaciones histórico-geográficas de la gobernación de Yucatán* (Mexico City: Universidad Autónoma Nacional de México, 1983); Antonio Arellano Moreno (ed.), *Relaciones geográficas de Venezuela* (Caracas: Academia Nacional de la Historia, 1964); Francisco del Paso y Troncoso (ed.), *Relaciones geográficas de la diócesis de Michoacán, 1579–1580*, 2 vols. (Guadalajara, 1958). An inventory of replies was published in Howard F. Cline, "*Relaciones geográficas*: Revised and Augmented Census of *Relaciones geográficas* of New Spain, 1579–1585," in *Handbook of Middle American Indians* 12 (Washington, DC: Library of Congress, 1966).

32. Barbara M. Mundy, *The Mapping of New Spain: Indigenous Cartography and the Maps of the Relaciones Geográficas* (Chicago: University of Chicago Press, 1996); and Raquel Álvarez Peláez, *La conquista de la naturaleza americana* (Madrid: CSIC, 1993).

33. Five instructions were sent to the Indies, in 1577–78, 1581, 1582, 1584, and 1588. The theory behind using eclipses to calculate longitude had been known since antiquity. The lunar eclipse serves as a global synchronizing event. Observers on one part of the globe record the time of the eclipse and compare it to the time observed elsewhere. The difference in time corresponds to the difference in longitude, since one hour equals fifteen degrees. This simple measurement, however, was impossible at the time given the inaccuracy of contemporary timepieces.

34. López de Velasco's eclipse project has been studied in Clinton R. Edwards, "Mapping by Questionnaire and Early Spanish Attempts to Determine New World Geographical Positions," *Imago Mundi* 23 (1969), 17–28; María Luisa Rodríguez-Sala, *El eclipse de Luna: Mision científica de Felipe II en Nueva España* (Huelva, Spain: Universidad de Huelva, 1998); and María M. Portuondo,

"Secret Science: Spanish Cosmography and the New World" (PhD diss., Johns Hopkins University, 2005).

35. María Isabel Vicente Maroto and Mariano Esteban Piñeiro, *Aspectos de la ciencia aplicada en la España de la Edad de Oro*, 2nd ed. (Valladolid, Mexico: Server-Cuesta, 2006), 70; and Catherine Wilkinson-Zerner, *Juan de Herrera: Architect to Philip II of Spain* (New Haven, CT: Yale University Press, 1993), 14.

36. Nicolas García Tapia and María Isabel Vicente Maroto, "Juan de Herrera, un científico en la corte española," in *Instrumentos científicos del siglo XV: La corte española y la Escuela de Lovaina* (Madrid: Fundación Carlos de Amberes, 1997), 42, 45.

37. Francisco J. Sánchez Cantón, *La libreria de Juan de Herrera* (Madrid: CSIC, 1941).

38. For more on Hernández, see Simon Varey, et al. (eds.), *Searching for the Secrets of Nature: The Life and Worlds of Dr. Francisco Hernández* (Stanford, CA: Stanford University Press, 2001); Jesús Bustamante García, "De la naturaleza y los naturales americanos en el siglo XVI: Algunas cuestiones críticas sobre la obra de Francisco Hernández," *Revista de Indias* 52 (1992), 297–328; José María López Piñero and José Pardo Tomás, *Nuevos materials y noticias sobre la historia de las plantas de Nueva España de Francisco Hernández* (Valencia, Spain: Universitat de València: CSIC, 1994); and from the same authors, *La influencia de Francisco Hernández (1515–1587) en la consititución de la botánica y la material médica moderna* (Valencia: CSIC, 1996).

39. "Relación y apuntamiento enviado al Consejo de Indias sobre lo que debe hacer el cosmógrafo Jaime Juan" (2 December 1582), AGS, GA-155, fols. 150–151v; reprinted in Vicente Maroto and Esteban Piñeiro, *Aspectos*, 420–23.

40. AGI, *Mapas y planos*, Mexico 34. The viceroy of New Spain urged Juan to send duplicates of his observations to the council before departing for the Philippines: "Carta del virrey Arzobispo Pedro de Moya y Contreras al rey . . . trata de tablas de Francisco Domínguez y eclipse" (Mexico, 22 January 1585), AGI, Mexico 336B, ramo 4, no. 176.

41. Historians Esteban Piñeiro and Vicente Maroto have investigated the origins and evolution of this institution in great detail and maintain its principal objective was to train mathematical practitioners, including cosmographers. See their *Aspectos*, 63–219.

42. Juan de Herrera, *Institución de la Academia Real Mathemática: Edición y estudios preliminares de José Simón Díaz y Luis Cervera Vera* (1584; Madrid: Instituto de Estudios Madrileños, 1995), 24.

43. Ibid., 13.

44. Vicente Maroto and Esteban Piñeiro, *Aspectos*, 74.

45. Alfonso de Ceballos-Escalera y Gila, "Una navegación de Acapulco a Manila en 1611: El cosmógrafo mayor Juan Bautista Labaña, el inventor Luis de Fonseca Coutinho, y el problema de la desviación de la aguja," *Revista de Historia Naval* 17, no. 65 (1999), 7–42.

46. Vicente Maroto and Esteban Piñeiro, *Aspectos*, 172–73.

Chapter Four

1. This is practically a generalized rule in the Anglo-American historiography of science. The quasi-sole exception is H. Floris Cohen, *The Scientific Revolution: A Historiographical Inquiry* (Chicago: University of Chicago Press, 1994). Cohen was a student of Hooykaas, also a lonely figure in the history of science in the English-speaking world, and one of the only scholars to pay attention to the Portuguese case. Hooykaas knew the Portuguese language, and that made a difference. Another author familiar with the Portuguese navigations was W. G. L. Randles, although he paid little attention to the issues here raised. See Randles, *The Unmaking of the Medieval Christian Cosmos, 1500–1760: From Solid Heavens to Boundless Aether* (Aldershot, England: Ashgate, 1999). Alfred W. Crosby was familiar with the Portuguese side of the maritime explorations but he also does not seem aware of what was going on in Portugal during the period he covered in *The Measure of Reality: Quantification and Western Society, 1250–1600* (Cambridge: Cambridge University Press, 1997). Another example, out of many, of authors who make claims that reflect lack of awareness of prior advancements in particular areas, and who deem these advancements as novel in the authors they study, is Anthony Grafton (with April Shelford and Nancy Siraisi), *New Worlds, Ancient Texts: The Power of Tradition and the Shock of Discovery* (Cambridge, MA: Belknap Press of Harvard University Press, 1992).

2. One specific example could be Toby E. Huff, *The Rise of Early Modern Science: Islam, China, and the West* (Cambridge: Cambridge University Press, 1993).

3. Francis Bacon, *Advancement of Learning; Novum Organum; New Atlantis* (Chicago: Encyclopaedia Britannica, 1952).

4. I have experienced students interpreting *Novum Organum* as a synthesis year after year in a course I teach at Brown on science during the period of Portuguese maritime discoveries, titled "On the Dawn of Modernity." *Novum Organum* is the last book read in the course after a full semester of readings, mostly written during the sixteenth century in Portugal.

5. Ibid., 121.

6. Ibid., 135.

7. Unfortunately, many of the works produced by these travelers are not available in English. See, however, Duarte Pacheco Pereira, *Esmeraldo de situ orbis*, vol. 79, ed. and trans. George H. T. Kimble (London: Hakluyt Society, 1937).

8. Nunes's famous work is *Coloquios dos simples e drogas he cousas medicinais da India*, published in Goa in 1563. English edition: *Colloquies of the Simples and Drugs of India* (London: Henry Sotheran & Co., 1913).

9. The best work on the concept of experience in Fernando Oliveira's work is Francisco Contente Domingues, "Experiência e conhecimento na construção naval portuguesa do século XVI: Os tratados de Fernando Oliveira," *Revista da Universidade de Coimbra* 33 (1985), 339–64. See also by the same author, "Science and Technology in Portuguese Navigation: The Idea of Experience in the Sixteenth Century," in *Portuguese Oceanic Expansion, 1400–1800*, ed. Fran-

cisco Bethencourt and Diogo Ramada Curto (Cambridge: Cambridge University Press, 2007), 460–79.

10. Francisco Sanches is the only one of these authors who did not participate directly in the process of the discoveries. Having emigrated very young with his parents, he spent most of his life in France, where he taught philosophy. I have attempted to demonstrate that his then-famous book *Quod nihil scitur,* a treatise attempting to show that we can never be sure of our knowledge, was in part prompted by the revolution in knowledge that the overseas discoveries brought about in the European mind. See O. T. Almeida, "Francisco Sanches: O 'elo perdido' entre os descobrimentos e a ciência moderna," *Cultura—Revista de História e Teoria das Ideias* 12, 2nd series (Spring 2001), 221–29. For an English translation of Sanches's text, see *That Nothing Is Known,* Introduction, notes, and bibliography by Elaine Limbrick; Latin text established, annotated, and translated by Douglas F. S. Thomson (Cambridge: Cambridge University Press, 1988).

11. Pereira, *Esmeraldo de situ orbis,* 134.

12. Ibid., 135.

13. Onésimo T. Almeida, "Portugal and the Dawn of Modern Science," in *Portugal, the Pathfinder: Journeys from the Medieval toward the Modern World, 1300–ca. 1600,* ed. George D. Winius (Madison, WI: Hispanic Seminary of Medieval Studies, 1995), 341–61.

14. See my "R. Hooykaas and His 'Science in Manueline Style': The Place of the Works of D. João de Castro in the History of Science," *Ibero-Americana Pragensia* 31 (1998), 95–101.

15. The best study of the works of D. João de Castro and their significance in the transition of the Medieval to the Modern European mind was done by the Dutch historian of science R. Hooykaas. Unfortunately it was published as an appendix in the Portuguese edition of Castro's collected works by Armando Cortesão and Luís de Albuquerque (eds.), *Obras Completas de D. João de Castro,* 4 vols. 1968–1982 (Coimbra: Academia Internacional de Cultura Portuguesa, 1982), vol. 4. This explains why such a superb piece of scholarship may have received such little attention.

16. Ernan McMullin is the author of a considerable body of work in the areas of history and philosophy of the sciences. Among his most notable books are *Newton on Matter and Activity* (Notre Dame, IN: University of Notre Dame Press, 1978) and the coordination of volumes such as *Galileo, Man of Science* (New York: Basic Books, 1968) and *Construction and Constraint: The Shaping of Scientific Rationality* (Notre Dame, IN: University of Notre Dame Press, 1988).

17. These quotes will be long, but they are important in that they allow me to more rigorously support the argument made in this chapter. As far as I know, they are from works that have never been translated from Portuguese into any another language, this being the first time they appear in English.

18. Prince D. Luís was the brother of King D. João III and Pedro Nunes's most beloved student.

19. Armando Cortesão and Luís de Albuquerque, _Obras Completas de D. João de Castro_, 4 vols. 1968–1981 (Coimbra: Academia Internacional de Cultura Portuguesa, 1968), 1:127–28.

20. Ibid., 129.

21. Ibid., 129–30.

22. Ibid., see p. 130.

23. Ibid., 140–41.

24. Ibid., 154.

25. Ibid., 181–84.

26. Ibid., 268.

27. Ibid., 190. Cortesão and Albuquerque note that the model of the instrument invented by Pedro Nunes, utilized by D. João de Castro, had been constructed by João Gonçalves (Ibid., 190n129).

28. Ibid., 202.

29. Ibid., and Cortesão and de Albuquerque, _Obras Completas de D. João de Castro_, 4 vols. 1968–1981 (Coimbra: Academia Internacional de Cultura Portuguesa, 1976), 3:42.

30. Ibid., 45.

31. See Luís de Albuqueruqe, "Sobre as prioridades de Pedro Nunes," in _Estudos de História de Ciência Náutica_, ed. Maria Emília Madeira Santos (Lisbon: Centro de Estudos de História de Cartografia Antiga, 1994), 595.

32. After a judicious analysis, Albuquerque concludes that, "in any case, we can say that the instrument and the process divulge much more due to the teachings of Pedro Nunes and the diligence of D. João de Castro in testing them, than does the treatise of Francisco Faleiro, which exerted little or no influence on the nautical science of the 16th century." (595)

33. Pedro Nunes, _"Tratado sobre certas duvidas"_ (1537), in A. Fontura da Costa (ed.), _A marinharia dos descobrimentos_ (Lisbon: Agência Geral do Ultramar, 1960), 219–20.

34. My transcription of the facsimile edition of the original of the work of Pedro Nunes inserted in Joaquim Bensaúde, _Opera omnia_, vol. 5 (Lisbon: Academia Portuguesa da História, 1995), 231.

35. Ibid.

36. Ibid., 215s.

37. I did not have access to this text, and could only resort to Francisco Teixeira Gomes' excellent analysis of it, and of all the works of Pedro Nunes in his _História das Matemáticas em Portugal_ (Lisboa: Academia das Ciências, 1934). The fact that Pedro Nunes published a work in Latin does not necessarily mean that it was not intended for sailors. Rather, it shows that the author meant to reach a wider public ignorant of the Portuguese language and that still used Latin as the lingua franca.

38. Ibid., 111–12. The impossibility of access to _De arte atque ratione navigandi_ of Pedro Nunes obliged us to make secondhand use of this long quote.

39. Ibid., 231.

40. Karl Popper writes in *Conjectures and Refutations: The Growth of Scientific Knowledge* (New York and Evanston: Harper Torchbooks, 1968), vii:

> The way in which knowledge progresses, and especially our scientific knowledge, is by unjustified (and unjustifiable) anticipations, by guesses, by tentative solutions to our problems, by *conjectures.* These conjectures are controlled by criticism; that is, by attempted *refutations,* which include severely critical tests . . . Criticism to our conjectures is of decisive importance: by bringing out our mistakes it makes us understand the difficulties of the problem which we were trying to solve.

41. Ernan McMullin, "The Goals of Natural Science," *Proceedings and Addresses of the American Philosophical Association* 58, no. 1 (September 1984), 54. Science-D is demonstrative science, which is different from science-P, or prediction.

42. We could criticize McMullin for not having been concerned with testing some of his assertions, but Aristotle was in fact an experimentalist, and the intensity of his constant preoccupation with the most minute details of the natural order should amaze us.

43. Ibid., 55.

44. Ibid.

45. McMullin reacts against the historians of science who assert that the term cannot be used appropriately before Newton and Galileo.

46. Ibid., 53.

47. Ibid., 54.

48. John R. C. Martyn (ed.), *Pedro Nunes (1502–1578): His Lost Algebra and Other Discoveries* (New York: Peter Lang, 1996), 8. Pedro Nunes, like all of the other figures connected with science in the period of the Portuguese discoveries, is rarely referred to in Anglo-Saxon historiography. Even in books where it would be natural for him to appear—for example, Alfred W. Crosby, *The Measure of Reality: Quantification and Western Society, 1250–1600* (Cambridge: Cambridge University Press, 1997)—he is alluded to only once (p. 236). In Salomon Bochner, *The Role of Mathematics in the Rise of Science* (Princeton, NJ: Princeton University Press, 1966), as in so many others, there is no reference made to him whatsoever. Curiously, it is in a book not strictly about the history of science—Patricia Seed, *Ceremonies of Possession in Europe's Conquest of the New World, 1492–1640* (New York: Cambridge University Press, 1995)—that an entire chapter is dedicated to the scientific activity of the Portuguese navigators, with Pedro Nunes repeatedly mentioned (see Chapter 4, "A New Sky and New Stars: Arabic and Hebrew Science, Portuguese Seamanship, and the Discovery of America"). In this chapter—just as in her later "Jewish Scientists and the Origin of Modern Navigation," in *The Jews and the Expansion of Europe to the West, 1450–1800,* ed. Paolo Bernardini and Norman Fiering (New York: Berghahan Books, 2001), 73–85—Seed stresses the Jewish mark of all the scientific activity realized in Portugal during the period in question, disregarding the work of key figures such as Duarte Pacheco Pereira or D. João de Castro.

49. I wish to thank Alexandra Montague for translating parts of this paper from the Portuguese original, as well as Robert Newcomb for his editorial help.

Chapter Five

1. Lorraine Daston, "The Nature of Nature in Early Modern Europe," in *The Scientific Revolution as Narrative*, ed. S. Harris and M. Biagioli, reprinted in *Configurations* 6, no. 2 (1998), 26–41. This essay begins by criticizing the "preformationist" view embedded in the term "early modern." Daston argues that the very term "early modern" implies that the past contains, prefigures, and announces the present.

2. The two terms "science" and "baroque" seem to hold some sort of congenital antipathy. The literature on baroque science is scarce, and it is limited to a few cases like Athanasius Kircher, Olaus Rudbeck, or Carlos Sigüenza. Apart from this, it is worth pointing out to English-language readers that "still lifes" can be expressed in Spanish either as *bodegones* or as *naturalezas muertas* (dead natures). In both cases time and changes in nature are evoked, as is the play between permanence and the expiration of things. My use of the term "baroque natures" seeks to evoke this analogy between the work of the painter of natures that are dead or about to die, and the historian of cultures that have been buried or have yet to be rescued.

3. The term "baroque" comes originally from the field of art history and was coined a century ago by Heinrich Wölfflin and Werner Weisbach, who used it in opposition to classicism and in its relation to the Counter-Reformation, respectively. Since then, the concept has colonized many other fields, but history of science has remained notably resistant to it.

4. See J. Simón Díaz, *El Colegio Imperial de Madrid*, 2 vols. (Madrid: Instituto de Estudios Madrileños, 1952); Víctor Navarro Brotóns, "El Colegio Imperial de Madrid: El Colegio de San Telmo de Sevilla," in *Historia de la ciencia y la técnica en la Corona de Castilla*, vol. 3, ed. J. M. López Piñero (Salamanca: Junta de Castilla y León, 2002), 53–81.

5. For the educational and scientific context of Madrid and its peculiar role as the center of operations of the court and Hispanic monarchy, see Antonio Lafuente and Javier Moscoso (eds.), *Madrid, ciencia y corte* (Madrid: Gráficas Palermo, 1999).

6. H. Didier, *Vida y pensamiento de Juan E. Nieremberg* (Madrid: Universidad Pontificia de Salamanca, 1976), 40–71.

7. P. Alonso de Andrade, *Varones ilustres de la Compañía de Jesús*, vol. 5, *Vida del muy espiritual y erudito P. Juan Eusebio Nieremberg* (Madrid, 1667).

8. See E. Zepeda-Henriquez, *Obras escogidas del R. P. Juan Eusebio Nieremberg*, vol. 2, *De la diferencia entre lo temporal y lo eterno: Crisol de desengaños con la memoria de la eternidad, postrimerías humanas y principales misterios divinos*, Biblioteca de Autores Españoles (Madrid: Atlas, 1957), 1–291.

9. Ibid., 44.

10. The reference is of course to *La vida es sueño* ("Life Is a Dream"), first performed in 1635, the year in which Lope de Vega died and Nieremberg published his *Historia naturae, maxime peregrinae*.

11. See Andrade, *Varones ilustres de la Compañía de Jesús*, vol. 5 and Didier, *Vida y pensamiento de Juan E. Nieremberg*, 40–71.

12. For recent work on Nieremberg, see R. Álvarez, "La historia natural de los animales," in *Historia de la ciencia y la técnica en la Corona de Castilla*, vol. 3, ed. J. M. López Piñero (Salamanca: Junta de Castilla y León, 2002), 573–95 and, above all, Luis Millones and Domingo Ledezma (eds.), *El saber de los jesuitas, historias naturales y el Nuevo Mundo* (Frankfurt-Madrid: Vervuert-Iberoamericana, 2005). This book contains two chapters directly related to the theme of my essay: L. Millones, "La intelligentsia jesuita y la naturaleza del Nuevo Mundo" (pp. 27–53), and D. Ledezma, "Una legitimación imaginativa del Nuevo Mundo: La *Historia naturae, maxime peregrinae* del jesuita Juan Eusebio Nieremberg" (pp. 53–85). In general terms, the argument of my essay follows the line of Ledezma's article.

13. See Bruno Latour and Steve Woolgar, *Laboratory Life: The Social Construction of Scientific Facts* (Princeton, NJ: Princeton University Press, 1986).

14. Juan Eusebio Nieremberg, *Curiosa y oculta filosofía: Primera y segunda parte de las maravillas de la naturaleza, examinadas en varias questiones naturales* (Alcalá, Spain: Imprenta de María Fernández, 1649). The following words from the cover page are worth reproducing here: "Containing very notable histories. Secrets and problems of nature are revealed with new Philosophy. Thorny parts of Scripture are explained. A very useful work, not only for the curious, but for learned jurists, philosophers, and doctors."

15. Juan Eusebio Nieremberg, *Historia naturae, maxime peregrinae* (Amberes: Imprenta palatina de Baltasar Moreto, 1635).

16. Lorraine Daston and Katharine Park, *Wonders and the Order of Nature, 1150–1750* (New York: Zone Books, 2001), 126.

17. G. Schott, *Physica curiosa sive mirabilia naturae et artis* (Herbipoli: Sumptibus Haeredum, 1662); J. Jonstonius, *Thaumatographia naturalis* (Amsterdam: Blaeu, 1632); P. Mexía, *Silva de varia lección* (Valladolid: Imprenta de Juan Villaquirán, 1550–51); A. de Torquemada, *Jardín de flores curiosas* (Salamanca: Imprenta de Juan Bautista de Terranova, 1570).

18. G. Correa, "El conceptismo sagrado en Alonso de Ledesma," in *Thesaurus* (Boletín del Instituto Caro y Cuervo) 30, no. 1 (1975), 4–80.

19. Nieremberg, *Historia naturae*, 1:2.

20. F. Rodríguez de la Flor, *Barroco: Representación e ideología en el mundo hispánico (1580–1680)* (Madrid: Cátedra, 2002), 240.

21. Nieremberg, *Curiosa y oculta filosofía*, 1:2.

22. Michel Foucault, *Las palabras y las cosas* (Barcelona: Planeta, 1984), 26–42.

23. Ibid., 304.

24. A. Kircher, *Iter extaticum coeleste*, cited in Rodríguez de la Flor, *Barroco*, 237.

25. The quote from Kircher is taken from Athanasius Kircher, *Magnes sive de magnetica arte* (Kalcoven, 1643). For Kircher as polymath, see Paula Findlen (ed.), *Athanasius Kircher: The Last Man Who Knew Everything* (New York: Routledge, 2004).

26. Nieremberg, *Curiosa y oculta filosofía*, 1:286.

27. Nieremberg, *Historia naturae*, 1:1.

28. Ibid.

29. David R. Olson, *El mundo sobre el papel* (Barcelona: Gedisa, 1994), 194–203.

30. Nieremberg, *Curiosa y oculta filosofía*, vol. 1, *De la mudanza de la naturaleza*.

31. Ibid., 35. The same analogy is used in Nieremberg, *De la diferencia entre lo temporal y lo eterno*, 114. The correspondence between the decadence of the monarchies and the corruption of animate matter could work both ways. Thus, Martín González de Cellorigo, one of the best-known treatise writers and *arbitristas* of the baroque period, believed that republics grew and withered like natural bodies.

32. Nieremberg, *Curiosa y oculta filosofía*, vol. 1, chap. 8.

33. Ann Blair, "The *Problemata* as a natural philosophy genre," in *Natural Particulars: Nature and the Disciplines in Renaissance Europe*, ed. A. Grafton and N. Siraisy (Cambridge MA: MIT, 1999), 171–205.

34. "The essential inclusiveness of the Jesuit scientific tradition consisted, however, in the way alternative theories, concepts, opinions, and hypotheses were always represented by the author of a text, even when criticized, censured, or rejected by the Jesuit author himself or by the Jesuit community." Rivka Feldhay, "The Cultural Field of Jesuit Science," in *The Jesuits: Cultures, Sciences, and the Arts, 1540–1773*, ed. John W. O'Malley, Gauvin Alexander Bailey, Steven J. Harris, and T. Frank Kennedy (University of Toronto Press, 1999), 117.

35. Thomas Browne, *Pseudodoxia Epidemica: Enquiries into Very Many Commonly Received Tenets and Commonly Presumed Truths* (London, 1646).

36. Nieremberg, *Historia naturae*, vol. 2, chaps. 2, 3, 9, 10.

37. Ibid., vol. 3, chap. 1.

38. Ibid., vol. 3.

39. For the slow assimilation of America in European minds, see John Elliott, *The Old World and the New 1492–1650* (Cambridge: Cambridge University Press, 1970). For the development of the language of marvels, see Stephen Greenblatt, *Marvelous Possesions: The Wonder of the New World* (Chicago and Oxford: University of Chicago Press, 1991). A review of mythological projection on the new continent in Juan Gil, *Mitos y utopías del descubrimiento*, 3 vols. (Madrid: Alianza, 1989).

40. José María López Piñero and José Pardo Tomás, *Nuevos materiales y noticias sobre la historia de las plantas de Nueva España de Francisco Hernández* (Valencia: CSIC, 1994), 129–33; *La influencia de Francisco Hernández (1515–1587) en la constitución de la botánica y la materia médica moderna* (Valencia: CSIC, 1996), 163–65.

41. Nieremberg, *Curiosa y oculta filosofía*, 1:96.

42. José Antonio Maravall, *Antiguos y modernos* (Madrid: Alianza, 1986), 25–113.

43. José de Acosta, *Historia natural y moral de las Indias*, Biblioteca de Autores Españoles, vol. 73 (Madrid: Ediciones Atlas, 1954), 3.

44. Baltasar Gracián, *Agudeza y arte de ingenio* (Madrid: Castalia, 1988), 1:140.

45. José Antonio Maravall, *La cultura del barroco* (Barcelona: Ariel, 2000), 458–70.

46. This historian was Luis Cabrera de Córdoba, author of *De historia, para entenderla y escribirla* (1611), cited in ibid., 459.

47. Nieremberg, *Historia naturae*, vol. 8, chap. 1.

48. As in other places, the subject of monsters has attracted a great deal of interest in Hispanic circles in recent years. In the field of history of science, see Antonio Lafuente and Javier Moscoso (eds.), *Monstruos y seres imaginarios en la Biblioteca Nacional* (Aranjuez, Spain: Doce Calles, 2000). For literature, see Elena del Río Parra, *Una era de monstruos: Representación de lo deforme en el Siglo de Oro español* (Madrid: Iberoamericana, 2003). Related to the history of culture and court practice, see Fernando Bouza and José Luis Betrán, *Enanos, bufones, monstruos, brujos y hechiceros* (Barcelona: Mondadori, 2005).

49. Gasparus Peucerus, *Commentarius de praecipuis generibus divinationum* (Wittenberg, Germany, 1553); Conrad Lycosthenes, *Prodigiorum ac ostentorum chronicon* (Basel, 1557); Ambroise Paré, *Des monsters* (Paris, 1573); Arnaud Sorbin, *Tractatus de monstris* (Paris, 1570); Cornelio Gemma, *Cosmocritica* or *De Naturae divinis Characterismis* (Antwerp, 1575).

50. Here I follow María José Vega, *Los libros de prodigios en el Renacimiento* (Barcelona: Universitat Autónoma de Barcelona, 2002), 11–25.

51. Johannes Jonstonius, *Thaumatographia naturalis*, 10 vols. (Amsterdam, 1632); Gaspar Schott, *Physica curiosa sive mirabilia naturae et artis* (Amsterdam, 1662); Fortunio Liceti, *De monstrorum natura, caussis, et differentiis* (Padua, 1616).

52. Nieremberg, *Historia naturae*, vol. 4, chap. 9.

53. Ibid., vol. 9.

54. Ibid., vol. 4, chap. 1.

55. León Pinelo (1589–1675), the Peruvian historian educated by Jesuits in Lima was a judge and historian of the Indies in the Casa de la Contratación, a leading expert on the New World and the author, among many other books, of *El Paraíso en el Nuevo Mundo* (1656).

56. Javier Malagón and José M. Ots Capdequí, *Solórzano y la política indiana* (Mexico City: Fondo de Cultura Económica, 1983), 50.

57. Nieremberg, *Historia naturae*, book 5, chaps. 1, 2, 27; Nieremberg, *Curiosa y oculta filosofía*, vol. 1, chap. 9.

58. Gregorio García, *Origen de los indios del Nuevo Mundo e Indias Occidentales* (1607).

59. Fabián Alejandro Campagne, *Homo catholicus: Homo superstitiosus; El discurso antispersticioso en la España de los siglos XV a XVIII* (Buenos Aires: Miño y Dávila, 2002), 275.

60. Nieremberg, *Curiosa y oculta filosofía*, vol. 1, chap. 17.

61. Domingo Ledezma, "Una legitimación imaginativa del Nuevo Mundo: La *Historia naturae, maxime peregrinae* del jesuita Juan Eusebio Nieremberg," in *El saber de los jesuitas, historias naturales y el Nuevo Mundo*, ed. Luis Millones and Domingo Ledezma (Frankfurt-Madrid: Vervuert-Iberoamericana, 2005), 66.

62. The olive tree was also chosen by Kircher to illustrate Ignatian distribution throughout the world in the *Horoscopium catholicum*, which figures in *Ars Magna Lucis et Umbrae* (The Great Art of Light and Darkness), a motif used by Steven J. Harris to explain the geography of the corporation: "Mapping Jesuit Science: The Role of Travel in the Geography of Knowledge," in O'Malley, et al. (eds.), *The Jesuits*, p. 220.

63. Nieremberg, *Historia naturae*, vol. 5, chap. 2.

64. Ibid., vol. 14. Ledezma, "Una legitimación imaginativa del Nuevo Mundo," 70, states that Nieremberg's description of the passionflower is taken from the *Cultura ingeniorum* (1610) by fellow Jesuit Antonio Posevino.

65. There is abundant use of eucharistic symbology and iconography in the art and literature of the Counter-Reformation. It can be seen in Calderón, Zurbarán, or Lope de Vega. It is therefore only to be expected that it should manifest itself in philosophical and scientific ways of thinking. I would like to thank Fernando Rodríguez de la Flor for lending me a copy of his unpublished text on this subject, "De lo abyecto a lo sublime: Trayectorias del alimento en la cultura hispana."

Chapter Six

1. See the essays collected in Bruce Moran (ed.), *Patronage and Institutions: Science, Technology, and Medicine at the European Court, 1500–1700* (Rochester, NY: Boydell Press, 1991); and Mario Biagioli's study of the Galileo affair through the lens of the absolutist court culture in Mario Biagioli, *Galileo, Courtier: The Practice of Science in the Culture of Absolutism* (Chicago: University of Chicago Press, 1993).

2. One of the most remarkable studies in the new history of science is Steven Shapin and Simon Schaffer's analysis of the dispute between Boyle and Hobbes over the nature of scientific knowledge in which they argue that the Royal Society provided an "ideal polity" through civil debate, in *Leviathan and the Air-Pump: Hobbes, Boyle, and the Experimental Life* (Princeton, NJ: Princeton University Press, 1985), 341. This study has been followed by essays extending the analysis of civility and etiquette in early modern scientific communities in Mario Biagioli, "Etiquette, Interdependence, and Sociability in Seventeenth-Century Science," *Critical Inquiry* 22, no. 2 (1996) and Lorraine Daston, "Baconian Facts, Academic Civility, and the Prehistory of Objectivity," in *Rethinking Objectivity*, ed. Allan Megill (Durham, NC: Duke University Press, 1994).

3. Pierre Bourdieu, "The Peculiar History of Scientific Reason," *Sociological Forum* 6, no. 1 (1991), 6. Bourdieu explicitly acknowledges Shapin and Schaffer's study as the basis for his analysis and opposes this to older "idealist" models of scientific change such as Thomas Kuhn's, in which paradigms appeared to shift under the weight of increasing contradictions or even a divergence with objective reality.

4. Bourdieu distinguishes his notion of "field" both from a history or sociology of institutions and from Michel Foucault's emphasis on discourse independent from social and political contexts. Thus while the "field of scientific reason" is not institutionally bound, it has a structuring effect on institutions

and on the discourse with which it shares cultural space. Bourdieu, "The Peculiar History of Scientific Reason," 11. Likewise, the notion that science could model a public sphere or civil society contrasts Jurgen Habermas's idea that the European public sphere developed from the field of aesthetics. See Jürgen Habermas, *The Structural Transformation of the Public Sphere*, trans. Thomas Burger (Cambridge, MA: MIT Press, 1989), 33–43.

5. Bourdieu, "The Peculiar History of Scientific Reason," 6.

6. Ibid. Bourdieu's characterization agrees with Shapin and Schaffer when they write, "either by decision or tacit processes, the space was restricted to those who gave their assent to the legitimacy of the game being played within its confines" in *Leviathan and the Air-Pump*, 336.

7. Bourdieu, "The Peculiar History of Scientific Reason," 7.

8. Ibid., 8. Here, Bourdieu suggests that the commonplace idea that scientific knowledge is disinterested has led to the assumption that the scientific field, its practitioners, societies, and institutions, are a refuge from political conflict.

9. As Lorraine Daston writes, "civility was often in woefully short supply in the new natural philosophy, as well as in the old scholasticism. Seventeenth-century science was a battlefield where rivals and factions stopped at nothing to scientifically discredit and personally abuse (the two were seldom distinguished) one another." Daston, "Baconian Facts," 52.

10. In this, the scientific community could be likened to Norbert Elias's description of the "court society." See Elias, *The Court Society*, trans. Edmund Jephcott (New York: Pantheon, 1983).

11. The fundamental study of the scholarly universe of the Habsburg court is R. J. W. Evans, *The Making of the Habsburg Monarchy, 1550–1700: An Interpretation* (Oxford: Clarendon Press, 1979). See also William B. Ashworth, "The Habsburg Circle," in Moran, *Patronage and Institutions*, for a further study of the culture of science and patronage of the court in the seventeenth century. For the Jesuit influence on colonial science in Mexico, see Jorge Cañizares-Esguerra, "Spanish America: From Baroque to Modern Colonial Science," in *The Cambridge History of Science*, vol. 4, *Eighteenth-Century Science*, ed. Roy Porter (Cambridge: Cambridge University Press, 2003), 720.

12. This perspective has been particularly advanced by the studies of Elías Trabulse who has focused on the "development" of science as a field in New Spain against the perception of colonial and Iberian backwardness. See Trabulse, "La obra científica de Carlos de Sigüenza y Góngora: 1667–1700," in *Ciencia colonial en América*, ed. José Antonio Lafuente and José Sala Catalá (Madrid: Alianza Editorial, 1992), reproduced in Elías Trabulse, "La obra científica de Don Carlos de Sigüenza y Góngora (1667–1700)," in *Carlos de Sigüenza y Góngora, homenaje 1700–2000*, ed. Alícia Mayer (Mexico City: UNAM, 2000), esp. 93–94, and the collected essays in Elías Trabulse, *Los orígenes de la ciencia moderna en México (1630–1680)* (Mexico City: Fondo de Cultura Económica, 1994). For a critique of Trabulse's categorization of Sigüenza's precursor Diego Rodríguez as "modern," see Cañizares-Esguerra, "Spanish America," 727.

13. Although there are no extant copies of this broadside, Sigüenza published it in its entirety as the first section of his *Libra astronómica*, in Carlos

de Sigüenza y Góngora, *Libra astronómica y filosófica*, 2nd ed. (1690; Mexico City: UNAM, 1984), 9–17.

14. Eusebio Kino, *Exposición astronómica de el cometa que el año de 1680 por los meses de noviembre, y diziembre, y este año 1681 por los meses de enero y febrero se ha visto en todo el mundo y le ha observado en la ciudad de Cadiz el P. Eusebio Francisco Kino de la Compañía de Jesús* [Astronomical exposition of the comet that was seen during the months of November and December in the year 1680 and January and February of this year 1681 in all the world and that the Father Eusebio Francisco Kino of the Company of Jesus has seen in the city of Cadiz] (Mexico City: Francisco Rodriguez Lupercio, 1681).

15. For the purposes of this chapter, I will cite José Gaos's latest edition (1959; revised 1984) of Sigüenza y Góngora, *Libra astronómica y filosófica*. The full title of Sigüenza's treatise is *Libra astronómica y philosófica en que D. Carlos de Siguenza y Góngora cosmographo, y mathematico regio en la Academia Mexicana, examina no solo lo que a su "Manifiesto philosophico contra los cometas" opuso el R. P. Eusebio Francisco Kino de la Compañía de Jesus, sino lo que el mismo R. P. opinó y pretendio haver demostrado en su "Exposición astronómica del cometa del año de 1681"* [Astronomical and Philosophical Balance in which D. Carlos de Siguenza y Góngora, Royal Cosmographer and Mathematician in the Mexican Academy Examines Not Only What in His *Philosophical Manifesto Against Comets* the Reverend Father Eusebio Francisco Kino of the Company of Jesus Opposed but Also What the Same Reverend Father Presented and Attempted to Demonstrate in His *Astronomical Exposition of the Comet of 1681*]. Sigüenza's allusion to Grassi's title in his own is suggestive and will be analyzed further below.

16. In his introduction, Guzmán y Córdova explains that a new comet in 1689 prompted him to publish Sigüenza's treatise in 1690 "in order to dissipate the panicked terror with which all who have seen the comet have been agitated." Ibid., 15.

17. In accordance with his argument that Sigüenza showed a pre-Enlightenment sensibility, Irving Leonard reads the dispute as fight over the ideals of reason and science, in Irving A. Leonard, *Don Carlos de Sigüenza y Góngora: A Mexican Savant of the Seventeenth Century* (Berkeley: University of California Press, 1929), 60, 70, 72–73.

18. Sigüenza was expelled from the Jesuit order in 1667 after having completed his initial studies. He petitioned at least twice to be readmitted, once in 1668 and again in 1676, and both times was denied. Although the exact cause for the expulsion is unknown, Ernest Burrus notes that it appears to have been a breach in decorum and discipline rather than morality. See Ernest J. Burrus, "Sigüenza y Góngora's Efforts for Readmission into the Jesuit Order," *Hispanic American Historical Review* 33 (1953), 389.

19. Sigüenza's attempts to separate himself from popular culture are clear in his yearly almanacs, in which he increasingly argued against astrology as superstition, referring to it as *bagatela* (hogwash), in José Miguel Quintana, *La astrología en la Nueva España en el siglo XVII (de Enrico Martínez a Sigüenza y Góngora)* (Mexico City: Bibliofilos Mexicanos, 1969), 195. Sigüenza's letter

to Admiral Pez following the 1692 riot in Mexico City explicitly uses scientific reason to measure the distance between himself and plebeian commoners. For instance, Sigüenza contrasts his telescopic observations of the solar eclipse to popular reactions: "Since all this was unexpected, at the moment that the light disappeared the birds flying overhead fell to earth, the dogs howled, women and children cried out, the Indian women left their stalls in which they had been selling fruits and vegetables and other items in the square and took refuge at full speed in the Cathedral; and since the bells rang out at that same moment not only in that but in all churches in the city, there was such an immediate confusion and mayhem that it caused horror" and "I, in the meantime, was contemplating the sun with my quadrant and long-range spyglass, extremely happy and giving repeated thanks to God for having provided me the opportunity to see what occurs in a given place only once in a long while and of which there are so few registered observations." Carlos de Sigüenza y Góngora, "Alboroto y motín de los indios de México," in *Seis obras*, ed. William G. Bryant (Caracas: Ayacucho, 1984), 108.

20. The most recent studies of Sigüenza's patriotism are Anthony Pagden, *Spanish Imperialism and the Political Imagination: Studies in European and Spanish-American Social and Political Theory 1513–1830* (New Haven, CT: Yale University Press, 1990), 91–97; and D. A. Brading, *The First America: The Spanish Monarchy, Creole Patriots, and the Liberal State, 1492–1867* (Cambridge: Cambridge University Press, 1991), 362–72. Scholars have tended to attribute Sigüenza's patriotism to Creole resentment and thus have downplayed the context of Spanish imperial decline. Nor has Sigüenza's defense of scientific reason been explicitly linked to his patriotism. For instance, Brading dedicates several pages to the dispute but states that "oddly enough, it was his scientific studies that occasioned a polemic in which his patriotism found its most embittered expression." Brading, *The First America*, 367.

21. Kino never seems to have responded to Sigüenza's *Libra astronómica*. He did, however, make note of it in 1695, at which point he denied ever having read the *Manifiesto*. Tellingly, he also claims that he was occupied in much more urgent occupations "such as those which brought me to the Indies." Cited in Herbert Eugene Bolton, *Rim of Christendom: A Biography of Eusebio Francisco Kino, Pacific Coast Pioneer* (New York: Russell & Russell, 1960), 82.

22. All translations from Spanish are mine. I'd like to thank Bryan Greene for his astute suggestions on many of these. In his preface to the *Libra*, Sebastián de Guzmán y Córdova describes the *Belerofonte* but states that it had never been published, a claim that Sigüenza validates in the *Libra*. Sigüenza y Góngora, *Libra astronómica*, prologue, 14, 147. It seems likely that the *Belerofonte* had also been written to publish Sigüenza's observations for the benefit of his erudite readers, a task that was subsequently taken over by the *Libra*. The title of Sigüenza's treatise alludes to the mythological battle between Bellerophon and the monster Chimera.

23. Despite Torre's criticism of the author's defense of astrology, in another note in the *Libra*, Sigüenza qualifies him as "the erudite mathematician don Martín de la Torre." Sigüenza y Góngora, *Libra astronómica*, 119. This apparent contradiction sheds light on Sigüenza's dispute with Kino. In both cases, the

authors' mathematical abilities prompted Sigüenza's initial respect, even if he differed with their overall interpretations of celestial phenomena.

24. Ibid., 19.

25. Ibid., 2.

26. Ibid., 4.

27. For gift exchange as a form of manifesting relations of obligation between patrons and scientists, see Paula Findlen, "The Economy of Scientific Exchange in Early Modern Italy," in *Patronage and Institutions: Science, Technology, and Medicine at the European Court, 1500–1750*, ed. Bruce Moran (Rochester, NY: Boydell Press, 1991). As I would like to show, reciprocity and exchange also formed the basis of "intellectual friendships" between scientific practitioners. The contradictions involved in this type of friendship will be further discussed below.

28. Sigüenza insists that anyone who had read the two treatises would see clearly that Kino's argument was a point-by-point refutation of the one he had made in the *Manifiesto* but that "no one knows better where the shoe pinches than the one who wears it." Sigüenza y Góngora, *Libra astronómica*, 151.

29. Ibid., 3.

30. Ibid.

31. Ibid., 7.

32. Ibid. Kino himself mentions the case of a "literary duel" on the topic of the 1677 comet, written and then publicly defended by the Jesuit Andres Waybel in the University of Ingolstadt. Kino, *Exposición astronómica*, 11.

33. Sigüenza y Góngora, *Libra astronómica*, 7.

34. Ibid., prologue, 18. Sigüenza also cites a classical text that says: "If these things be considered false, it is better not to praise them but if they are believed to be true they should be judged publicly." Ibid., 7–8.

35. Leonard, *Don Carlos de Sigüenza y Góngora*, 75.

36. Sigüenza y Góngora, *Libra astronómica*, 6.

37. Ibid., 2.

38. Ibid.

39. Ibid., 88.

40. Ibid., 119.

41. Ibid., 6.

42. Ibid., 85–86.

43. Ibid., 151.

44. Ibid., 68.

45. Víctor Navarro Brotóns, "*La libra astronómica y filosófica* de Sigüenza y Góngora: La polémica sobre el cometa de 1680," in *Carlos de Sigüenza y Góngora: Homenaje 1700–2000*, ed. Alicia Mayer (Mexico City: UNAM, 2000), 152.

46. Peter Dear, "The Church and the New Philosophy," in *Science, Culture, and Popular Belief in Renaissance Europe*, ed. Paolo Rossi and Maurice Slawinski Stephen Pumfrey (Manchester, UK: Manchester University Press, 1991), 134. See also Peter Dear, *Discipline and Experience: The Mathematical Way in the Scientific Revolution* (Chicago: University of Chicago Press, 1995).

47. As Dear explains, the term was a polemic attempt to overcome the Aristotelian disciplinary framework in order to authorize mathematics: "physico-mathematics was a bid for disciplinary authority over knowledge of nature." *Discipline and Experience,* 167–68.

48. Riccioli, whose *Almagestum Novum* Sigüenza cites often, in particular provided an argument for mathematics as a means to certify truth. In the preface to his encyclopedic work, Riccioli argues that mathematics provides a method free from sensual confusion and that together with metaphysics was the only means to arrive at truth that provided no further proof. Alfredo Dinis, "Giovanni Battista Riccioli and the Science of His Time," in *Jesuit Science and the Republic of Letters* (Cambridge, MA: MIT Press, 2003), 200.

49. Sigüenza y Góngora, *Libra astronómica,* 10.

50. Ibid. For fideism in the work of Pierre Gassendi, see Margaret Osler, "Divine Will and Mathematical Truth: Gassendi and Descartes on the Status of the Eternal Truths," in *Descartes and His Contemporaries: Meditations, Objections, and Replies,* ed. Roger Ariew and Margorie Grene (Chicago: University of Chicago Press, 1995).

51. Sigüenza y Góngora, *Libra astronómica,* 14. For the association between comets and monsters, see Lorraine Daston, "Marvelous Facts and Miraculous Evidence in Early Modern Europe," *Critical Inquiry* 18, no. 1 (1991), 103; and Sara Schechner Genuth, *Comets, Popular Culture, and the Birth of Modern Cosmology* (Princeton, NJ: Princeton University Press, 1997), 31. For debates around interpretations of monsters as portends, see Lorraine Daston and Katharine Park, *Wonders and the Order of Nature, 1150–1750* (New York: Zone Books, 2001), 173–214.

52. Sigüenza y Góngora, *Libra astronómica,* 61, 22–23.

53. Ibid., 40–41.

54. Ibid., 35. Here, as Paula Findlen has written, Sigüenza had something of a home field advantage, as his access to his extensive library allowed him to take advantage of Kircher's eclecticism and find a contradiction to beat Kino at his own game. Findlen, "A Jesuit's Books in the New World: Athanasius Kircher and His American Readers," in *Athanasius Kircher: The Last Man Who Knew Everything,* ed. Paula Findlen (New York: Routledge, 2004), 348.

55. As Noel Malcolm notes, Kircher was not an accepted member of the Franco-Dutch axis of the Republic of Letters. Malcolm, "Private and Public Knowledge: Kircher, Esotericism, and the Republic of Letters," in Findlen, *Athanasius Kircher,* 299. R. J. Evans has shown, however, that he was the central figure of the wide intellectual circle of the Habsburg empire. Evans, *The Making of the Habsburg Monarchy,* 316–45.

56. After his expulsion from the Jesuit order, Sigüenza attempted at least two times to be readmitted. See Burrus, "Sigüenza y Góngora's Efforts for Readmission into the Jesuit Order," 368.

57. The full sentence is worth citing, as it indicates the emotional force with which Sigüenza treated his relationship to the order: "When innate duty reminds me of all that I owe such a learned, exemplary, and sacred religion my most tender age when, out of the kindness of the reverend fathers of this

Mexican province, my friends, my teachers, and my fathers, I received such singular favors which I always recognize publicly and which I would wish to repay even with blood from my veins and since all of this is true and all know it to be so, for the same reason all will realize that in this controversy, to which I am compelled to enter for the reasons I will express, I speak to the reverend father not as part of such a venerable whole but rather as an individual subject and mathematician." Sigüenza y Góngora, *Libra astronómica*, 1.

58. Ibid., 151.

59. Höpfl declares that self-defense was the "only natural right which Jesuit theologians regarded as entirely uncontentious." Harro Höpfl, *Jesuit Political Thought: The Society of Jesus and the State, c. 1540–1630* (Cambridge: Cambridge University Press, 2004), 292.

60. Sigüenza's desire to emphasize his association with the Jesuit order and his advocacy of reasoned public debate seems the most likely reason for his choosing the title of Grassi's attack on Galileo for his own treatise. Sigüenza sought to sanction his conclusions and present Kino as a renegade, in much the same way that Grassi's Jesuit-sponsored attack on Galileo succeeded in doing the earlier controversy. Thus he defends his position in the "literary duel" with Kino as analogous to that of Grassi in the dispute with Galileo:

> And if not in this (which is not just), at least in titling this work *Astronomical and Philosophical Balance* I wished to imitate the reverend father Horatio Grassi who with the same epigraph labeled the book he published against what Mario Guiducio and Galileo Galilei wrote of the same comet of 1618; and if this was not a censurable action in that father, who provoked, how could it be in me, who was provoked, unless one tramples upon reason and justice? [Sigüenza y Góngora, *Libra astronómica*, 7]

For the Grassi and Galileo controversy from the perspective of the politics of court patronage and the Jesuit order, see Biagioli, *Galileo, Courtier*, 268–308.

61. Sigüenza y Góngora, *Libra astronómica*, 15.

62. This agrees with the position Sigüenza will take on the 1692 riot of Mexico City, which he attributes to divine punishment for the collective sin of pulque drinking. In his letter to the Admiral Pez he states: "if there is no reform, He will perfect his justice." Sigüenza y Góngora, "Alboroto y motín," 134 and in his 1693 astrological almanac: "it has already been some time since like a good Father He warned us that He is guarding a greater punishment for the future if there is no reform." Quintana, *La astrología en la Nueva España*, 228.

63. If, as Findlen has argued, American scholars' "dreaming of Kircher was one of the great intellectual fantasies of Baroque Mexico," it is just as true that European scholars also dreamed of the exotic. Findlen, *A Jesuit's Books in the New World*, 337. Bolton states that, from a young age, Kino "dreamed of China." Bolton, *Rim of Christendom*, 37.

64. The phrase comes from Virgil's *Aeneid*, book 9, line 641 and translates "thus you shall go to the stars." In the text he wrote to accompany his 1680 triumphal arch for the entrance of the viceroy to Mexico City, Sigüenza explains the meaning of the Pegasus: "it represents man who reveals his soul almost

always turned toward that which is sublime, for the benefit of his homeland." Sigüenza y Góngora, "Teatro de virtudes políticas que constituyen a un príncipe," in *Seis obras,* ed. William G. Bryant (Caracas: Ayacucho, 1984), 174.

65. Citing Biagioli's study of astronomy and court culture, Victor Navarro has argued that the dispute between Sigüenza and Kino should be understood in the context of patronage politics in the viceregal court. Navarro, "La Libra astronómica y filosófica de Sigüenza y Góngora," 169–73.

66. In what follows, I have been particularly influenced by Derrida's discussion of friendship in *The Politics of Friendship.* In this, he examines two notions of friendship intertwined in Michel de Montaigne's essay "On Friendship." In one, derived primarily from Cicero and the Stoic tradition, friendship is a rare and singular experience of "one soul in two bodies" in which calculation, exchange, and even division by death is not possible. In the other, derived from Aristotle, friendship may be divided into three types, that based on virtue, that based on pleasure, and that based on utility. As Derrida shows, the Aristotlean divisions point toward the contamination of ideas of usefulness and virtue in the first two forms of friendship. See Jacques Derrida, *Politics of Friendship,* trans. George Collins (London: Verso, 1997), esp. 195–226.

67. For the expectation of hospitality and reciprocity in a virtuous friendship, see ibid., 178–80.

68. In particular, Derrida finds this contradiction in Montaigne's espousal of the Ciceronian ideal of a singular and perfect friendship. According to Derrida, Montaigne's emphasis on the absolute uniqueness of the "sovereign friendship" creates an abyss of unsatisfiable calculation. This is because friendship, unlike love, is inherently based on a reciprocity and exchange inimicable to the classical idea of virtue. Derrida, *Politics of Friendship,* 215–16.

69. Sigüenza y Góngora, *Libra astronómica,* 5. I would like to thank Bryan Green for pointing out that Sigüenza's phrase "starry bleariness" *(astrosas lagañas)* refers to the Spanish refrain, *ojos hay que de lagañas se enamoran.*

70. Leonard, *Don Carlos de Sigüenza y Góngora,* 56.

71. Sigüenza y Góngora, *Libra astronômica,* 5.

72. Octavio Paz, *Sor Juana Inés de la Cruz o Las trampas de la fe,* 3rd ed. (Mexico City: Fondo de Cultura Económica, 1983), 206–7.

73. As for instance when Sigüenza writes that the "love that one owes to his *patria* [homeland]" has led him to look to local history for the motif of his 1680 triumphal arch. Sigüenza y Góngora, "Teatro de virtudes," 172.

74. For the history of the term "Republic of Letters" and its seventeenth-century use to describe the cosmopolitan network of authors linked through epistolary correspondence, see Paul Dibon, "Communication in the Respublica Literaria of the 17th Century," *Res Publica Litterarum* 1 (1978); Françoise Waquet, "Qu'est-ce que la République des Lettres? Essai de sémantique historique," *Bibliothèque de l'École des Chartes* 147 (1989); and Dena Goodman, *The Republic of Letters: A Cultural History of the French Enlightenment* (Ithaca, NY: Cornell University Press, 1994), 1–11. While Pierre Bayle could write that the literary republic was an "empire of truth and reason," the field of scientific reason, as described by Bourdieu, establishes truth rather than

recognizing an agreed-upon body of knowledge. Although Sigüenza uses the term "literary republic" to describe the audience he hopes to attain through his *Libra astronómica*, he appears to have in mind the structure of scientific reason. In his defense of "literary duels," for instance, these will decide upon truth: "they almost always provide many truths to the literary republic." Sigüenza y Góngora, *Libra astronómica*, 151. Bayle's comments are cited in Waquet, "Qu'est-ce que la République des Lettres?" 484.

75. See the literature summarized in note 2 of this chapter.

76. Sigüenza calls Caramuel "my great friend and most excellent correspondent" and José de Zaragoza a "very singular friend of mine" in Sigüenza y Góngora, *Libra astronómica*, 34 and 141, respectively. For Favián's epistolary relationship with Athanasius Kircher, see Findlen, *A Jesuit's Books in the New World*, 337–43.

77. Likewise, it is distinct from the context of exchange that Paula Findlen has discussed in the context of scientific patronage. See Findlen, "The Economy of Scientific Exchange in Early Modern Italy."

78. Sigüenza y Góngora, *Libra astronómica*, 121.

79. Giovanni Francesco Gemelli Careri, *Viaje a la Nueva España*, trans. Francisca Perujo (Mexico City: UNAM, 1983), 119.

80. Sigüenza's contradictory expectations for both exchange of information and hospitality outside of calculation and exchange correspond to the paradoxes of what Derrida has called "ethical friendship" or "comradeship" that establishes a political community while maintaining the ideal of friendship as virtue:

> When, on the other hand, the parties leave the matter to each other's discretion, in a sort of trust without contract, credit becoming an act of faith, then friendship 'wants to be' moral, ethical [*ethike*] and of the order of comradeship [*etarike*]. Why is it that in this case recriminations and grievances abound? Because this ethical friendship is against nature [*para phusin*]. Indeed, those who associate themselves in this way wish to have both friendships at once, one in the service of interest (based on usefulness) and one appealing to virtue (the reliability of the other), friendship of the second type and primary friendship. [*Politics of Friendship*, 204–5]

81. Steven J. Harris, "Mapping Jesuit Science: The Role of Travel in the Geography of Knowledge," in *The Jesuits: Cultures, Sciences, and the Arts, 1540–1773*, ed. John O'Malley, Gauvin Alexander Bailey, Steven J. Harris, and T. Frank Kennedy (Toronto: University of Toronto Press, 1999), 213–26.

82. Sigüenza y Góngora, *Libra astronómica*, 141–42, 146.

83. In this sense, Sigüenza's espousal of scientific reason should be differentiated from the exoticism of the Jesuit natural histories produced over the course of the sixteenth and seventeenth centuries. While it is true that he shared many of the Neoplatonic sources that informed what has been called the "emblematic worldview" of these histories, Sigüenza's mathematical skepticism tempers his interest in "emblems" as a source of truth or certainty. This did not preclude his use of analogies, etymologies, and other hermeneutic instruments common

to the baroque for purposes other than ascertaining physical properties. In fact, it appears that Sigüenza tended toward fideism and simply cordoned off what reason could access and what it could not, leaving what it could not to more Neoplatonic structures of knowledge. In this sense, he could be considered an interesting exception to Cañizares-Esguerra's discussion of the Neoplatonic tradition in colonial science, in Cañizares-Esguerra, "Spanish America," 722–29. For the "emblematic worldview" of early modern natural histories, see William B. Ashworth, Jr., "Natural History and the Emblematic World View," in *Reappraisals of the Scientific Revolution*, ed. David C. Lindberg and Robert S. Westman (Cambridge: Cambridge University Press, 1990); and the collected essays in Luis Millones Figueroa and Domingo Ledezma (eds.), *El saber de los jesuitas, historias naurales y el nuevo mundo* (Frankfurt: Iberoamericana, 2005).

84. Sigüenza y Góngora, *Libra astronómica*, prologue, 15.

85. Ibid., 181.

86. Bernardino de Sahagún's *Códice Florentino* and the Colegio de Santa Cruz de Tlatelolco with which he was involved are examples of a sixteenth-century "missionary humanism" in which indigenous knowledge was understood to be active and valid, if ultimately misdirected. By the time Sigüenza wrote the *Libra astronómica*, he and other Creoles had established a clear distinction between the knowledge and virtue of an indigenous nobility, then in sharp decline, and the superstitious knowledge of the indigenous popular class. For Sahagún, see J. Jorge Klor de Alva, H. B. Nicholson, and Eloise Quiñones Keber (eds.), *The Work of Bernardino de Sahagún: Pioneer Ethnographer of Sixteenth-Century Aztec Mexico* (Austin: University of Texas Press, 1988).

87. Michel Foucault uses the term "governmentality" to describe the increasing rationalization of governance in seventeenth and eighteenth centuries. See Foucault, "Governmentality," in *Power*, vol. 3, *Essential Works of Foucault, 1954–1984*, ed. James D. Faubion, trans. Robert Hurley et al. (New York: New Press, 2000).

88. The riot of 1692 in Mexico City was preceded by severe flooding and a corn shortage due to heavy rains. During this crisis the viceroy convened a body of advisors to deliberate on how to best drain the lake. Sigüenza participated in this meeting and advised the viceroy to dredge the city canals, scoffing at suggestions that there was a drain hole in the lake that could be unplugged to let out the excess water. Although projects to drain Mexico's lake were not new, Sigüenza linked his advice to a patriotic dedication and throughout his letter emphasizes the benefits of Creole knowledge for governance. Sigüenza y Góngora, "Alboroto y motín," 104–5.

89. Sigüenza himself makes this association in his letter describing the 1692 riot, which he opens by referring to the contemporaneous rebellions among the Tarahumara and other northern nations. These groups are "so absolutely barbarous and bestial and for this reason so impossible for this reason to subjugate that even though their encampments are found less than thirty leagues from this court until now it has not been possible to rein them in nor were the Mexicans able to do so when their empire flourished." Ibid., 100.

Chapter Seven

1. The issues surrounding the politics of reproduction and pregnancy analyzed in this essay are explored further in my forthcoming book, tentatively titled "Colonial Medicine and Local Healing Cultures in Guatemala, 1680–1830."

2. Santiago de Guatemala (today Antigua) was the capital of the Audiencia of Guatemala, part of the viceroyalty of New Spain, until its partial destruction from an earthquake in 1773. The capital then transferred to its present location in today's Guatemala City.

3. For a good introduction to formal medical practices dealing with women's reproductive health, see John Tate Lanning, *The Royal Protomedicato: The Regulation of the Medical Professions in the Spanish Empire*, ed. John TePaske (Durham, NC: Duke University Press, 1985), esp. chap. 13, "Government and Obstetrics."

4. For more on female curing and midwifery as it related to colonial understandings of sorcery practices, supernatural illnesses, and community conflict, see Martha Few, *Women Who Live Evil Lives: Gender, Religion, and the Politics of Power in Colonial Guatemala* (Austin: University of Texas Press, 2002), esp. chap. 4, "Illness, Healing, and the Supernatural World."

5. I have also written about monstrous births as a distinct category within difficult pregnancies in "Atlantic World Monsters: Monstrous Births and the Politics of Pregnancy in Colonial Guatemala," in *Gender and Religion in the Atlantic World (1600–1800)*, ed. Lisa Vollendorf and Daniela Kostroun (Toronto: University of Toronto Press, in press).

6. On acculturation, see the pioneering work of Gonzalo Aguirre Beltrán, *Medicina y magia: El proceso de aculturación en la estructura colonial* (Mexico City: Instituto Nacional Indigenista, 1963); on colonialism and the professionalization of medicine, see the equally pioneering work of Lanning, *The Royal Protomedicato*.

7. I am grateful to Lisa Vollendorf for encouraging me to elaborate about this process.

8. I chose the word "diagnosis" deliberately, to emphasize that formal medical doctors in colonial Guatemala did not have an ideological or professional monopoly on diagnosing the causes of difficult pregnancies. Illness in colonial Guatemala was a familial and community event and not the kind of individual experience that we think about today. The evidence here shows that those involved constructed arguments about how to catalogue the symptoms and diagnose the causes of the illness. See Steve Feierman, "Explanation and Uncertainty in the Medical World of Ghaambo," *Bulletin of the History of Medicine* 74 (2000), 317–44.

9. See, for example, Lanning, *The Royal Protomedicato*.

10. For more on the early establishment of hospitals and formal medicine in New Spain, see Guenter B. Risse, "Medicine in New Spain," in *Medicine in the New World: New Spain, New France, and New England*, ed. Ronald L. Numbers (Knoxville: University of Tennessee Press, 1987), 38.

11. Gonzalo Fernández de Oviedo, *Historia general de Indias* (1535), tomo 3, libro 33, cap. 46, p. 505, cited in José Luis Gómez Ratón, "Capitulos médicos

en la obra de los historiadores de Indias," *Cuadernos de historia de la medicina Española* 2 (1963), 60.

12. Bartolomé de Las Casas, *Historia de las Indias* (1561), cap. 215, p. 563; cited in Gómez Ratón, "Capítulos médicos," 60.

13. For an analysis of abortifacients from the New World and the lack of circulation of the medical plants and knowledge about their uses to Europe, see Londa Schiebinger, *Plants and Empire: Colonial Bioprospecting in the Atlantic World* (Cambridge, MA: Harvard University Press, 2004), esp. chap. 3, "Exotic Abortifacients."

14. Las Casas, *Historia de las Indias* (1561), cap. 39, p. 101; cited in Gómez Ratón, "Capitulos médicos," 60.

15. For a good introduction to the official practice of formal medicine in New Spain, see Lanning, *The Royal Protomedicato.*

16. This type of community consensus about illness diagnosis can also be seen in cases of witchcraft accusations. See Few, *Women Who Live Evil Lives.*

17. In previous research, I argued that in colonial Guatemala, community members categorized illness into two exclusive categories: supernatural illnesses and natural illnesses, the latter also described as "from God." See ibid., 68–99. As I have looked more deeply into these issues, I have found other illness categories besides these two, and that illness categories as explanatory frameworks for a certain set or sets of symptoms were not necessarily mutually exclusive. See also Feierman, "Explanation and Uncertainty in the Medical World of Ghaambo."

18. Those working on illness and healing among contemporary indigenous Maya cultures have tended to argue for the diagnosis of illness causation into two categories, natural and supernatural. However, it appears that illness categories could overlap in the colonial period. For contemporary studies of illness categories in Mesoamerica, see for example Azzo Ghidinelli, "El sistema de ideas sobre la enfermedad en Mesoamérica," *Tradiciones de Guatemala* 26 (1986), 69–89; Alfredo Méndez Domínguez, "Illness and Medical Theory Among Guatemalan Indians," in *Heritage of Conquest: Thirty Years Later,* ed. Carl Kendall, John Hawkins, and Laurel Bossen; preface by Sol Tax (Albuquerque: University of New Mexico Press, 1983), 267–98; Sheila Cosminsky, "Medical Pluralism in Mesoamérica," in Kendall et. al., *Heritage of Conquest*; and Richard N. Adams and Arthur J. Rubel, "Sickness and Social Relations," in *Handbook of Middle American Indians: Social Anthropology* (Austin: University of Texas Press, 1967), 6:333–55.

19. Archivo General de Centro América (AGCA), Guatemala City, Guatemala, A1-4929-42045. While no specific ethnic group is listed in the documents, in the seventeenth century this town was made up of Kakchikel Maya (Christopher Lutz, personal communication). In fact, Marta de la Figueroa's *defensor* (lawyer) who represented her after the case transferred to Santiago de Guatemala, characterized Pinula as "a pueblo without any Spaniards" (fols. 27v–29). While here I emphasize the medical-related aspects of this case, I have previously analyzed this document within the context of gendered religious cultures and the Inquisition in *Women Who Live Evil Lives*, 96–99.

20. AGCA, A1-4929-42045, fol. 1v.

21. Ibid., fol. 4–4v. Maldonado is listed as being able "to speak and understand Spanish."

22. Ibid., fols. 2v–3. Andrés Yos, husband of Juana Candelera, has the surname of Andrés Gonzalez in other documents (see ibid., fol. 39). For consistency and to avoid confusion, I will use the name Andrés Yos in this analysis.

23. AGCA, A1-4929-42045, fol. 39.

24. Ibid.

25. Ibid., fol. 6. Maeda spoke through an interpreter.

26. Ibid., fol. 1v.

27. Ibid., fols. 27v–29, from the testimony of her lawyer, Pedro de San Juan y Prado.

28. For more on Mesoamerican medical cultures and practices, see Beltrán, *Medicina y magia* and Alfredo López Austin, *Cuerpo humano e ideología: Las concepciones de los antiguos nahuas*, 2 vols., 2nd ed. (Mexico City: UNAM, Instituto de Investigaciones Antropológicas, 1984).

29. AGCA, A1-4929-42045, fol. 39r–v.

30. Ibid., fol. 12. The scribe is named Lázaro de Verganza.

31. Archivo General de la Nación, Mexico City, Ramo de Inquisición (hereafter AGN, Inq.), vol. 644, exp. 2, fols. 196–347: fol. 222.

32. Arriola did not name the woman.

33. Pedro Mariano Iturbide, *Breve y diminuto compendio de la obligación que hay de bautizar los fetos* (Nueva Guatemala: Oficina de D. Ignacio Beteta, 1788), fol. 23 (archived at Museo del Libro, Antigua, Guatemala).

34. Francisco Antonio Fuentes y Guzmán was born in 1643 in Santiago de Guatemala. He served as *alcalde* of Santiago de Guatemala and *alcalde mayor* of Totonicapán (1661) and Sosonate (1699).

35. Santo Domingo Sinacoa is located near the present-day Guatemalan town of Sumpango. Francisco Antonio de Fuentes y Guzmán, *Historia de Guatemala: Ó Recordación florida* (Madrid: L. Navarro, 1882–83), vol. 2, book 13, chap. 6, p. 978 (archived at Newberry Library, Chicago).

36. Lorraine Daston and Katharine Park, *Wonders and the Order of Nature, 1150–1750* (New York: Zone Books, 2001), 176.

37. The word *miembro* here literally translates into "member," a colonial-era term that in this context signifies a penis.

38. Fuentes y Guzmán, *Historia de Guatemala*, 2:978.

39. Ibid.

40. Francisco J. Santamaría (ed.), *Diccionario de mejicanismos* (Mexico City: Editorial Porrua, 1992).

41. I am continuing to investigate the etymology of the word *nannoso*.

42. I have not yet been able to locate this painting of conjoined twins.

43. AGN, Inq., vol. 830, exp. 7, fols. 100–28. The documentation that specifically refers to the deformed birth considered here comes from fols. 108f–v and 124f–v.

44. Charles Wisdom, "The Supernatural World and Curing," in *The Heritage of Conquest: The Ethnology of Middle America*, ed. Sol Tax (Glencoe, IL: Free Press, 1952), 128.

45. Edward Geoffrey Parringer, *Witchcraft: European and African* (London: Faber & Faber, 1963), 146.

46. Francisco Sunzin de Herrera, *Consulta práctico-moral en que se pregunta si los fetos abortivos se podran bautizar a lo menos debaxo de condición, a los primeros días de concebidos* (Guatemala City: Imprenta Nueva de Sebastián de Arébalo, 1756), n.p. (archived at John Carter Brown Library, Providence, RI). See the section "Argumento sexto" for the quoted excerpt.

47. AGN, Inq., vol. 680, exp. 1, fols. 1–16 (1694). All of the information regarding Doña María Limón is located on fol. 9v.

48. Dr. Fernandéz's position at the university is listed as Cathedrático de Prima de Medicina.

49. Real Academia Española, *Diccionario de autoridades* (1732; facsimile edition, Madrid: Editorial Gredos, 1964), 532.

Chapter Eight

1. Alberto Saladino García, *Dos científicos de la ilustración hispanoamericana: J. A. Alzate y F. J. Caldas* (Mexico City: UNAM, 1990).

2. *La Gazeta de Literatura de México*, 3 vols. (Mexico City: Imprenta de D. Felipe de Zúñiga y Ontiveros, 1788–95). Two later editions appeared in 1831 and 1892. Most research undertaken on Alzate's work tends to be based on the 1831 Puebla edition. However, its format is problematic due to the omission of all details providing the text with its original sense of periodicity and its inclusion of work not included in the original *Gazeta de Literatura (GLM)*. All references in this study are to the original publication. The reader should be advised of the following problems in the copy from which this work has been taken. The page numbers in volume 1 do not follow a strictly chronological order: for issue 12, a new numbering sequence begins when the printing house is changed, followed by another new sequence with issue 13. Also in volume 1, the first issue of the second subscription is numbered page 1 of issue 1, instead of page 1 of issue 25, and so on. In order to avoid confusion and to differentiate between the two subscriptions within the same volume, I have chosen to number each issue of the second subscription as 1b, 2b, and on up to 24b, which is the final issue of volume 1. The page numbering for volume 2 remains steady throughout, while volume 3 skips ten numbered pages between issues 33 and 34. Again it is hoped that any references will be easily accessed through numbering and dates. All quotations have been translated into English by the author.

3. *Gazeta de México* 9, no. 28 (3 April 1799), 223. This *Gazeta* will be referred to as *GM* in these notes.

4. See Bernabé Navarro, "Alzate, símbolo de la cultura ilustrada mexicana," *Memorias y Revista de la Academia Nacional de Ciencias* 54 (1952), 85–97 and José Luis Peset, "La naturaleza como 'símbolo' en la obra de José Antonio de Alzate," *Asclepio* 39, no. 2 (1987), 285–95.

5. *GLM* 3, no. 44 (22 October 1795), 343.

6. See Roberto Moreno, "José Antonio de Alzate y los Virreyes," *Cahiers du Monde Hispanique et Luso-Brézilien* 12 (1969), 104.

7. Anne Goldgar, *Impolite Learning: Conduct and Community in the Republic of Letters, 1680–1750* (New Haven, CT: Yale University Press, 1995); and Françoise Waquet and Hans Bots, *La République des Lettres* (Berlin: De Boeck, 1997).

8. Goldgar, *Impolite Learning*, 3.

9. Waquet and Bots, *La République*, 14–15, 60.

10. Goldgar, *Impolite Learning*, 12–19.

11. Ibid., 29.

12. Ibid., 48.

13. *GLM* 2, no. 2 (21 September 1790), 18.

14. *GLM* 2, no. 32 (17 January 1792), 258: "Advertencia relative al Suplemento de la Gazeta de Literatura publicado antes de esta."

15. The list of Alzate's reports and activities is too extensive to include in this study. See also Ramón Aureliano, A. Buriano, and S. López (eds.), *Índice de las Gacetas de Literatura de México de José Antonio Alzate y Ramírez* (Mexico City: Instituto de Investigaciones Mora, 1996); Francisco de las Barras y de Aragón, "Noticia de la vida y obras de D. José Antonio Alzate y Ramírez," *Boletín de la Real Sociedad Española de Historia Natural* 48, no. 3 (1950), 339–53; Francisco Fernández del Castillo, *Apuntes para la bibliografía del presbítero bachiller José Antonio Félix de Alzate y Ramírez Cantilla* (Mexico City: Talleres Gráficos de la Nación, 1927).

16. For information on this process within the academy, see Rafael Aguilar y Santillán, "Una carta interesante de Alzate," *Memorias de la sociedad científica "Antonio Alzate,"* vol. 23 (Mexico City: Imprenta del Gobierno Federal, 1905), 87.

17. Patrice Bret provides an excellent in-depth study of the correspondence maintained between Alzate and the Academy and discusses certain possible reasons for the exclusion of his name from their lists after 1786. See Patrice Bret, "Alzate y Ramírez et l'Académie Royale des Sciences de Paris: La reception des travaux d'un savant de nouveau monde," in *Periodismo científico en el siglo XVIII: José Antonio Alzate y Ramírez*, ed. Patricia Aceves Pastrana (Xochimilco, Mexico: Universidad Autónoma Metropolitana, 2001), 123–205.

18. *Journal des Sçavans* (October 1771), 117–38: "Observaciones Meteorológicas de los últimos nueve meses de el año 1769, hechas en esta Ciudad de Mexico, por D. Joseph Antonio de Alzate y Ramírez. Eclypse de Luna del 12 de Diciembre de 1769 [o]bservado en la Impérial ciudad de Mexico. Impresso en Mexico en la imprenta del Lic. D. Joseph de Jauregui en la Calle de S. Bernardo. 1770"; *Journal des Sçavans* (June 1773), 238–44: "Observations tirées d'une lettre écrite de Mexico à l'Académie Royale des Sciences, par Dom de Alzate y Ramyres"; *Journal de Physique* (February 1773), 221–23: "Observations tirées d'une lettre écrite de Mexico à l'Académie Royale des Sciences, par Dom de Alzate y Ramyres sur des poissons vivipares, & quelques autres objets d'histoire naturelle."

19. *Memorial Literario Instrutivo y Curioso de Madrid* 14 (May 1788), 87–98: "Historia de la Nueva España, por el viajero francés, alias el Abate Delaporte." This was first printed in *GLM* 1, no. 2 (31 January 1788), 9–20. Later, in *GLM* 1, no. 7b (9 September 1789), 53, Alzate mentions that it has come to his attention that the article was reprinted in the Spanish periodical *Memorial Literario*.

20. His remit seems to have been to send various botanical and mineralogical samples from New Spain to the Royal Cabinet. However, there is no record of the promised samples. Letter from Alzate to Pedro Franco Dávila (26 April 1777), Archivo del Museo Nacional de Ciencias Naturales, Real Gabinete, doc. 418.

21. Alzate was requesting a copy of Maquer's *Chemical Dictionary* for use in his own saltpeter factory. In his zeal he goes so far as to furnish Gómez Ortega with instructions on the most effective means of postage and packaging. Included with his letter were samples of the herb known as *yerba del pollo* (chicken's herb), and two issues of the *GM*. Letter from Alzate to Casimiro Gómez Ortega (16 July 1784), Real Jardín Botánico, Legajo 20, 1, doc. 3.

22. Letter from Peris to Casimiro Gómez Ortega (15 January 1785), ibid., doc. 4. González Bueno notes that it is unlikely that Alzate would have received the publication he was seeking but that, in a move as politically strategic as scientific, Gómez Ortega puts his name forward as correspondent to the Botanical Garden on 8 January 1785. This letter was read, along with a presentation of the samples sent by Alzate, before the general assembly of the Botanical Garden on the same day that Alzate was named correspondent. See Antonio González Bueno, "Las relaciones de José Antonio de Alzate y Ramírez con los Reales Gabinetes de la Metrópoli," in Aceves Pastrana, *Periodismo científico*, 107–22.

23. Yolanda Agudín, *Historia del periodismo en México, desde el virreinato hasta nuestros días* (Mexico City: Panorama, 1987), 17.

24. *Diario Literario de México, dispuesto para la utilidad pública a quien se dedica* (12 March–10 May 1768), Imp. de Biblioteca Mexicana, 8 issues; the final issue was suspended by order of the Viceroy Marquis de Croix. *Asuntos Varios sobre Ciencia y Artes* (26 October 1772–4 January 1773), Imp. de José de Jáuregui, 12 issues, until finally being prohibited by the government.

25. *Mercurio Volante con Noticias Importantes i Curiosas sobre Varios Asuntos de Física i Medicina* (Mexico City: D. Felipe de Zúñiga y Ontiveros, Calle de la Palma, October 1772–February 1773). The *Mercurio Volante* appeared regularly during its short run.

26. Yves Aguila, "El periodismo científico en la Nueva España: Alzate y Bartolache, 1768–1773," in *La América española en la época de las luces* (Madrid: Cultura Hispánica, 1988), 303.

27. José Antonio Alzate y Ramírez, *Asuntos Varios sobre Ciencia y Artes* 11 (28 December 1772); *Obras*, vol. 1, *Periódicos*, ed. Roberto Moreno (Mexico City: UNAM, 1980).

28. John D. Browning, "The Periodical Press: Voice of the Enlightenment in Spanish America," *Dieciocho* 5, no. 1 (1980), 10.

29. D. Manuel Antonio Valdés, *Gazetas de México, compendio de noticias de Nueva España, desde principios del año 1784*, con Licencia y Privilegio (Mexico City: D. Felipe de Zúñiga y Ontiveros, Calle del Espíritu Santo, n.d.).

30. Yves Aguila, "El periodismo en la Nueva España: Alzate y Bartolache, 1768–1773," in *La América española en la época de las luces* (Madrid: Cultura Hispánica, 1988), 309. Alberto Saladino García has also presented an excellent overview of the scientific content of the most important publications in the

late-colonial Spanish American periodical press in his study, *Ciencia y prensa durante la Ilustración latinoamericana* (Mexico City: UNAM, 1996).

31. José Antonio Alzate y Ramírez, *Observaciones sobre la Física, Historia Natural y Artes Útiles* (21 March 1787–30 July 1788). Fourteen issues were printed by José Francisco Rangel.

32. See Antonello Gerbi, *The Dispute of the New World: The History of a Polemic, 1750–1900*, trans. Jeremy Moyle (Pittsburgh: University of Pittsburgh Press, 1973).

33. See David Brading. *The First America: The Spanish Monarchy, Creole Patriots, and the Liberal State, 1492–1867* (Cambridge: Cambridge University Press, 1991).

34. See Jorge Cañizares-Esguerra. *How to Write the History of the New World: Histories, Epistemologies, and Identities in the Eighteenth-Century Atlantic World* (Stanford, CA: Stanford University Press, 2001).

35. For further discussion of this aspect of Alzate's work, see Rosalba Cruz Soto, "El Nacionalismo de José Antonio de Alzate en el periódico científico, *Gaceta de Literatura*," in Aceves Pastrana, *Periodismo científico*, 618–50; Rafael Moreno, "La ciencia y la formación de la mentalidad nacional en Alzate," *Quipu* 6, no. 1 (1989), 93–107 and "Creación de la nacionalidad mexicana," *Historia Mexicana* 12 (1963), 531–51; Roberto Moreno, *Un eclesiástico criollo frente al estado Borbón: Discurso* (Mexico City: UNAM, 1980), "El indigenismo de Clavijero y de Alzate," in *Estudios sobre política indigenista española en América* (Valladolid, Spain: Universidad de Valladolid, 1977), 43–52, "Las notas de Alzate a la historia antigua de Clavijero," *Estudios de Cultura Nahuatl* 10 (1972), 359–92, and his Introduction to Alzate y Ramírez in *Obras*, vol. 1, *Periódicos*; and Alberto Saladino García and Juan José Saldaña (eds.), *José Antonio Alzate y Ramírez: Homenaje en el bicentenario de su fallecimiento* (Mexico City: Universidad Autónoma de México, 1999).

36. According to Alzate, two other individuals were involved with him in the original planning of the *GLM*. See *GLM* 1, no. 15 (16 December 1788), 39. He was joined in the first volume by two main contributors: José Moziño (who would, in 1789, leave to participate in the Botanical Expedition under Martín de Sessé) and the lesser known Mariano Castillejo, a young Mexican lawyer. This stands in stark contrast to contemporary periodicals in Spanish America that were either edited by, or acted as the voice pieces of, learned societies such as the Sociedades Económicas de Amigos del País. Periodicals launched within this role include: *Gazeta de Guatemala* (1796); *Mercurio Peruano* (1791); *El Papel Periódico de la Havana* (1790); *Papel Periódico de Santafé de Bogotá* (1791); and *Primicias de Quito* (1792).

37. Few correspondents used pseudonyms. Many of those who contributed, some of whom are linked through family ties—for example, Mariano Castillejo, his father, and his brother-in-law—appear to come from a variety of professions and social backgrounds: priests, doctors, surgeons, members of the Mining Tribunal, royal officials, poets, and lawyers as well as those whose actions or knowledge are reported indirectly, such as native peoples and sometimes children. In geographical terms we find references to Guadalajara, Tonalá, Veracruz, Puebla, and Oaxaca, among others.

38. See Tables 8.1, 8.2, 8.3, and 8.4 for the contrasting number of articles available in each volume. The decrease in articles in volume 3 was, to a large extent, due to Alzate's lengthy publication of his study of the cochineal beetle that appeared between February and September of 1794. Added to this was the fact that no issues were published in July or August of that year.

39. Alzate y Ramírez, *Observaciones*.

40. François Rozier, Jean-André Mongez, and Jean-Claude de La Métherie, *Observations sur la physique, sur l'histoire naturelle et sur les arts* (Paris: Journal de Physique, 1773–1793).

41. See Alzate, *Obras*, 153–63.

42. Quotation taken from correspondence sent to Rozier by the Italian mathematician, Pistoi, and published in the *Journal de Physique* (November 1777). *GLM* 1, no. 15b (12 April 1790), 115–17.

43. As is frequently the case when various subjects are dealt with within one issue, the print layout fails to clearly mark divisions between the different articles. Alzate will regularly jump from one subject to another without clearly defining the end of each section or, at times, even distinguishing between the words of the original author and his own comments and additions.

44. The tables in this chapter are based on the articles that have, to date, been verified as belonging to the cited journals. A few additional references exist that I have not yet located, at times despite the fact that Alzate names the title of the journal to which they belong. It is not yet clear why this dichotomy exists.

45. *Bibliothèque Physico-Économique, Instructive et Amusante . . . contenant des mémoires et observations-pratiques sur l'économie rurale,—sur les nouvelles découvertes les plus intéressantes . . .* (Paris: Buisson, Libraire, Hôtel de Mesgrigny, 1782–1830).

46. *Le Journal des Sçavans* (Paris: Jean Cusson, 1665–1797).

47. Pierre Rousseau, *Journal Encyclopédique ou Universel* (Liège; Boullion: L'Imprimerie du Journal, 1756–1794).

48. For our present purposes the many articles linking the *Gazeta de Literatura* with the *Gazeta de México* have been omitted, as they form the subject of a different and more complex discourse within the Mexican periodical press. See Fiona Clark, "The *Gazeta de Literatura de México* (1788–1795): The Formation of a Literary-Scientific Periodical in Late-Viceregal Mexico," *Dieciocho: Hispanic Enlightenment* 28, no. 1 (2005), 7–30.

49. *Memorial literario instructivo y curioso de la Corte de Madrid* (Madrid: Imprenta Real, 1784–1791); *Diario curioso, erudito, económico, y comercial* (Madrid, 1786–1787); Miguel Germonymo Suárez y Núñez, *Memorias instructivas, útiles, y curiosas, sobre agricultura, comercio, industria, economía, medicina, química, botánica, histora natural, etc.* (Madrid: P. Marín, 1778–1791). Alzate provides other references to the *Gazeta de Madrid*; however, I have not been able to locate the stated articles on any search to date.

50. *Mercurio Peruano* (Lima: Imp. Real de los Niños Huérfanos, 1791–95); *Papel Periódico de la Havana* (Havana, 1790–1805).

51. *GM* 4, no. 13 (6 July 1790), 131 announces opportunities to subscribe to the twenty-four issues of the 1791 publication run of the *Memorial Literario Instrutivo y Curioso* and provides further opportunities to buy previous issues.

52. *GM* 8, no. 11 (10 May 1796), 92 announces that all twenty-one volumes of the *Diario de Madrid* (previously titled *Diario Curioso*) will be available for sale at Manuel Lopez de Luna in Veracruz.

53. *GM* 3, no. 34 (23 June 1789), 342 announces that the Inquisition has ruled against numbers 269 and 270 of the *Diario Curioso*, as well as certain numbers of *El Censor* and the *Correo de Madrid*. *GM* 8, no. 23 (11 November 1796), 186 announces that an edict has been pronounced banning numbers 71–100 of volume 7 of the Suarez *Memorias* for containing doctrine that went against church teaching.

54. *GLM* 1, no. 1 (15 January 1788), 3.

55. Ibid., 2.

56. *GLM* 2, no. 9 (30 December 1790), 67:

The author of this Gazeta has been successful in vindicating the Nation against the impostures heaped upon it by certain foreigners . . . Continually driven forward by love for the truth, I waste no opportunity to resist the daring way in which some foreigners insult us . . . I will only present some articles correcting them through the use of a few additional notes to demonstrate the level of deception evident among many individuals in Europe who are reputed to be intellectuals of the highest order.

57. Ibid., 70.

58. GLM 2, no. 13 (22 February 1791), 105.

59. *GLM* 2, no. 14 (8 March 1791), 111nC.

60. *GLM* 1, no. 9 (28 June 1788), 78.

61. *GLM* 2, no. 19 (17 May 1791), 147–53.

62. In the case of the lightning rod, for example, a closer study of the translations of the articles would show that he omits from his text much of the material relating to the debate over the preferential use of rounded or pointed tips.

63. *GLM* 2, no. 11 (8 April 1788), 94.

64. *GLM* 3, no. 15 (7 March 1793), 114.

65. *GLM* 3, no. 13 (28 May 1793), 97. See also *GLM* 3, no. 37 (17 February 1795), 289.

66. *GLM* 2, no. 15 (22 March 1791), 115n.

67. Filippo Salvatore Gilij, *Saggio di storia americana, o sia storia naturale, civile e sacra de' regni e delle provincie spagnuole di Terra-ferma nell'America meridionale*, 4 vols. (Rome: Luigi Perego Salvioni, 1780–1784).

68. *GLM* 1, no. 7b (9 December 1789), 53 and no. 8b (23 December 1789), 58.

69. Fiona Clark, "Lost in Translation: The *Gazeta de Literatura de México* and the Epistomological Limitations of Colonial Travel Narratives," *Bulletin of Spanish Studies* 75, no. 2 (2008), 151–73.

Chapter Nine

1. J. C. Pinto de Sousa, *Biblioteca histórica de Portugal e seus domínios ultramarinos* (Lisbon: Typografia Chalcographica do Arco Cego, n.d.), 44 (archived at Biblioteca Pública de Évora).

2. Minas Gerais was the name of the captaincy that was created in 1720 in Brazil's southwest where the Portuguese found gold and diamonds in the eighteenth century.

3. José Rodrigues Abreu, *Historiologia médica, fundada e estabelecida nos princípios de George Ernesto Stahl,* 4 vols. (Lisbon: Oficina de Antonio de Sousa da Silva, 1733–1752), vol. 2 (1745), part 2.

4. At that time, Minas Gerais was part of a very large captaincy named Repartição Sul, which comprised the entire colony to the south: Rio de Janeiro, São Paulo, Minas do Ouro, and Espírito Santo. After the Emboabas War, São Paulo and Minas do Ouro became a separate captaincy and Antônio de Albuquerque was appointed its first governor.

5. A. J. Russell-Wood, "Identidade, etnia e autoridade nas Minas Gerais do século XVIII: Leituras do Códice Costa Matoso," *Vária Historia* 21 (Belo Horizonte: Editora UFMG, July 1999), 100–18; Adriana Romeiro, *Um visionário na corte de D. João V: Revolta e milenarismo nas Minas Gerais* (Belo Horizonte: Editora UFMG, 2001), 195–97.

6. The "Brazilica," or general language, was created by the Jesuits in order to make comunication with the natives easier. Luciano Raposo and Maria Veronica Campos, *Códice Costa Matoso* (Belo Horizonte: Fundação João Pinheiro, 1999), 1:202, 206.

7. Laura de Mello e Souza, "Os nobres governadores de Minas, mitologias e histórias familiares," in *Norma e conflito: Aspectos da história de Minas no século XVIII* (Belo Horizonte: Editora UFMG, 1999), 185.

8. José Joaquim da Rocha, *Geografia histórica da capitania de Minas Gerais: Estudo crítico,* ed. Maria Efigênia Lage de Resende (Belo Horizonte: Fundação João Pinheiro, 1995), 88.

9. Raposo and Campos, *Códice Costa Matoso,* 1:213, 235.

10. Mello e Souza, "Os nobres governadores de Minas," 185.

11. Arquivo Público Mineiro, Seção Colonial, Belo Horizonte, códice 7, fol. 89.

12. Abreu, *Historiologia médica.*

13. Ibid., 2:514–29. The simples were the natural elements that when combined with other elements resulted in the medicaments.

14. Diogo Barbosa Machado, *Biblioteca lusitana* (Lisbon: Oficina de Inácio Rodrigues, 1747), 2:895–96.

15. Arquivos Nacionais da Torre do Tombo (ANTT), Chancelaria de Dom Pedro II (Lisbon, 1699), livro 49, fol. 358.

16. This post was an important honor. Francis A. Dutra, "The Practice of Medicine in Early Modern Portugal," in *Libraries, History, Diplomacy, and the Performing Arts,* ed. Israel J. Katz (New York: Pendragon Press, 1991), 135–37.

17. Machado, *Biblioteca lusitana,* 2:895–96.

18. Joaquim Veríssimo Serrão, *História de Portugal—1640–1750,* 2nd ed. (Lisbon: Editorial Verbo, 1982), 5:254.

19. Since the eighteenth century, "the noble class was being broadened to include, among others, important physicians and doctors." Dutra, "The Practice of Medicine," 140.

20. ANTT, Ministério do Reino, Decretamentos de Serviço, maço 1, doc. 39.

21. Abreu, *Historiologia médica*, 1:a1.

22. ANTT, "Real Mesa Censória," *Gazeta de Lisboa*, caixa 465, no. 7 (17 February 1750); no. 40 (quinta feira, 9 November 1752); no. 36 (4 October 1755). I thank André Belo for this reference.

23. Machado, *Biblioteca lusitana*, 2:895–96.

24. Manoel Cardozo, "The Internationalism of the Portuguese Enlightenment: The Role of the Estrangeirado, c. 1700–c. 1750," in A. O. Aldridge, *The Ibero-American Enlightenment* (Urbana: University of Illinois Press, 1971), 153–67; Kenneth Maxwell, *Marquês de Pombal: Paradoxo do Iluminismo* (São Paulo: Paz e Terra, 1996), 14–19; Kenneth Maxwell, *Pombal, Paradox of Enlightenment* (Cambridge: Cambridge University Press, 1995).

25. Overseas Council, or Conselho Ultramarino, was a body that served as the primary institution for Portuguese overseas politics.

26. ANTT, Chancelaria de Dom João V, livro 122, fol. 36r–36v.

27. Abreu, *Historiologia médica*, 1:b4v.

28. The collection is very rare. It can be found in Biblioteca Nacional and Biblioteca da Ajuda, both in Lisbon.

29. At that time (1517–1768), the books to be printed in Portugal had to be licensed in a triple system: by the Inquisition; by the Ordinários—parochial ecclesiastical judges; and by the Desembargo do Paço—a superior tribunal.

30. Abreu, *Historiologia Médica*, 1:b–b5.

31. Anthony Grafton's comments on Renaissance astrology are also valid for the study of medicine during this period: medicine was at the same time marked by rationalism and irrationality, by tradition and its novel contents, based both in ancient sources and in its modern social role. Anthony Grafton, *Cardano's Cosmos: The Worlds and Works of a Renaissance Astrologer* (Cambridge, MA: Harvard University Press, 1999), 15.

32. The philosophers of that time believed that the universe was created by God and was governed by perfect laws, as a mirror of the Creator. It was their mission to discover the rules that the Creator had established and that governed the world, compounding the Book of Nature.

33. Followers of Paracelsus also held that effervescence, fermentation, and putrefaction were the processes of life, and denied the existence of an incorruptible element. Mary Lindemann, *Medicine and Society in Early Europe* (Cambridge: Cambridge University Press, 1999), 77–83.

34. C. S. Lewis, *The Discarded Image: An Introduction to Medieval and Renaissance Literature* (Cambridge: Cambridge University Press, 1964), 75–76.

35. Anthony Stevens, *Ariadne's Clue: A Guide to the Symbols of Humankind* (Princeton, NJ: Princeton University Press, 1998), 132–34.

36. A. Tavares Sousa, *Curso de história da medicina: Das origens aos fins do século XVI* (Lisbon: Fundação Calouste Gulbenkian, 1996), 377.

37. Anthony Grafton and Nancy Sirasi (eds.), *Natural Particulars: Nature and the Disciplines in Renaissance Europe* (Cambridge, MA: MIT Press, 1999); Grafton, *Cardano's Cosmos*.

38. Alfred Gierer, "Organisms-Mechanisms: Stahl, Wolff, and the Case against Reductionist Exclusion," *Science in Context* (Cambridge University Press) 9, no. 4 (1996), 515.

39. "The soul, in Stahl, is an active being: intelligent, autonomous, architectural, capable of varying the vital movements, which it produces according to the ends of life. The soul has an ontological reality." Paul Hoffman, "La controverse entre Leibniz et Stahl sur la nature de l'âme," *Studies on Voltaire and the Eighteenth Century* 199 (1919), 239.

40. George Ernest Stahl, *Philosophical Principles of Universal Chemistry* (London: John Osborn, 1730), 9.

41. Ibid., front cover, 319–21, 393–424.

42. Ibid., 394, 415, 405–6.

43. Abreu, *Historiologia médica*, 2:b3, b1v, 514.

44. Stahl, *Philosophical Principles of Universal Chemistry*, 291.

45. Ibid., 319.

46. Solar medicine was based on the effect of gold (brilliant like the sun) in curing illness in the human body.

47. Abreu, *Historiologia médica*, 2:528–29.

48. Ibid., 514–15.

49. Ibid., 520.

50. Ibid., 515.

51. Ibid., 584, 598, 515.

52. Ibid., 527.

53. In the eighteenth century, the Portuguese Inquisition tried to eliminate from the popular culture the idea of an earthly paradise, considering it a heretical superstition.

54. Ibid., 515–16.

55. Ibid., 516.

56. Ibid.

57. Ibid., 517, 525.

58. Ibid., 523–24.

59. Plínio Freire Gomes, *Um herege vai ao paraíso: Cosmologia de um ex-colono condenado pela Inquisicão (1680–1744)* (São Paulo: Companhia das Letras, 1997), 111; Romeiro, *Um visionário na corte de Dom João V*, 162–67.

60. André Thevet, *As singularidades da França Antártica* (1557-1558; Belo Horizonte, Brazil: Itatiaia, 1978), 121; Gabriel Soares de Sousa, *Tratado descritivo do Brasil* (1587; Recife: Editora Massangana, 2000), 27.

61. Simão de Vasconcelos, *Chronica da Companhia de Jesus do Estado do Brasil e do que obrarão seus filhos nesta parte do Novo Mundo* (Lisbon: Oficina de Henrique Valente de Oliveira, 1663), 1:31. I thank Guita and José Mindlin for allowing me to consult their precious library, where I found this and other books used in this chapter.

62. Abreu, *Historiologia médica*, 2:519.

63. Sérgio B. Holanda, *Visão do Paraíso*, 6th ed. (São Paulo: Brasiliense, 1994), 8; Freire Gomes, *Um herege vai ao paraíso*, 112–13.

64. Alfredo Pinheiro Marques, *A cartografia dos descobrimentos* (Lisbon:

Elo, 1994), 65; Sérgio B. Holanda, "Um mito geopolítico: A ilha Brasil," in *Tentativas de mitologia* (São Paulo: Perspectiva, 1979).

65. Abreu, *Historiologia médica*, 2:518.

66. Jaime Cortesão, *História do Brasil nos velhos mapas*, 2 vols. (Rio de Janeiro: Instituto Rio Branco, 1957); Sérgio B. Holanda, *O extremo oeste* (São Paulo: Brasiliense, 1986), 92–93.

67. Afonso Taunay (ed.), *Relatos monçoeiros* (São Paulo: Martins Fontes Editora, 1976).

68. Sérgio B. Holanda, *Monções* (São Paulo: Brasiliense, 1990).

69. Hans Staden, *A verdadeira história dos selvagens nus e ferozes devoradores de homens encontrados no Novo Mundo, a América* (1557; Rio de Janeiro: Dantes, 1999).

70. Jacques Cartier, *Voyages de découverte au Canada entre les années 1534 et 1542: Suivis d'une biographie de Jacques Cartier par René Maran* (Paris: Éditions Anthropos, 1968).

71. Girolamo Benzoni, *La Historia del Mondo Novo* (Venice: Pietro e Francesco Tini, 1572); Gonzalo Fernandes de Oviedo y Valdés, *Historia general y natural de las Indias, islas y tierra-fierme del mar océano*, 5 vols. (1535–1548; Madrid: Real Academia Española, 1959).

72. Abreu, *Historiologia médica*, 2:517.

73. Ibid., 526.

Chapter Ten

1. David N. Livingstone, *The Geographical Tradition: Episodes in the History of a Contested Enterprise* (Oxford and Cambridge, MA: Blackwell, 1992), 102ss.

2. Richard Helgerson, "The Land Speaks: Cartography, Chorography, and Subversion in Renaissance England," *Representations* 16 (1986), 50–85.

3. Anne Godlewska, "Traditions, Crisis, and New Paradigms in the Rise of Modern French Discipline of Geography 1760–1850," *Annals of the Association of American Geographers* 79, no. 2 (1989), 192–213; Anne Godlewska, *Geography Unbound: French Geographic Science from Cassini to Humboldt* (Chicago and London: Chicago University Press, 1999).

4. Although the empire developed instruments for the monitoring of the geographic evolution of cities from the time of Felipe II, this information would generally be linked to the economic (and historical) management of the territory, and it would not receive an overall graphical form, partly because many of the maps sent by the colonies came from the natives and were, therefore, useless for the purpose of obtaining a consistent overall image. In the case of New Spain, 65 percent of the maps were commissioned from natives, because "images were the province of the indios." See Barbara E. Mundy, *The Mapping of New Spain: Indigenous Cartography and the Maps of the Relaciones Geográficas* (Chicago and London: University of Chicago Press, 1996), 30.

5. Godlewska, *Geography Unbound*, 44.

6. We have explored the biopolitical dimension of the expeditionary enter-

prise and how metropolitan management of natural wealth clashes with local interests in A. Lafuente and N. Valverde, "Linnaean Botany and Spanish Imperial Biopolitics," in *Colonial Botany: Science, Commerce and Politics in the Early Modern World*, ed. Londa Schiebinger and Claudia Swan (Philadelphia: University of Pennsylvania Press, 2005), 134–47.

7. Letter from Cruz Cano to the Marqués de Grimaldi (31 March 1770), in Ricardo Donoso, "El mapa de la América meridional de la Cruz Cano y Olmedilla," *Revista Chilena de Historia y Geografía* 131 (1963), 142; letter from Cruz Cano to the Marqués de Grimaldi (8 December 1767), in ibid., 141, 136; Thomas R. Smith, "Cruz Cano's Map of South America, Madrid, 1775: Its Creation, Adversities, and Rehabilitation," *Imago Mundi* 20 (1966), 55, 57.

8. Donoso, "El mapa de la América meridional," 143.

9. Letter from Hipólito Ricarte to Francisco Manuel Mena (8 December 1775), in ibid., 154.

10. Ibid., 155.

11. On the geographical myths included in the map, see Ken Mitchell, "Science, Giants and Gold: Juan de la Cruz Cano y Olmedilla's Mapa Geografico de la América Meridional," *Terra Incognita* 31 (1999), 25–41.

12. Donoso, "El mapa de la América meridional," 144.

13. Orders from Ceballos to Vertiz (12 June 1778), in ibid., 123–26.

14. Report of Francisco Requena on Cruz's map (9 December 1802), in ibid., 167.

15. Less demand for ethnographic information does not imply a lack of interest in native or local cultures on the part of the metropolis. For centuries, but significantly throughout the 1700s, data would be requested in geographical accounts on the physical, cultural, and economic environment of the colonies. Gerhard compares the evolution of the accounts from 1523 to 1825 and shows the loss of interest in pre-Hispanic culture as of 1648–50, while showing a more constant monitoring of the economy and culture and that of infrastructures from then onward. See Paul Gerhard, *Geografía histórica de Nueva España 1519–1821* (Mexico City: UNAM, 1986), 30.

16. *Mapa corográfico del Nuevo Reyno de Granada, que comprehende desde los cuatro grados de latitud Norte hasta la costa del mar del Norte. Construido sobre las mejores observaciones astrónomicas, modernas noticias y operaciones trigonométricas por D. Vicence Talledo y Rivera, teniente coronel del Real Cuerpo de Ingenieros. Por disposición del excelentísmo seño don Antonio Amar y Bortón, virrey, gobernador y capitán general de dicho reyno. Año de 1808.*

17. What follows is based fundamentally on Michel Antochiw, "La visión total de la Nueva España: Los mapas generales del siglo XVIII," in *México a través de los mapas*, ed. Héctor Mendoza Vargas (Mexico City: UNAM, 2000); and Elías Trabulse, "La cartografía en la historia de la ciencia en México," in *Cartografía mexicana tesoros de la nación: Siglos XVI a XIX* (Mexico City: Archivo General de la Nación, 1983).

18. Antochiw, "La visión total de la Nueva España," 81.

19. Trabulse, "La cartografía," 26.

20. Gerhard, *Geografía histórica*, 33.

21. Trabulse, "La cartografía," 25.

22. Alzate's map is no exception. From the same period comes, for example, the *Mapa corográfico de la provincia que propiamente se llama Sonora . . . situada en la America septentrional* (1768) of Nicolás Medina y Cabrera, advocate in the royal court of Mexico, with a border depicting agricultural and stockbreeding activities (152 x 151 cm, Museo Naval, Madrid).

23. *Tercera parte del mapa que comprende la frontera de los dominios del rey en la America septentrional copiado por Dn. Luis de Surville según el original que hizo Dn Jph. De Urrutia sobre varios puntos observados por él y el capitán de yngenieros D. Nicolas Lafora* (Servicio Geográfico del Ejército, Madrid, 1769).

24. This same type of information appears, for example, in Surville's map, the anonymous *Map of Sonora dedicated to D. José Tienda de Cuervo, governor of the provinces of Sinaloa, Sonora . . .* , 1770 (Museo Naval, Madrid), which also includes watering holes and springs; the *Mapa Coro-grafico de la Nueva Andalucia de D. Luis Surville*, produced by order of José Gálvez in 1778 and published in the *Historia coro-grafica, natural y evangelica de la Nueva Andalucia* (Madrid, 1779) of Father Antonio Caulin.

25. Luis Navarro García, *José Gálvez y la comandancia general de las provincias internas* (Seville: CSIC, 1964), 118. A series of forts in the North area of New Spain, along the border of the Internal Provinces, was built to protect the new settlements and control the indigenous forces. The exhaustion of resources, movements of populations, and other circumstances registered in maps justified the creation or suppression of forts along the imaginary line that connected them.

26. The foundations of this idea had been laid by the Real Ordenanza de Intendentes (Royal Ordinance of Governors), particularly articles 57 and 58, which Revilla Gigedo, with insufficient men to carry out the work, solved by commissioning several officers in 1791 to "draw up plans of the places which they consider worthy of this operation, extending their reports and observations, in order to give some idea of their local situation and of the advantages they may yield or defects that should really be attended to." Revilla Gigedo to Carlos Urrutia (30 October 1793), cited in Carlos Urrutia, "Noticia geográfica del reino de Nueva España y estado de su población, agricultura, artes y comercio (1794)," in E. Florescano and I. Gil, *Descripciones económicas generales de Nueva España 1784–1817* (Mexico City: Instituto Nacional de Antropología e Historia, 1973), 74.

27. Navarro García, *José Gálvez*, 135.

28. Ibid., 216–17.

29. This was the case of the town and encampment of San Juan de Sonora, the first capital of the province, which disappeared due to the incursions of the Seri Indians. Ibid., 121.

30. Ibid., 187.

31. The encampment at Basis, for example, had 594 people in 1778 but in the following year was down to only 289. Ibid., 413.

32. This was the case in Sonora around 1777, when the disappearance of many settlements left the forts disconnected from each other and the units of each fort were obliged to perform other duties, such as providing supplies of

provisions and mail for the fort, and military support to frontier towns. The military force was thus fragmented, and the strategic value of a chain of forts that worked as a wall was lost. Ibid., 387.

33. See Richard Helgerson, "The Land Speaks: Cartography, Chorography, and Subversion in Renaissance England," *Representations* 16 (1986), 50-85.

34. For example, in the pacification of Nueva Vizcaya, soldiers, business and mining representatives, and individuals were consulted about the relocation of the forts. Navarro García, *José Gálvez*, 109-10.

35. One of the more important collective contributions was without a doubt that of the Spanish trade delegation of Jalapa; the consulates and merchants of Mexico, Puebla, and Veracruz; the ecclesiastical chapters of Durango and Oaxaca; and some merchants, miners, and neighbors of the missions and the border, who financed most of the campaign of Gálvez. Ibid., 149-50. The authorities were confident of the economic commitment to stabilize the territories, and for that reason after the pacification of Sonora, Gálvez published a "Plan of a company of Shareholders to foment the activity of benefit of the rich Sinaloa and Sonora mines, and to restore the Pearl Fisheries in the Gulf of California" (1771), although the project did not have the awaited endorsement. Ibid., 200, 253.

36. This form of participation was particularly encouraged due to its political impact. For that reason Colonel Domingo Elizondo in 1770 informed Bucareli that the Seris and Suaquis of Pitic had constructed a corral and an irrigation ditch to irrigate their crops, and had a dam under construction. Navarro considers this information as the veritable announcement of the pacification of the zone (ibid., 205). Beneath these ideas of collective responsibility in the construction of the territory lies the internalization of biopolitics, the notion that "the more citizens there are in a Kingdom, the more the taxpayers and the greater the number of hands available for all the enterprises and resources to increase the general wealth." Urrutia, "Noticia geográfica del Reino de Nueva España," 101.

37. The 1770 discovery of the extraordinarily productive La Cieneguilla mine in Sonora, for example, had to confront the problem of water shortage in the area. Navarro García, *José Gálvez*, 205-6.

38. *Carta esférica de las costas de la América meridional desde el paralelo de 36° 30' de latitud S. hasta Cabo de Hornos levantada de orden del Rey en 1789, 90, 94 y 95 por varios Oficiales de la Marina* (1798), in Julio Guillén Tato, *Monumenta chartographica indiana* (Madrid: Ministerio de Asuntos Exteriores, 1942), map 95.

39. Brian Harley, "Silences and Secrecy: The Hidden Agenda of Cartography in Early Modern Europe," in *The New Nature of Maps: Essays in the History of Cartography*, ed. Paul Laxon (Baltimore and London: Johns Hopkins University Press, 2001), 98.

40. The Dirección Hidrográfica (Hydrographic Office) set up in 1797 to ensure accurate Spanish hydrographic mapmaking, possessed in 1808 a collection of 80,000 charts, plans, and pictures. Ursula Lamb, *Martín Fernández de Navarrete Clears the Deck: The Spanish Hydrographic Office (1802-1824)*, Centro de Estudos de Cartografía Antiga, Série Separatas 81 (Coimbra: Junta de Investigaçoes Científicas do Ultramar, 1980), 31.

41. Monica L. Smith, "Networks, Territories, and the Cartography of Ancient States," *Annals of the Association of American Geographers* 95, no. 4 (2005), 832–49.

42. Philip E. Steinberg, "Lines of Division, Lines of Connection: Stewardship in the the World Ocean," *Geographical Review* 89, no. 2 (1999), 254–64.

43. On the vicissitudes of the British about-face in matters of foreign policy and its influence on American expansion, see the classic article by Felix Gilbert, "The English Background of American Isolationism in the Eighteenth Century," *William and Mary Quarterly* 1 (1944), 138–60, reproduced in *American Empire in the Pacific: From Trade to Strategic Balance, 1700–1922 (The Pacific World)*, ed. Arthur Power Dudden (Aldershot, England: Ashgate Variorum, 2004).

44. John R.Gillis, "Islands in the Making of an Atlantic Oceania, 1400–1800." Paper presented at Seascapes, Littoral Cultures, and Trans-Oceanic Exchanges, Library of Congress, Washington DC, February 12–15, 2003, http://www.historycooperative.org/proceedings/seascapes/gillis.html (Accessed on 14 May 2008).

45. Juan Pimentel has analyzed how this change affected the perception of overseas territory and the strategies of exploration in the cases of Quirós—for whom the discovery of a coast was sufficient grounds to claim a continent—and of Cook, where in spite of surveying 3,000 km of coastline, there was still a great void. Juan Pimentel, *Testigos del mundo: Ciencia, literatura y viajes en la Ilustración* (Madrid: Marcial Pons, 2003), 71–109.

46. Lauren Benton, "Oceans of Law: The Legal Geography of the Seventeenth-Century Seas" (paper presented at Seascapes, Littoral Cultures, and Trans-Oceanic Exchanges, Library of Congress, Washington, DC, February 12–15, 2003; see http://www.historycooperative.org/proceedings/seascapes/benton.html, accessed on May 14, 2008).

47. A good summary of the events surrounding this crisis is in Salvador Bernabeu Albert, "El tratado de Tordesillas y su Repercusión en el Tratado de Límites de 1792," in Juan Francisco Bodega y Quadra, *Nutka 1792: Viaje a la costa noroeste de la América septentrional por Don Juan Francisco de la Bodega y Quadra, capitán de navío*, ed. Mercedes Palau, Freeman Tovell, Pamela Sprätz, and Robert Inglis (Madrid: Ministerio de Asuntos Exteriores, 1998), 47–59. On the complex diplomatic negotiations over the crisis, see Howard V. Evans, "The Nootka Sound Controversy in Anglo-French Diplomacy, 1790," *Journal of Modern History* 46, no. 4 (1974), 609–40.

48. Only in 1580 had Britain expressed doubts about the Spanish king's sovereign rights over the Indias, based on the principle of occupation. See James Simsarian, "The Acquisition of Legal Title to Terra Nullius," *Political Science Quarterly* 53, no. 1 (1938), 113.

49. "Reflexiones sobre la convención hecha con Inglaterra en 28 de octubre de 1790," Archivo Historico Nacional (AHN), Estado, Legajo 4291; quoted in Bernabeu, "El Tratado de Tordesillas," 58.

50. Juan Francisco Bodega y Quadra to George Vancouver (13 September 1792), in Bodega y Quadra, *Nutka 1792*, 162.

51. "Tratado del Escorial," in Juan Francisco Bodega y Quadra, *El descubrimiento del fin del mundo (1775–1792)*, ed. Salvador Bernabeu (Madrid: Alianza Editorial, 1990), 259.

52. Bodega y Quadra, *Nutka 1792*, 173.

53. Ibid., 146.

54. Even those maps that refused to get into open argument on questions about property and jurisdiction were part of this new iconographic strategy. However, some of the maps, such as the *Carta General de Cuanto Hasta Hoy se Ha Descubierto y Examinado por los Españoles en la Costa Septentrional de California* (1791) of Bodega y Quadra, showed the dates of discovery of and visits to the different settlements, thus fully capturing the new role of charts in the field of international law.

55. Alejandro Malaspina, "Examen político de las colonias inglesas en el mar Pacífico" (1793), in Juan Pimentel, *En el panóptico del Mar del Sur: Orígenes y desarrollo de la visita australiana de la Expedición Malaspina (1793)* (Madrid: CSIC, 1992), 135.

56. Antonio Lafuente and Nuria Valverde, "The Emergence of Early Modern Commons: Technology, Heritage and Enlightenment" (preprint, 2004) in Digital CSIC, http://hdl.handle.net/10261/2854.

Chapter Eleven

1. I discuss further the emergence of empirical practices in sixteenth-century Spain in *Experiencing Nature: The Spanish American Empire and the Early Scientific Revolution* (Austin: University of Texas Press, 2006). On the history of science in Spain, see Jorge Cañizares-Esguerra, "Iberian Science in the Renaissance: Ignored How Much Longer?" *Perspectives on Science* 12 (2004), 86–124; David Goodman, "The Scientific Revolution in Spain and Portugal," in *The Scientific Revolution in National Context*, ed. Roy Porter and Mikuláš Teich (Cambridge: Cambridge University Press, 1992); David C. Goodman, *Power and Penury: Government, Technology, and Science in Philip II's Spain* (Cambridge: Cambridge University Press, 1988); Luis García Ballester (ed.), *Historia de la ciencia y de la técnica en la Corona de Castilla*, 4 vols. (Salamanca: Junta de Castilla y León y Consejería de Educación y Cultura, 2002), vol. 3; José María López Piñero, *Ciencia y técnica en la sociedad española de los siglos XVI y XVII* (Barcelona: Editorial Labor, 1979).

2. Deb Harkness is producing a very interesting work on the artisans' activities in the streets of London; see her "'Strange' Ideas and 'English' Knowledge: Natural Science Exchange in Elizabethan London," in *Merchants and Marvels: Commerce, Science, and Art in Early Modern Europe*, ed. Pamela H. Smith and Paula Findlen (New York: Routledge, 2001), 137–60.

3. Fernand Braudel, *The Mediterranean and the Mediterranean World in the Age of Philip*, 2 vols. (New York: Harper, 1966), 1:372.

4. On this subject, see John Law, "On the Methods of Long-Distance Control: Vessels, Navigation and the Portuguese Route to India," *Sociological Review Monograph* 32 (1986), 234–63.

5. On the mapping activities of the *Casa*, see the pioneering work of Ursula Lamb, "Cosmographers of Seville: Nautical Science and Social Experience," in *First Images of America: The Impact of the New World on the Old*, ed. Fredi Chiappelli (Berkeley: University of California Press, 1976); "Science by Litigation: A Cosmographic Feud," *Terrae Incognitae* 1 (1969); and "The Spanish Cosmographic Juntas of the Sixteenth Century," *Terrae Incognitae* 6 (1974). See also Alison Sandman, "Mirroring the World: Sea Charts, Navigations, and Territorial Claims in Sixteenth-Century Spain," in Smith and Findlen, *Merchants and Marvels*, 83–108.

6. José Cervera Pery, *La Casa de Contratación y el Consejo de Indias: Las razones de un superministerio* (Madrid: Ministerio de Defensa, 1997); Ana Crespo Solana, *La Casa de Contratación y la Intendencia General de la Marina en Cádiz, 1717–1730* (Cádiz: Servicio de Publicaciones Universidad de Cádiz, 1996); José Pulido Rubio, *El piloto mayor de la Casa de la Contratación de Sevilla* (Seville: Publicaciones de la Escuela de Estudios Hispano-Americanos, 1950); Joseph de Veitía Linage, *Norte de la Contratación de las Indias Occidentales* (Seville, 1672; Buenos Aires: Publicaciones de la Comisión Argentina de Fomento Interamericano, 1945); José Manuel Piernas y Hurtado, *La Casa de la Contratación de las Indias* (Madrid: Librería de don Victoriano Suárez, 1907); Manuel de la Puente y Olea, *Los trabajos geográficos de la Casa de Contratación* (Seville: Escuela Tipográfica y Librería Salesianas, 1900); Manuel Ruiz del Solar y Uzuriaga, *La Casa de Contratación*, vol. 1, *El retablo y sus retratos*; vol. 2, *Los trabajos geográficos* (índice o breve resumen); vol. 3, *La celebración de su IV centenario en 1903* (Seville: Escuela Tipográfica y Librería Salesianas, 1900); Edward L. Stevenson, "The Geographical Activities of the Casa de la Contratación," *Annals of the Association of American Geographers* 27, no. 2 (1927), 39–59.

7. See the royal decree of 11 March 1532 (Ocaña, Spain), Archivo General de Índias (AGI), Indiferente 422, libro 15, fols. 18v–19r.

8. Cédula real otorgada a Diego Ribeiro (Granada, 26 November 1526), in Germán Latorre, *Diego Ribero: Cosmógrafo y cartógrafo de la Casa de la Contratación de Sevilla* (Seville: Centro Oficial de Estudios Americanistas de Sevilla, Tipografia Zarzuela, 1919), 17. This is my own translation.

9. Cédula real a los oficiales de la Casa de la Contratación (Avila, 17 July 1531), AGI, Indiferente 1961, libro 2, fols. 84v–85; and cédula real a los oficiales de la Casa de la Contratación (Medina del Campo, 13 October 1531), ibid., 99v.

10. Cédula real a los oficiales de la Casa de la Contratación (Medina del Campo, 4 November 1531), AGI, Indiferente 1961, libro 2, fols. 105v–106r.

11. See Latorre, *Diego Ribero*, 18ff.

12. Armando Cortesão, *Cartografia e cartógrafos portugueses dos séculos XV e XVI*, 2 vols. (Lisbon: Seara Nova, 1935), 2:137.

13. Roger Hahn describes a similar process of vetting information in the case of the French Académie des Sciences a century or so later. See Roger Hahn, *The Anatomy of a Scientific Institution: The Paris Academy of Sciences (1666–1803)* (Berkeley: University of California Press, 1986).

14. Consulta del Consejo (Madrid, 4 May 1570), AGI, Indiferente 738, no.

119. The Council made that statement in the context of the mining invention made by Benito de Morales.

15. See Latorre, *Diego Ribeiro*, 16.

16. Carta del Consejo de Indias a los oficiales de la Casa de la Contratación (Madrid, 28 June 1535), AGI, Indiferente 1961, libro 3, fols. 293v–294r.

17. Cédula real a Vicente Barreros (Madrid, 22 January 1536), AGI, Indiferente 1962, libro 4, fols. 31v–32v.

18. I discuss Villasante's case in more detail in Antonio Barrera-Osorio, "Local Herbs, Global Medicines: Commerce, Knowledge, and Commodities in Spanish America," in *Merchants and Marvels*, 163–81.

19. Cédula real a varios médicos y cirujanos (Madrid, 5 April 1530), AGI, Indiferente 422, libro 14, fols. 73r–74v.

20. I have not been able to locate these reports: perhaps they ended up in the hands of Villasante and his partners, and never made it into the royal archives.

21. Cédula real de la Reina a las justicias de Sus reinos (Madrid, 5 April 1530), ibid., 68r.

22. Cédula real a los visitadores del Hospital del Cardenal de la ciudad de Toledo (Madrid, 5 April 1530), ibid., 72v.

23. Cédula real a los visitadores de varios hospitales (Madrid, 5 April 1530), ibid.

24. Cédula real al bachiller Andrés de Jodar médico, vecino de Baeza (Madrid, April 5, 1530), ibid., 73r–74v.

25. Cédula real a varios médicos y cirujanos (Madrid, April 5, 1530), ibid., 73r–74v.

26. Cédula real a Pedro Benito de Basniana y Franco Leardo para que puedan subir los salarios asignados a los médicos que contribuyen a la propaganda del bálsamo (Madrid, 12 July 1530), ibid., 102r–103r.

27. Cédula real a los oficiales de Cuéllar (Madrid, 16 October 1532), AGI, Indiferente 422, libro 15, fols. 197v–198r.

28. Cédula real a los oficiales de Cuéllar (Madrid, 16 October 1532), ibid.; cédula real a Diego de la Haya para que pague a Melchor de Angulo (Madrid, 27 November 1532), ibid., 199v.

29. Cédula real a Juan de Vargas para que venga a la corte (Madrid, 21 November 1532), ibid., 199r; and mandamiento a Diego de la haya para que pague a Juan de Vargas por haber estado en la corte (Madrid, 27 February 1533), ibid.

30. Mandamiento a Diego de la Haya para que pague a Juan de Vargas por haber estado en la corte (Madrid, 27 February 1533), ibid.; cédula real a Diego de la Haya para que pague cierta suma a Juan de Vargas (Monzón, 3 October 1533), ibid., libro 16, fol. 43v; cédula real a Juan de Vargas (Toledo, 18 April 1534), ibid., 75v.

31. Cédula real a los alcaldes ordinarios de la villa de Amusco (Toledo, 23 May 1539), AGI, Indiferente 423, libro 19, fols. 247–48.

32. Eric H. Ash discusses the role mediators played between artisans and court officials in the English setting. For more, see his *Power, Knowledge, and Expertise in Elizabethan England* (Baltimore: Johns Hopkins University Press, 2004).

33. German miners were all over the Spanish empire from the Caribbean to Peru. For their dominance as experts in mining technology and practices, see ibid., esp. chap. 1.

34. Modesto Bargalló, *La minería y la metalurgia en la América española durante la época colonial* (Mexico City and Buenos Aires: Fondo de Cultura Económica, 1955), 56f.

35. "Petición de la Ciudad de México sobre el repartimiento general y perpetuo de la Nueva España: La presentó Juan Velázquez de Salazar al licenciado Juan de Ovando en Madrid, a 6 de junio de 1571," in *Epistolario de Nueva España*, ed. Francisco del Paso y Troncoso (Mexico City: Antigua Librería Robredo, de José Porrúa e Hijos, 1940), 11:118, doc. 659; Bargalló, *La minería y la metalurgia*, 91.

36. Muro, "Bartolomé de Medina," 524.

37. Troncoso, *Epistolario*, 11:118, doc. 659.

38. Juan Manuel Menes Llaguno, *Bartolomé de Medina: Un sevillano pachuquero* (Pachuca, Mexico: Universidad Autónoma del Estado de Hidalgo, 1989), 47ff.; Manuel Castillo Martos and Mervyn Francis Lang, *Metales preciosos: Unión de dos mundos* (Seville and Bogotá: Muñoz Moya y Montraveta, 1995), 96.

39. Merced, ca. November 1554. Mexico City, Archivo General de la Nación, Mexico, ramo de Mercedes, 5, fol. 87r–v, transcribed in Luis Muro, "Bartolomé de Medina, introductor del beneficio de patio en Nueva España," *Historia Mexicana* 13, no. 4 (1964), 517–31, 518ss.

40. Castillo and Lang, *Metales preciosos*, 99ss.

41. Vannocio Biringucci described the process in his *De la Pirotechnia* (Venice, 1540), book 9, chap. 11; see Bargalló, *La Minería y la metalurgia*, 107ff. English translation: Cyril Stanley Smith and Martha Teach Gnudi, *The Pirotechnia of Vannocio Biringucci: The Classic Sixteenth-Century Treatise on Metals and Metallurgy* (New York: Dover, 1990).

42. On this point, see Castillo and Lang, *Metales preciosos*, 97.

43. Nothing is known about this German expert save that Medina tried to take him to New Spain, but the German did not obtain a license. Medina went alone and worked for almost a year in the new method. See ibid., 97; Eli de Gortari, *La ciencia en la historia de México* (Mexico City: Editorial Grijalbo, 1979), 198f.

44. Muro, "Bartolomé de Medina," 522; Bargalló calculated that around 126 people were using Medina's technology; see Bargalló, *La Minería y la metalurgia*, 112.

45. Don Francisco de Mendoza, administrator of the Guadalcanal mines, in Andalucia, was ordered to implement the amalgamation process in those mines in instructions dated October 30, 1557. After some unsuccessful experiments, the *indiano* (as those who became rich from the New World were called) Mosén Boteller was called from New Spain to Guadalcanal to implement the new method. See Bargalló, *La minería y la metalurgia*, 121, 111.

46. Peter Bakewell, "Technological Change in Potosi: The Silver Boom of the 1570s," *Jahrbuch für Geschichte von Staat, Wirtschaft und Gesellschaft lateinamerikas* 14 (1977), 57.

47. Ibid., 58. On the *mita*, see Peter Bakewell, "Mining," in *Colonial Spanish America*, ed. Leslie Bethell (Cambridge: Cambridage University Press, 1987), 203–49, esp. 221f; Guillermo Lohmann Villena, "El Virreinato del Perú," in *El descubrimiento y la fundación de los reinos ultramarinos hasta fines del siglo XVI*, ed. Manuel Lucena-Salmoral (Madrid: Ediciones Rialp, 1982), 525–61, esp. 532ff.

48. About this junta, see Raquel Alvarez Peláez, "La obra de Hernández y su recuperación ilustrada," *La Real Expedición Botánica a Nueva España, 1787–1803* (Madrid: Real Jardín Botánico/Consejo Superior de Investigaciones Científicas, 1987), 147–58, quote on p. 147; Raquel Álvarez Peláez, *La conquista de la naturaleza americana* (Madrid: CSIC, 1993), 131ss; Demetrio Ramos-Pérez, "La Junta Magna y la nueva política," in Lucena-Salmoral, *El descubrimiento*, 437–53.

49. Carta del Virrey Toledo a Juan de Ovando (Cuzco, 10 June 1572), AGI, Lima 28b, libro 4, fols. 302–4.

50. Bakewell, "Mining," 222f; on the colonial economy of Peru, see S. J. Stern, *Peru's Indian Peoples and the Challenge of Spanish Conquest* (Madison: University of Wisconsin Press, 1993), 8off.

51. Bargalló, *La minería y la metalurgia*, 130f; Muro, "Bartolomé de Medina," 530n30.

52. This account is based on Menes-Llaguno, *Bartolomé de Medina*, 105–7 and Muro, "Bartolomé de Medina," 530n30.

53. Carta del Virrey don Martín Enríquez, 1576, in Bargalló, *Minería*, 131s; Menes-Llaguno, *Bartolomé de Medina*, 107–8.

54. Menes-Llaguno, *Bartolomé de Medina*, 108–9.

55. Juan de Cárdenas, *Problemas y secretos maravillosos de las Indias* (Mexico City, 1591; facsimile edition, Madrid: Ediciones Cultura Hispanica, 1945); José de Acosta, *Historia natural y moral de las Indias* (Mexico City: Fondo de Cultura Económica, 1962).

56. Nicolás Bautista Monardes, *Primera y segunda y tercera partes de la historia medicinal de las cosas que se traen de nuestras Indias occidentales que sirven en medicina* (Seville: Casa de Alonso Escrivano, 1574), fols. 29r–30r. Daniela Bleichmar discusses this case in her essay, "Books, Bodies, and Fields: Sixteenth-Century Transatlantic Encounters with New World Materia Medica," in *Colonial Botany: Science, Commerce, and Politics in the Early Modern World*, ed. Londa L. Schiebinger and Claudia Swan (Philadelphia: University of Pennsylvania Press, 2005), 83–99; see also Francisco Esteve Barba, *Cultura virreinal: Historia de América y de los pueblos americanos* (Barcelona: Salvat, 1965), 717f.

57. Monardes, *Historia medicinal*, fol. 30r.

58. Pedrarias de Benavides, *Secretos de chirurgia, especial de las enfermedades de Morbo galico y Lamparones y Mirrarchia, y asimismo la manera como se curan los Indios de llagas y heridas y otras passiones en las Indias, muy util y provechoso para en España y otros muchos secretos de chirurgia hasta agora no escritos* (Valladolid, 1567), fols. 21v–23r.

59. Esteve Barba, *Cultura virreinal*, 700.

60. Monardes received his degree in medicine from the Colegio-Universidad de Santa María de Jesús de Sevilla in 1547; see Berta Ares-Queija, *Tomás López Medel: Trayectoria de un clérigo-oidor ante el Nuevo Mundo* (Guadalajara: Institución Provincial de Cultura "Marqués de Santillana," 1993), 26; Juan Jiménez Castellanos, "Prólogo," in Monardes, *Historia medicinal* (1574; Seville: Padilla Libros, 1988), viii; Esteve Barba, *Cultura virreinal*, 715.

61. Monardes, *Historia medicinal*, fol. 30r–v.

62. Ibid., 31v.

63. Ibid., 71r.

64. Ibid., 31r–v.

65. I am following Raquel Álvarez Peláez in her description of Santa Cruz's memorial; see her *La conquista de la naturaleza americana*, 176–77. Santa Cruz's proposal belongs to the history of the *Relaciones geográficas de Indias* (1577). For an overview of the *Relaciones*, see Howard F. Cline, "The Relaciones Geográficas of the Spanish Indies, 1577–1648," in *Handbook of Middle American Indians*, ed. Howard F. Cline (Austin: University of Texas Press, 1972), 183–242; for a history of the *Relaciones*, see my article, "Empire and Knowledge: Reporting from the New World," *Colonial Latin American Review* 15 (2006), 39–55.

66. On Sahagún and his questionnaire, see José M. López Piñero et al., *Diccionario histórico de la ciencia moderna en España* (Barcelona: Ediciones Península, 1983), under the entry, "Sahagún, Bernardino de." The questionnaire is missing, but from his book it is clear he used one to collect information.

67. Cédula real a Marco de Ayala (Madrid, 1 November 1562), AGI, Mexico, 2999, libro 2, fols. 6r–7r.

68. Cédula real al gobernador de la provincia de Yucatán (El Escorial, 25 June 1565), ibid., 34r–v.

69. Monardes, *Historia medicinal*, fol. 74v.

70. Ibid., 73v.

71. Lorraine Daston and Katharine Park, *Wonders and the Order of Nature, 1150–1750* (New York: Zone Books, 1998); Paula Findlen, *Possessing Nature: Museums, Collecting, and Scientific Culture in Early Modern Italy* (Berkeley, Los Angeles, London: University of California Press, 1994); Stephen Greenblatt, *Marvelous Possessions* (Chicago: University of Chicago Press, 1991); Joy Kenseth, *The Age of the Marvelous* (Hanover, NH: Hood Museum of Art, 1991).

72. In my discussion of marvels I am following Paula Findlen, *Possessing Nature*, and Daston and Park, *Wonders*, esp. chap. four.

73. Ruth Pike, *Aristocrats and Traders: Sevillian Society in the Sixteenth Century* (Ithaca, NY: Cornell University Press, 1972), 83–84.

74. Carta del licenciado Maldonado al rey (La Española, 30 August 1554), AGI, Contratación, 5103.

75. Francisco Iñiguez-Almech, *Casas reales y jardines de Felipe II* (Madrid: Consejo Superior de Investigaciones Científicas, 1952), 129.

76. Consulta del Consejo (Madrid, 20 November 1578), AGI, Chile, 1, no. 4.

77. Monardes, *Historia medicinal*, fol. 45r.

78. Germán Somolinos D'Ardois, *El doctor Francisco Hernández y la primera*

expedición científica en América (Mexico City: Secretaría de Educación Pública, 1971), 28.

79. Simon Varey (ed.), *The Mexican Treasury: The Writings of Dr. Francisco Hernández*, trans. R. Chabran, C. L. Chamberlin, and S. Varey (Stanford, CA: Stanford University Press, 2001), 46; "Instrucciones al Dr. Francisco Hernández" (Madrid, 1 November 1570), AGI, Indiferente, 1228; Francisco Hernández, *Historia natural de Nueva España* (Mexico City: Universidad Nacional de México, 1959). On Hernández, see R. Álvarez Peláez, "La historia natural en la segunda mitad del siglo XVI: Hernández, Recchi y las relaciones de Indias," in *Nouveau Monde et renouveau de l'histoire naturelle*, 3 vols., ed. Marie-Cécile Bénassy et al. (Paris: Presses de la Sorbonne Nouvelle, 1986), vol. 3; R. Álvarez Peláez, "El doctor Hernández, un viajero ilustrado del XVI," in *Ciencia y contexto histórico nacional en las expediciones ilustradas a América*, ed. F. del Pino Diaz (Madrid: Consejo Superior de Investigaciones Científicas, 1988); R. Álvarez Peláez, "La obra de Hernández y su recuperación ilustrada," *La real expedición botánica a Nueva España, 1787–1803* (Madrid: Real Jardín Botánico/Consejo Superior de Investigaciones Científicas, 1987); Enrique Beltrán, "Una polémica sobre Francisco Hernández y su obra," *Anales de la sociedad mexicana de la historia de la ciencia y de la tecnología* 5 (1979), 49–73; Eli de Gortari, *La ciencia en la historia de México* (Mexico City, Barcelona, Buenos Aires: Editorial Grijalbo, 1979); Germán Somolinos D'Ardois, *Vida y obra de Francisco Hernández* (Mexico City: Universidad Nacional de México, 1960).

80. Varey, *Mexican Treasury*, 46.

81. Carta del Dr. Hernández (Mexico City, 1 September 1574), AGI, Mexico, 69, ramo 6, no. 98.

82. Daston and Park, *Wonders*, 136f.

83. Varey, *Mexican Treasury*, 53.

Chapter Twelve

1. *Certificación de Juan Rodríguez de Salamanca* (Acapulco, 29 December 1606), Archivo General de Índias (AGI), Patronato 51-3-8, fols. 289v–296r. An English translation of the contents is found in Celsus Kelly O.F.M., ed., *La Austrialia del Espíritu Santo: The Journal of Fray Martín de Munilla, O.F.M. and Other Documents Relating to the Voyage of Pedro Fernández de Quirós to the South Sea (1605–1606) . . .*, 2 vols. (Cambridge: Hakluyt Society, 1966), 2333–40.

2. Pedro Fernández de Quirós, *Relación*, Museo Naval Madrid (MNM), ms. 951, fols. 89v–90r. The best modern edition of this work was published in 1876 by Justo Zaragoza and has recently been reprinted. See *Historia del descubrimiento de las regiones austriales hecho por el general Pedro Fernández de Quirós* (Madrid: DOVE, 2000).

3. J. C. Beaglehole, *The Exploration of the Pacific*, 3rd ed. (Stanford, CA: Stanford University Press, 1966), 105. See also p. 81 for Beaglehole's reflections on the importance of Cook and his achievement.

4. This traditional assessment of Spanish achievement is represented in ibid., 324, and Beaglehole, *The Life of Captain James Cook* (Stanford, CA: Stanford

University Press, 1992), 126–27. These sources provide a succinct summary of Beaglehole's opinion of the "scientific" character of the voyages of Cook. Other authors have been less willing to adopt the model proposed by Beaglehole. See Glyndwr Williams, "The *Endeavour* Voyage: A Coincidence of Motives," in *Science and Exploration in the Pacific: European Voyages to the Southern Oceans in the Eighteenth Century*, ed. Margarette Lincoln (Boydell and Brewer, 2001), 15–16: "In more general terms, only in retrospect does the *Endeavour* voyage appear as part of a coherent official strategy of Pacific exploration. It is doubtful, indeed, if there was any coherent strategy."

5. A widely circulated contemporary map illustrates the extent of this continent. See Abraham Ortelius, *Maris Pacifici (quod vulgo Mar del Zur) cum regionibus circumiacentibus insulisque in eodem passim sparsis novissima descriptio* (Antwerp: Plantin Press, 1592).

6. Of these, only one is known to have survived in the form of a rough sketch of the Pacific. Pedro Fernández de Quirós, *Terra Australis incognita* (1598), in *Portvgaliae monvmenta cartographica*, 6 vols., ed. Armando Cortesão and Avelino Teixeira da Mota (Lisbon: Comemorações do V Centenário da Morte do Infante D. Henrique, 1960), vol. 5, plate 525. The original is found in the Newberry Library, Chicago. This chart shows evidence of having been based upon that of Juan López de Velasco, *Demarcación y División de las Indias* (1575). The original map is to be found at the beginning of Velasco's *Geografía*. It is conserved in the John Carter Brown Library. Contemporary editions of Velasco's do not reproduce this map. See Juan López de Velasco, *Geografía y descripción universal de las Indias*, ed. Marcos Jiménez de la Espada (Madrid: Atlas, 1971).

7. Sir Clements Markham (ed. and trans.), *The Voyages of Pedro Fernandez de Quiros 1595 to 1606* (London: Hakluyt Society, 1904), 1:xix. Such views are not confined to early-twentieth-century historiography. H. E. Maude describes events associated with the 1606 possession ceremony on the island of Espíritu Santo orchestrated by Quirós in terms of a "tragicomedy." See his "Spanish Discoveries in the Central Pacific: A Study in Identification," *Journal of the Polynesian Society* 68 (1959), 318. The most intellectually stimulating and archivally based study on this period in the Pacific is found in the work of the Spanish historian Juan Gil. Nevertheless, Gil provides a no less critical and presentist view of Quirós's activities on Espíritu Santo. See his *Mitos y utopias del descubrimiento*, vol. 2, *El Pacífico* (Madrid: Alianza Editorial, 1989), 117–19. An essential guide to the vast array of documentary sources associated with the Spanish voyages in the South Pacific is provided by Celsus Kelly, O.F.M., *Calendar of Documents: Spanish Voyages in the South Pacific from Alvaro de Mendaña to Alejandro Malaspina 1567–1794 and the Franciscan Missionary Plans for the Peoples of the Austral Lands 1617–1634* (Australia: Franciscan Historical Studies, and Madrid: Archivo Ibero-Americano, 1965). Despite a wealth of documentation available to researchers, there is virtually no new study on the voyage since that of Gil's.

8. Rather than a flight of fancy, these formalities were in fact a crucial part of establishing the fundamentals of Spanish law and governance in new settlements. For a contemporary account of their importance, see Pedro de Madriga,

"Description of the Government of Peru," in *The East and West Indian Miror, Being an Account of Joris Van Speilbergen's Voyage Round the World (1614–1617), and the Australian Navigations of Joacob le Maire*, ed. and trans. J. A. J. de Villiers (London: Hakluyt Society, 1906), 86.

9. For a text of these laws, see "Ordenanzas de Su Magestad hechas para los nuevos descubrimientos, conquistas y pacificaciones" (1573), in *Colección de documentos inéditos relativos al descubrimiento, conquista y organización de las antiguas posesiones españoles de América y Oceania*, 42 vols., ed. Joaquín F. Pacheco, Francisco de Cárdenas, Luis Torres de Mendoza, et al. (Madrid, 1864–1884), 16:142–87; see also 8:484–537.

10. Juan Gil is one of the few commentators on the voyage to draw a probable link between the events on Espíritu Santo and the intellectual inheritance of millenarianism and Joachimism. See his *Mitos y utopías*, 2:118–19.

11. Perhaps the greatest affirmation of Quirós's achievement is recorded by eighteenth-century advocate of English exploration in the Pacific, Alexander Dalrymple:

> The *discovery* of the Southern Continent, *whenever*, and by *whomsoever* it may be completely effected, is in *justice* due to this *immortal* name. . . . The voyages previous to *that* of Pedro Fernandez de Quiros, were not directed, at least immediately, to this great object; but Quiros formed his plan on the observations he himself had made in Mendana's voyage, in 1595: and, reasoning from principles of science and deep reflexion, he asserted the *existence* of a Southern Continent; and devoted, with unwearied, though contemned diligence, the *remainder* of his *life* to the prosecution of this *sublime conception*. [emphasis in original]

An Historical Collection of the Several Voyages and Discoveries in the South Pacific Ocean, 2 vols. (London, 1770–71), 1:95 (archived at Mitchell Library, Sydney).

12. This definition takes as its inspiration the research of Antonio Barrera-Osorio. See his *Experiencing Nature: The Spanish American Empire and the Early Scientific Revolution* (Austin: University of Texas Press, 2006), 12.

13. David Goodman, *Power and Penury: Government, Technology and Science in Philip II's Spain* (Cambridge: Cambridge University Press, 1988).

14. See Mariano Esteban Piñeiro, "La Academia de Matemáticas de Madrid," in *Felipe II, la ciencia y la técnica*, ed. Enrique Martínez Ruiz (Madrid: ACTAS, 1999), 113–32.

15. On this point, see Fray Gerónimo de Mendieta's comment from the 1560s: "Your Majesty is like a blind man who has excellent understanding but can only see exterior objects . . . through the eyes of those who describe them to you." Cited in Geoffrey Parker, *The Grand Strategy of Philip II* (New Haven, CT, and London: Yale University Press, 1998), 58.

16. Fray Gerónimo de Villacarrillo, O.F.M., *Memorial al rey Felipe II* (Guanuco, 25 July 1574), AGI, Lima 270, fol. 572r–v.

17. For a detailed monograph on Lerma's foreign policy, see Bernardo José García García, *La pax hispanica: Política exterior del Duque de Lerma* (Leuven, Belgium: Leuven University Press, 1996).

18. For a schematic presentation of the wars of the sixteenth century, see Parker, *The Grand Strategy of Philip II*, 2, fig. 1.

19. The above arguments feature strongly in deliberations in the Council of Indies, the Council of State, and the Council of Portugal during the first decade of the seventeenth century.

20. That Philip III sought a venture of this nature is argued in Patrick Williams, "Philip III and the Restoration of Spanish Government, 1598–1603," *English Historical Review* 88, no. 349 (October 1973), 751–69. See also Antonio Feros, *Kingship and Favoritism in the Spain of Philip III, 1598–1621* (Cambridge: Cambridge University Press, 2000), 53.

21. The documents referred to here date from March 1595. The first is a receipt for a copy of the *asientos* and *capitulaciones* (the contract) between Mendaña and Philip II dating from 1574. The receipt bears the signature of Quirós. A copy of this receipt is found in AGI, Patronato 51-3-8, fols. 104r–108r[mr: 1]. See also Kelly, *Calendar of Documents*, 157, no. 289. The documents covered by the receipt include the following: (1) Philip II, *Capitulación con Álvaro de Mendaña* (Madrid, 27 April 1574), AGI, Patronato 18, no. 10, ramo 8, libro 3, fols. 28v–35r, printed in Celsus Kelly, O.F.M. (ed.), *Austrialia Franciscana* (Australia: Franciscan Historical Studies, in collaboration with Madrid: Archivo Ibero-Americano, 1962), 5:62–72; and (2) Philip II, cédula dirigida a Álvaro de Mendaña (Aranjuez, 12 May 1574), AGI, Patronato 51-3-8, fols. 93v–97r. This *cédula* is contained within a memorial written by Mendaña to the viceroy of Peru: Álvaro de Mendaña, "Memorial al virrey" (Callao, April 1595), AGI, Patronato 51-3-8, fols. 93v–97r (cédula fols. 94r–97r). For further details on these documents, see Kelly, *Calendar of Documents*, 123, nos. 119, 120; 158, no. 292. The second source to mention Quirós is a letter written by Mendaña, appointing Quirós as both *piloto mayor* of the fleet and *capitán* and *maestre* of the *San Gernónimo*. Mendaña, *Carta a Pedro Fernández de Quirós* (Callao, 7 March 1595), AGI, Patronato 51-3-8, fols. 97r–99r; printed in Kelly, *Austrialia Franciscana*, 6:7–10. This letter is also important in that it contains Mendaña's assessment of Quirós's navigational skills and experience. He describes Quirós thus: "Perona de mucha satisfaçion y confiança y tener gran pratica y notiçia de las cossas de la mar, y que su persona hera mui nessesaria e inportante para la nauegaçion de la dicha jornada y descubrimiento" [A very agreeable and confident person who has great experience and knowledge of the things of the sea, and he is a most necessary and important individual to undertake the said voyage and discovery]. See AGI, Patronato 51-3-8, fol. 97r and Kelly, *Austrialia Franciscana*, 6:5. A further significance should be noted in the above-mentioned documentary material. The memorials indicate the type of information Quirós had at his disposal in the formulation of his future plans for the exploration of the South Sea.

22. Pedro Fernández de Quirós, *Memorial al virrey* (June 1597), Archivo de Universidad, Salamanca, ms. 337, fols. 121–123v.

23. Fernando Oliveira, *Planisphere I & II*, in *A cartografia portuguesa do Japão (séculos XVI–XVII): Catálogo das cartas portuguesas* [The Portuguese Cartography of Japan (XVI–XVII Centuries): A Catalogue of Portuguese

Charts], ed. Alfedo Pinheiro Marques, trans. Martin A. Kayman (Lisbon: Fundação Oriente, Comissão Nacional para as Comemorações dos Descobrimentos Portugueses, Imprensa Nacional—Casa da Moeda, 1996), 140–42, plates 43, 44. For the origin of these ideas, see Diodorus Siculus, *Bibliotheca historica*, ed. Fredericus Vogel (Lipsiae, Germany: B.G. Teubneri, 1888), vol. 1, libro 2, chaps. 55–57.

24. Quirós's plans would be modified radically through a variety of influences. Not the least was the contentious issue of the legal right to continue the rights of exploration and settlement originally granted to Álvaro de Mendaña by Philip II. Mendaña's widow, Isabel Barreto, had survived the 1596 voyage to the Philippines and married a resident of Manila, Fernándo de Castro. In 1599 Barreto and Castro claimed the rights to the region around the Solomons. See Fernánde de Castro, *Memorial al rey* (Mexico City, 25 March 1599), AGI, Patronato 18, no. 10, ramo 8. In order to avoid potential conflict with these claims, the focus of Quirós's proposed voyage was a probable mainland in the vicinity of the Marquesas. For a commentary on these circumstances, see Celsus Kelly, O.F.M., "Introduction," in *La Australia del Espíritu Santo: The Journal of Fray Martín de Munilla O.F.M. and Other Documents Relating to the Voyage of Pedro Fernández de Quirós to the South Sea (1605–1606) and the Franciscan Missionary Plan (1617–1627)*, ed. Celsus Kelly, O.F.M. (Cambridge: Hakluyt Society, 1966), 1:14–15. For Quirós's plan, see his *Memorial al virrey de Perú* (Cartagena de las Indias, August–October 1598), Archivo Historico de Loyola (AHL), Guipúzcoa (formerly the Santa Casa de Loyola), ms. *De Marina*, estante 10, pluteo 5, no. 14. At the time of researching this paper, inquiries at the AHL proved fruitless in locating this manuscript. A microfilm copy was consulted at the library of St. Paschal's College, Melbourne.

25. Don Luis de Velasco, *Carta al rey* (Callao, 16 April 1598), AGI, Lima 33.

26. Fray Diego de Soria, O.P., "Memorial al rey" (Madrid, 6 December 1600), AGI, Filipinas 79, ramo 4, no. 63, 2 fols.

27. Fray Diego de Soria, O.P., "Memorial al rey" (Madrid, 1602), Archivo General de Simancas (AGS), Estado 191, 2 fols.

28. Sesa held the position of ambassador from 1592 to 1603. See Luciano Serrano, O.S.B., *Archivo de la Embajada de España cerca de la Santa Sede*, vol. 1, *Indice analítico de los documentos del siglo XVI* (Rome: Palacio de España, 1915), xxx.

29. Pedro Fernández de Quirós, *Memorial al rey* (Madrid, 17 June 1602), AGS, Estado, K-1631, c. 37, doc. 243.

30. This information is supplied by Diogo Barbosa Machado, who writes:

> Passou a Roma no anno de 1600 em que com jubilo do mundo catholico se celebrava o Anno Santo, e como conheceste o seu grande talento o Duque de Sessa Embaixador de Castella em a Curia o admitio por familiar da sua Casa para instruir a seu filho na intelligencia dos Mapas do mundo, e cartas de marear.

> He went to Rome in the year 1600 in which the Holy Year was celebrated with a jubilee of the Catholic world, and as the Duke of Sesa, ambassador of Castile at the cruia came to know of his great talent, he admitted him as a member of his household to instruct his son in the understanding of world maps, and nautical charts.

Bibliotheca Lusitana: Historica, critica, e cronologica; Na qual se comprehende a noticia dos authores portuguezes, e das obras, que compuzeraõ desde o tempo da promulgaçaõ da Ley da Graça até o tempo presente (Lisbon: Ignacio Rodrigues, 1752), 3:578a. The author does not indicate the source of his information, and these details concerning Quirós's employment as a tutor are not found in any of the known documentation.

31. Duque de Sesa, *Carta al rey* (Rome, 2 February 1602), AGS, Estado, K-1631, c. 37, doc. 33. The meeting is described by Quirós in his *Relación*:

> Parecióle bien a Su Excelencia, e hizo juntar en su casa los mayores pilotos y matemáticos que se hallaban en Roma; y habiendo en su presencia hecho largo examen de mis papeles, discursos y cartas de marear, y quedando satisfechos de que todo lo que yo decía era probable y digno de ponerse en ejecución, me negoció el señor duque audiencia para con Su Santidad de Clemente VIII, la cual tuve a veintiocho de agosto, habiendo primero comido en la mesa de los pobres.

> It seemed well to his excellency, and he called together in his house the best pilots and mathematicians that were in Rome, and having in his presence made an extensive survey of my papers, arguments and sea charts, and they were satisfied that all that I was saying was probable and worthy of being undertaken. The lord Duke negotiated an audience for me with His Holiness Clement VIII, which took place on the 28th of August, having first eaten at the table of the poor.

32. Lorenzo Ferrer Maldonado, "Discurso del Capitán Lorenzo Ferrer Maldonado sobre la variación de la aguja," in *Papeles varios sobre Indias, sec. XVII*, Biblioteca Nacional de Madrid (BNM), Raros 17.270, fols. 245r–248v. Note the references in this pamphlet to a meeting in the residence of Conde de Salinas, in which Ferrer Maldonado was present, along with João Bautista Lavanha and Luis de Fonseca.

33. Quirós, *Relación*, MNM, ms. 951, libro 3, cap. 2, fol. 74v; Biblioteca del Palacio Real, Madrid (BPM), ms. 1686, c. 41, fol. 113r.

34. Christopher Clavius, S.J., *In Sphaeram Ioannis de Sacro Bosco Commentarivs* (Rome: Ex Officina Dominici Basae, 1581). Other works by Clavius dealing with nautical science include his *Astrolabivm* (Rome: Impensis Bartholomaei Grassi, 1593). See also *Compendivm brevissimvm describendorvm horologiorum horizontalium ac declinantium* (Rome: Apud Aloysium Zannettum, 1603). It is probable that he and Villalpando shared a similar late-Ptolomaic cosmology. Clavius's interpretation of Ptolomy is provided in James M. Lattis, *Between Copernicus and Galileo: Christoph Clavius and the Collapse of Ptolemaic Cosmology* (Chicago and London: University of Chicago Press, 1994), 64–85.

35. Further evidence of intellectual influences on Quirós during this period are seen in the similarities between his arguments concerning the sphericity of the earth and that employed by Benito Arias Montano in his *Natvrae historia* (Antwerp: Ex Officina Plantiniana, apud Ioannem Moretum, 1601), 182–83. Compare this to Quirós, *Memorial al virrey del Perú* (*Cartagena de Indias, 3 March 1599*), Archivo Historico de Loyola, MS *De Marina*, estante 10, pluteo 5, no. 14:

> Digo, Ecmo. Príncipe, que la forma que hacen la tierra y el agua sobre que los hombres andamos es redonda, y ya aprobada esta opinión por todos los antiguos

y modernos . . . hace la sombra conforme a su figura y así en las eclises (*sic*) de la luna causadas por la sombra de la tierra, en ella se ve claro esta verdad por sus extremos cuando va creciendo o menguando el tal eclise.

I maintain, most excellent sir, that the form taken by the earth and the water upon which we men dwell is round, and this opinion is already proven by all the ancients and moderns . . . [that] the shadow conforms to its figure, and thus in the eclipses of the moon caused by the shadow of the earth, this is seen clearly.

36. Jerónimo de Prado, S.J., and Juan Bautista Villalpando, S.J., *In ezechielem explanationes et apparatus vrbis, ac pempli hierosolymitani: Commentariis et imaginibvs illvstratvs,* 3 vols. (Rome: Carolus Vulliettus, Typis Illefonsi Ciacconii, vol. 1: 1596; vols. 2–3: 1604). At the time of research, the pertinent section of the Archivo de Asuntos Exteriores in Madrid housing the archives of the Spanish embassy in Rome was unavailable for consultation. The following source was used: Lucian Serrano, O.S.B., *Archivo de la Embajada de España cerca de la Santa Sede,* vol. 1, *Indice analítico de los documentos del siglo XVI* (Rome: Palacio de España, 1915). See, in particular, p. 121 for a description of *legajo* 49:

Obra del P. Juan B. Villalpando, jesuita, fol. 275–292. Cartas de Sessa y Condes de Miranda y Olivares acerca de la ayuda de costa y anticipos de dinero sobre Nápoles, que mandó el Rey dar al P. Villalpando para imprimir su obra del «Templo de Salomón», 1594–1599. - Recibos originales de dicho subsidio que pagará dicho Padre con los productos de la venta de la obra.

Work of Fr. Juan B. Villalpando, Jesuit, fols. 275–292. Letters of Sessa and the Counts of Miranda and Olivares regarding assistance for costs and advances in funds concerning Naples, that the King ordered to Fr. Villalpando to print his work on "The Temple of Solomon," 1594–1599. Original receipts of the aforesaid subsidy, which the aforesaid Father will pay with the proceeds of the sale of the work.

37. Juan Bautista Villalpando, S.J., *Carta al príncipe Felipe* (Rome, 1 January 1597), BNM, ms. 6.035, fols. 151r–154r, transcribed by René Taylor in Juan Antonio Ramírez, René Taylor, André Corboz, Robert Jan van Pelt, and Antonio Martínez Ripoll, *Dios arquitecto: J. B. Villalpando y el Templo de Salomón,* 2nd ed. (Madrid: Ediciones Siruela, 1995), 351–52.

38. Villalpando's emphasis on the transcendent meaning of Jerusalem is established in his initial exegesis of Ezekiel:

Vrbem Hierusalem in medio gentium sitam fuisse testatur Dominus per Ezechielem, dicens: *Ista est Hierusalem: in medio gentium posui eam, et in circuitu eius terras.* [Ez. 5:5]. Quem locum enarrans Hieronymus, sic habet: Hierusalem in medio mundi sitam, hic idem Propheta testatur, vmbilicum terrae eam esse demonstrans.

The city of Jerusalem in the middle of all peoples was declared by the Lord through Ezekiel, saying: Thus is Jerusalem: I have placed her in the midst of the peoples, and surrounded by lands. [Ez. 5:5]. Of which place Jerome describes, thus I have placed Jerusalem at the center of the world. The Prophet testifies to this in the same way, showing that it is the navel of the world.

Prado and Villalpando, *In ezechielem explanationes,* 3:13b.

39. Juan de Herrera, *Discurso sobre la figura cúbica*, ed. Edison Simons and Roberto Godoy (Madrid: Editora Nacional, 1976).

40. Tomás Carreras y Artau and Joaquín Carreras y Artau, *Historia de la filosofía española: Filosofía cristiana de los siglos XIII al XV* (Madrid: Real Academia de Ciencias Exactas, Físicas y Naturales, 1943), 2:257–58.

41. Philip II, "Real Provisión al licenciado Arias de Loyola" (San Lorenzo, 19 October 1591), AGI, Indiferente General 426, libro 28, fols. 110v–112r.

42. The Spanish ambassador at the Roman court, the Duke of Sesa, was intimately involved in the formation of this grand design, orchestrating meetings between Quirós, local astronomers, mathematicians, and cosmographers. D. Antonio de Cardona y Córdova, duque de Sesa, *Carta a Su Magestad* (Rome, 2 February 1602), AGS, Estado, K-1631, c. 37, doc. 33; Pedro Fernández de Quirós, *Relación*, MNM, ms. 951, fol. 74v.

43. Quirós, *Memorial al rey*, "Señor/Con liçensia de V. M. el Adelandado Aluaro de Mendaña" (El Escorial, 17 June 1602), AGS, Estado, K-1631, c. 37, doc. 244, 2ff.[mr: 4]. For the context, see Quirós, *Relación*, MNM, ms. 951, libro 3, cap. 1, fol. 72v.

44. The correspondence Quirós carried with him to Peru is also indicative of the company he sought during this period. With reference to this documentation he states: "A estas cedulas acompañaron muchas cartas que en la corte me dieron algunos grandes señores para el virrey del Peru." Quirós, *Relación*, MNM, ms. 951, libro 3, c. 2, fol. 76v. He gives further details in a memorial to Philip III:

> Sali de Valladolid, y de Aranjuez por Abril, y Mayo de 1603, con cuatro Cedulas de V. M. emanadas del Consejo de Estado, y con dos Cartas del Conde de Lemus y Andrade, Presidente del Consejo de las Indias, y otra Carta del Condestable de Castilla, otra del duque de Sesa, otra de don Juan de Ydiaguez, Comendador mayor de Leon, y otra de la Condesa de Monterrey, diciendome todas estas personas, y tan graves ministros de V. M. que en obra tan grandiosa y encaminada à tantos bienes, querian tener su parte.

> I departed from Valladolid and Aranjuez in April, and May of 1603, with four *cédulas* of His Majesty ammended by the Council of State, and with two letters of the Count of Lemos y Andrade, president of the Council of Indies, and another setter from the Constable of Castile, another of the Duke of Sesa, another of Don Juan de Idiaquez, Comendador Mayor of León, and another of the Countess of Monterrey, all of these persons, and very important ministers of Your Majesty, telling me that they wished to take their part in a work so grand and directed to so many benefits.

Quirós, *Memorial al rey* (Madrid, 14 December 1607), Biblioteca de la Real Academia de Historia, Madrid, *Papeles varios de Indias*, doc. 90 (estante 26, gr. 4, doc. 90), fol. 2r–v[mr: 10].

45. The sources indicate that the Council of State was far more sympathetic to Quirós's plans than that of the Indies. Following the chronology found within the narrative, it seems Quirós met with members of the Council of Indies in a

garden of one of the estates of the Count of Lemos sometime between March 31 and May 9, 1603:

> Auiendo acudido al Real consejo de las Yndias con los breves de su Santidad para refrendarlos, quiso el Conde de Lemos que era Presidente de aquel consejo y los demas señores de el enterarse de mi yntento y promesa, y me mandaron que lleuase vn Mapa, y les fuese a dar quenta de todo esto a vn jardin del Conde, donde se juntaron para este efecto y auiendome oydo, mostraron quedar satisfechos y avn embidiosos de que mi despacho se vbiese encaminado por el consejo de estado.

> Having turned to the Royal Council of the Indies with the briefs of His Holiness to endorse them, the Count of Lemos who was the president of that council and its other lords wished to learn of my intention and promise, and they summoned me to bring a map, and I went to give an account of all this to a garden of the Count where they met for this purpose, and having listened to me, they appeared satisfied and even envious that my dispatch had been generated by way of the Council of State.

Quirós, *Relación*, MNM, ms. 951, libro 3, c. 2, fol. 76v. There was little, in fact, that the Council of Indies could do. The negotiations were a fait accompli. Any obstruction or intervention by the Council of Indies would have been a direct challenge to the king's decision. This appears to have been the result of a deliberate attempt by Lerma to extend the power of the Council of State.

46. Quirós, *Memorial al rey* (Madrid, after 17 June 1602), AGS, Estado, K-1631, c. 37, doc. 243, 2 fols. ff. It seems in fact there were more than two memorials presented to the king in this period, as the narrative indicates: "me fue forzoso yr haziendo mas ynstanzia con su Magestad dandole cada dia nuevos memoriales y representando las razones que hauia en fauor de mi empresa y procurando satisfacer a las que se oponian en contrario" [I was forced to proceed with greater insistence with His Majesty, sending him new memorials each day, and presenting the reasons I had in favor of my enterprise, while seeking to satisfy those who were opposed in the contrary]. Quirós, *Relación*, MNM, ms. 951, libro 3, c. 2, fol. 73r–v.

47. "Yo que deseo los ocultos moradores de la parte antartica en la dotrina euangelica repastados aprouechados y sustentados y a V.M. senor conoçido obedeçido y servido como lo es de levante a poniente del uno hasta otro polo" [I who desire that the hidden dwellers of the antarctic part (be) pastured, benefited, and sustained in the evangelical doctrine, and that Your Majesty (be) the lord recognized, obeyed and served as you are from the rising of the sun to its setting, from one pole to the other]. Quirós, *Memorial al rey*, AGS, Estado, K-1631, c. 37, doc. 244, fol. 1v.

48. "Tengo praticado muchos años prior que los descubridores pasados tuuiesen una sola ora de semejantes negoçios y que soy testigo de vista y vengo de aquellas partes a estas a dar el auiso" [I have many more years' experience than past explorers who might have had only an hour of similar dealings, and I am an eyewitness and come from those regions to these to give testimony]. Quirós, *Memorial al rey* (Madrid, after 17 June, 1602), AGS, Estado, K-1631, c. 37, doc. 243.

49. Granfell Price (ed.), *The Explorations of Captain James Cook in the Pacific, as told by selections of his own journals, 1768–1779* (New York: Dover Publications, 1971), 21.

50. Beaglehole, *Exploration of the Pacific*, 237.

51. See Manuel Godinho de Herédia, *Planisfério*, in *Atlas Miscelânea*, c. 1615–c. 1622, reproduced in *A cartografia portuguesa do Japão (séculos XVI–XVII): Catálogo das cartas portuguesas* [The Portuguese Cartography of Japan (XVI–XVII Centuries): A Catalogue of Portuguese Charts], ed. Alfredo Pinheiro Marques, trans. Martin A. Kayman (Lisbon: Imprensa Nacional/Casa da Moeda, 1996), plate 65.

52. "Journal & Miroir de la Navigation australe du Jaques le Maire chef and conducteur de deux Navires Concorde & Horne," in Herrera, *Description des Indes occidentales* (Amsterdam, 1622), 177. Cited in Kelly, *La Austrialia*, 1:5.

53. Jorge Cañizares-Esguerra, "Iberian Science in the Renaissance: Ignored How Much Longer?" *Perspectives on Science* 12, no. 1 (2004), 91. Cañizares-Esguerra suggests that Bacon's description of the island utopia implied his high esteem for Spanish imperial institutions of learning. While inaccurate in some of its factual details—such as a reference to Joachim of Fiore as a Franciscan or a misreading of the nomenclature associated with the discoveries of Pedro Fernández de Quirós (Australia as opposed to the original Austrialia in honor of the Habsburg monarchy)—Cañizares-Esguerra nevertheless accurately assesses the likely origin of Bacon's ideas in the utopianism of Fernández de Quirós. More research needs to be done to see the intellectual matrix from which the ideas of Fernández de Quirós emerged. See Jorge Cañizares-Esguerra, *Nature, Empire, and Nation: Explorations of the History of Science in the Iberian World* (Stanford, CA: Stanford University Press, 2006), 22; also his *Puritan Conquistadores: Iberianizing the Atlantic, 1550–1700* (Stanford, CA: Stanford University Press, 2006), 50.

54. See also later seventeenth-century utopian views based in part on a reading of Quirós: Maria Teresa Bovetti Pichetto, "Gabriel de Foigny, utopista e libertino," in *Studi sull'Utopia*, ed. Luigi Firpo (Florence: Leo S. Olschki, 1977), 189–221.

55. Kelly, *Calendar of Documents*, 235–50.

56. "The Copie of a Petition presented to the King of Spaine, by Captaine Peter Ferdinand de Quir, touching the Discouerie of the fourth part of the World, called Terra Australis incognita: and of the great riches and fertilite of the same: Printed with license in Siuill, An. 1610," in Samuel Purchas, *Haklvytus Posthumus or Purchas His Pilgrimes, Contayning a History of the World, in Sea voyages & lande Trauells, by Englishmen & others. Wherein Gods Wonders in Nature & Prouidence, The Actes, Arts, Varieties, & Vanities of Men w[i]th a world of the Worlds Rarities, are by a world of Eyewitnesses-Authors, Related to the World* (London: Printed by William Stansby for Henrie Fetherstone, and are to be sold at his shop in Pauls Church-yard at the signe of the Rose, 1626), 4:1422–27. The English translation is followed by a Spanish version on pp. 1427–32.

Chapter Thirteen

Abbreviations used in the endnotes to this Chapter:

AHU   Arquivo Histórico Ultramarino (Portuguese Overseas Historical Archive); Lisbon, Portugal

ANTT   Arquivo Nacional da Torre do Tombo (National Archives of Portugal, Lisbon)

HAG   Historical Archive of Goa, India

MR   Livros do Monções do Reino (annual volumes of official state correspondence to the Estado da Índia)

cx   *caixa* (box)

doc   *documento* (document)

fol   *folho* (folio)

1. See, for example, HAG MR 181A, fols. 9–45; and 181B, fols. 370–98.

2. John Huyghen Van Linschoten, *The Voyage of John Huyghen Van Linschoten to the East Indies . . .* , ed. Arthur Coke Burnell and P. A. Tiele, 2 vols. (London: Hakluyt Society, 1885), 1:237.

3. Alberto C. Germano da Silva Correia, *La Vieille-Goa* (Bastorá, India, 1931), 274–75; F. P. Mendes da Luz, "Livro das Cidades," *Studia* 6 (1960), fol. 8. I am grateful to Michael Pearson for this reference.

4. HAG MR 181A, fols. 65, 194–201; and 212A, fol. 200v.

5. HAG MR 52, fol. 191r–v; and 115, fols. 88r–89r.

6. John M. de Figueiredo, "Ayurvedic Medicine in Goa According to European Sources in the Sixteenth and Seventeenth Centuries," *Bulletin of the History of Medicine* 58, no. 2 (Summer 1984), 225–35.

7. João Telles e Cunha, "Socio-Cultural Aspects of the Catholic Missionary Works in India," in *The Portuguese and the Socio-Cultural Changes in India* (Tellicherry, India: MESHAR and Fundação Oriente, 2001), 237–67.

8. Charles R. Boxer, *Portuguese Society in the Tropics* (Madison: University of Wisconsin Press, 1965), 25–26.

9. See, for example, HAG MR 46A, fols. 96r–97v (report of medicines and their prices sent from Goa to the Hospital Novo of Moçambique, 1681); HAG 7926, fol. 56r–v (report of medicines sent from Hospital Real of Goa to the Fortress of Diu, 1785); and HAG 1346, fol. 183 (report of medicines and their prices sent from Goa to the Hospital Publico Militar of the Islands of Soldar and Timor, 1838).

10. A *converso* is a person of Jewish heritage who has converted to Christianity.

11. Conde de Ficalho, *Garcia da Orta e o seu tempo*, 2nd ed. (Lisbon: Imprensa Nacional-Casa da Moeda, 1983), introduction; A. J. R. Russell-Wood, *A World on the Move: The Portuguese in Africa, Asia and America, 1415–1808* (Manchester: Carcanet Press, 1992), 83–84.

12. Ibid., 149.

13. Ibid, 149–50.

14. Richard H. Grove, *Green Imperialism: Colonial Expansion, Tropical Island Edens and the Origins of Environmentalism, 1600–1860* (New Delhi: Oxford University Press, 1995), 77–80.

15. The only known extant copy of this report may be found in the Bibliothèque Nationale de France, Manuscrits Occidentale, Fonds Portugais no. 59, fols. 29–77v.

16. Ibid., 2–79v.

17. Ibid., 79–155 (Chinese text on fols. 151v–152; French text on fols. 124–127v).

18. Treaties of 1513–1515, contained in Julio Firmino Judice Biker (ed.), *Collecção de tratados e concertos de pazes que o Estado da India Portugueza fez com os reis e senhores com quem teve relações nas partes da Asia e Africa Oriental desde o principio da conquista até ao fim do seculo XVIII* (Lisbon: Imprensa Nacional, 1881–1886), 1:10, 33. Cited in F. C. Danvers, *Report of the Portuguese Records Relating to the East Indies* (London: India Office, 1892), 8–9.

19. Manuel Godinho de Erédia, *Suma de árvores e plantas da Índia intra Ganges*, ed. John G. Everraert, J. E. Mendes Ferrão, and E. M. Cândida Liberato (1612; Lisbon: Comissão Nacional para as Comemorações dos Descobrimentos Portugueses, 2001).

20. Ibid., 9–26.

21. Van Linschoten, *Voyage of John Huyghen Van Linschoten*, 1:xxiii–xl.

22. Ibid, 1:61.

23. Ibid., 2:113–14.

24. Ibid., 2:115.

25. Magalhães Basto (ed.), *Viagem de Francisco Pyrard de Laval* (Oporto, Portugal: Civilização Editora, 1944), 2:87.

26. Philip Baldaeus, *A True and Exact Description of the Most Celebrated East-India Coasts of Malabar and Coromandel and also of the Isle of Ceylon* (1703, London; New Delhi: Asia Educational Services, 2000), 3:608.

27. Boxer, *Portuguese Society in the Tropics*, 25–26.

28. João Manuel Pacheco de Figueiredo, "The Practice of Indian Medicine in Goa during the Portuguese Rule, 1510–1699," *Luso-Brazilian Review* 4, no. 1 (June 1967), 51–52.

29. Ana Maria Amaro, *Introdução de medicina ocidental em Macao e as receitas de segredo da botica do Colégio de São Paulo* (Macau: Instituto Cultural de Macao, 1992), 7–11.

30. HAG 7795, fols. 25–27; Pacheco de Figueiredo, "The Practice of Indian Medicine in Goa," 53–54.

31. Amaro, *Introdução de Medicina Ocidental*, 101.

32. Ibid., 102.

33. Russell-Wood, *A World on the Move*, 83–84.

34. Eduardo de Castro Almeida, *Inventário dos documentos relativos ao Brasil existantes no Archivo de Marinha e Ultramar de Lisboa*, vol. 1, *Bahia, 1613–1762* (Rio de Janeiro: Officinas Graphicas da Biblioteca Nacional, 1913), doc. 2917, 255–56.

35. See HAG 7795, fols. 25–27; and HAG MR 59, fol. 305v.

36. HAG MR 178B, fol. 646r; and 175, fols. 222v–223v.
37. HAG MR 177A, fol. 212.
38. HAG MR 178A, fol. 272.
39. HAG MR 177A, fol. 212.
40. See HAG MR 173, fol. 168; 176B, fol. 436; 176B, fol. 448; and 177A, fol. 218.
41. HAG MR 175, fols. 219–30.
42. See Figueiredo, "Ayurvedic Medicine in Goa," 225–35. See also V. V. Sivarajan and Indira Balachandran, *Ayurvedic Drugs and Their Plant Sources* (New Delhi and Bombay: Oxford and IBH Publishing, 1994).
43. HAG MR 175, fols. 220r–221v.
44. Garcia da Orta, *Coloquios dos simples e drogas e cousas medicianais da Índia . . .* (Goa, 1563; facsimile edition, Lisbon: Academia das Ciências de Lisboa, 1963), Colloquy, 42.
45. Sebastião Rodolfo Dalgado, *Glossário luso-asiático* (1921, Lisbon; New Delhi: Asian Educational Services, 1988), 2:196–97.
46. Sivarajan and Balachandran, *Ayurvedic Drugs*, 185–86, 218.
47. Pedro Sousa Dias and Rui Pita, "A botica de S. Vicente e a farmácia nos mosteiros e conventos da Lisboa setecentista," in *A botica de São Vicente de Fora*, ed. P. Sousa Dias and R. Pita (Lisbon: Associação Nacional das Farmácias, 1994), 19.
48. ANTT, Ministério do Reino, maço 469 (n.d.); cited in José Pedro Felipe de Sousa Dias, "Inovação técnica e sociedade na farmácia da Lisboa setecentista" (doctoral diss., Universidade de Lisboa, Faculdade de Farmácia, 1991), 2:638–39. See also Sousa Dias and Pita, "A botica de S. Vicente," 20.
49. Sousa Dias and Pita, "Inovação técnica," 19–20.
50. Ibid., 18, 21.
51. Manoel Rodrigues Coelho, *Farmacopeia tubalense chimico-galenica* (Lisbon: Officina de Antonio de Sousa Sylva, 1735), 845–46.
52. Paula Basso and João Neto, "O Real Mosteiro de S. Vicente de Fora," in *A botica de São Vicente de Fora* (Lisbon: Associação Nacional das Farmácias, 1994), 14; João Neto, "A botica do real mosteiro de S. Vicente de Fora," *Medicamento, História e Sociedade* (Nova Série) 3, no. 4 (September 1994), 10–11.
53. Dom Caetano de Santo António, *Pharmacopea lusitana: Método prático de preparar, e compor os medicamentos na forma galenica com todas as receitas mais usuais* (Coimbra: Impressão de João Antunes, 1704) and *Pharmacopea lusitana reformada: Método prático de preparar os medicamentos na forma galenica e chimica* (Lisbon: Impressão no Real Mosteyro de São Vicente de Fóra, 1711); Neto, "A botica do real mosteiro," 10–11.
54. Sousa Dias, "Inovação técnica," 2:696–702.
55. Arquivo Histórico do Tribunal de Contas, *Junta da Inconfidência*, no. 112, fols. 58–73, cited in Sousa Dias, "Inovação técnica," 2:626–33.
56. Ibid., 697–99.
57. Ibid.
58. Russell-Wood, *A World on the Move*, 126–27.
59. *Lista da botica de São Roque*, Arquivo Histórico do Tribunal de Contas,

Junta da Inconfidência, no. 112, fols. 58–73, cited in Sousa Dias, "Inovação técnica," 2:626–33.

60. Ibid.

61. Maria Benedita Araújo, *O conhecimento empírico dos fármacos nos séculos XVII e XVIII* (Lisbon: Edições Cosmos, 1992), 20–21.

62. Ibid.

63. HAG MR 85, fol. 59v.

64. HAG MR 135B, fol. 489v.

65. Van Linschoten, *Voyage of John Huyghen Van Linschoten*, 2:113–14.

66. HAG 1736, fols. 9v–12r.

67. HAG 8031, fols. 4r–v.

68. HAG MR 46A, fols. 96r–97v.

69. Ibid.

70. HAG 1346, fol. 183.

71. Fátima da Silva Gracias, *Health and Hygiene in Colonial Goa, 1510–1961* (New Delhi: Concept Publishing, 1994), 105–6.

72. HAG 1429, fol. 229.

73. HAG 1436, fol. 11r–v.

74. Central Library of Panaji, Goa, ms. no. 18, fols. 2–58.

75. Ibid., 27–36.

76. Ibid., 28.

77. Ibid., 27.

78. Ibid., 32.

79. Eduardo de Castro Almeida, *Inventário dos documentos relativos ao Brasil existantes no Archivo de Marinha e Ultramar de Lisboa*, vol. 1, *Bahia, 1613–1762*, doc. 5018 (Rio de Janeiro: Officinas Graphicas da Biblioteca Nacional, 1913), 401.

80. Ibid.

81. Ibid.

82. ANTT, Ministério do Reino, *caixa* 555, maço 444.

83. Ibid., fol. 2.

84. AHU, São Tomé and Príncipe Collection, *caixa* 55, doc. 75.

85. Arquivo Histórico de São Tomé e Príncipe, *Alfandega* of São Tomé (1 February 1899). See also *Sociedade e emigração para São Thomé e Príncipe*, Relatorio da Direcção; Paracer do Conselho Fiscal; Lista dos Acionistas, anno 2 (Lisbon, 1914), 93.

86. HAG MR 178B, fol. 644.

87. Ibid., 644–64.

88. HAG MR 175, fols. 219–30.

89. HAG MR 178B, fols. 645–48.

90. HAG no. 646, fols. 28–29.

91. Ibid., 29.

92. See, for example, Mark Harrison, *Climates and Constitutions: Health, Race, Environmentalism and British Imperialism in India, 1600–1850* (New Delhi: Oxford University Press, 1999) and two collections of essays on the subject edited by David Arnold: *Imperial Medicine and Indigenous Societies* (New

Delhi: Oxford University Press, 1989) and *Warm Climates and Western Medicine: The Emergence of Tropical Medicine, 1500–1900* (Amsterdam: Editions Rodopi, 1996).

93. A. M. G. Rutten, *Dutch Transatlantic Medicine Trade in the Eighteenth Century under the Cover of the West India Company* (Rotterdam: Erasmus Publications, 2000); J. Heniger, *Hendrik Adriaan Van Reede Tot Drakenstein (1636–1691) and Hortus Malabaricus: A Contribution to the History of Dutch Colonial Botany* (Rotterdam: A. A. Balkema, 1986).

94. José Pedro Sousa Dias, "Bibliografia sobre a farmácia e a matéria médica da expansão e da colonização portuguesa (séculos XVI a XVIII)," in *Mare Liberum* 11–12 (January–December 1996), 165–207.

95. For rare examples of this literature in English, see articles by João Manuel Pacheco de Figueiredo, "The Practice of Indian Medicine in Goa," and his son, John M. de Figueiredo, "Ayurvedic Medicine in Goa According to European Sources in the Sixteenth and Seventeenth Centuries," in *Bulletin of the History of Medicine* 58, no. 2 (1984). See also Michael N. Pearson, "First Contacts between Indian and European Medical Systems: Goa in the Sixteenth Century," in *Warm Climates and Western Medicine: The Emergence of Tropical Medicine, 1500–1900*, ed. David Arnold (Amsterdam: Editions Rodopi, 1996), 20–41.

96. For example, see João Frada, "Contributos portugueses do período expansionista e da época colonial para as ciências médicas," in *Medicamento, História e Sociedade* 4, no. 6 (July 1995), 8–14.

97. Londa Schiebinger, *Plants and Empire: Colonial Bioprospecting in the Atlantic World* (Cambridge, MA: Harvard University Press, 2004).

98. Harold J. Cook, professorial inaugural lecture, University College London, 27 February 2003.

99. Harold J. Cook, *Global Commerce and the Rise of Science: Medicine and Natural History in the Dutch Golden Age* (New Haven, CT: Yale University Press, in press), 1–7; cited from correspondence and Dr. Cook's website at University College London: http://www.ucl.ac.uk/histmed/people/academics/cook .html (5 May 2006).

Chapter Fourteen

1. "Expediente sobre la remisión de 24 cajones de curiosidades de la Naturaleza y del arte, recogidas por el Obispo de Truxillo (hoi Arzobispo de Santa Fe) y remitidas por el Virrey de Lima, venidas en la Fragata Rosa" (Lima, 1789). Archivo General de Índias, Seville (AGI/S), Lima 798. Martínez Compañón was named Archbishop of Santa Fe in 1788, but did not arrive there to take up his duties until 1791. The tour of inspection took place between 1782 and 1785. His highly significant work is the current subject of a dissertation by Emily Berquist (University of Texas, Austin, 2007). For further bibliographic references, see Paz Cabello Carro, *Coleccionismo americano indigena en la España del siglo XVIII* (Madrid: Editiones de Cultura Hispánica, 1989), chap. 13.

2. *Relaciones geográficas* were compiled from periodic questionnaires sent to colonial officials in the Americas and the Philippines in order to gather

information on natural resources and local customs. I have argued elsewhere that the bureaucratic collections discussed here were part of that tradition. See Paula De Vos, "Natural History and the Pursuit of Empire in Eighteenth-Century Spain," *Eighteenth-Century Studies* 40, no. 2 (2007), 209–39 and "Research, Development, and Empire: State Support of Science in the Later Spanish Empire," *Colonial Latin American Review* 15, no. 1 (2006), 55–79. For studies of the *Relaciones*, see Francisco de Solano and Pilar Ponce (eds.), *Cuestionarios para la formación de las relaciones geográficas de Indias, siglos XVI–XIX* (Madrid: CSIC, 1988); Raquel Álvarez Peláez, *La conquista de la naturaleza americana* (Madrid: CSIC, 1993); Barbara Mundy, *The Mapping of New Spain: Indigenous Cartography and the Maps of the Relaciones Geográficas* (Chicago: University of Chicago Press, 1996); and Antonio Barrera-Osorio, *Experiencing Nature: The Spanish American Empire and the Early Scientific Revolution* (Austin: University of Texas Press, 2006).

3. De Vos, "Natural History and the Pursuit of Empire." In that essay, I focus on the "useful" natural history specimens gathered and shipped to Spain, whereas in this essay I focus on and try to explain the presence of an important minority of "curious" items also included in the shipments. Works that deal more generally with the relationship between science, economic reform, and social improvement in eighteenth-century Spain are Richard Herr, *The Eighteenth-Century Revolution in Spain* (Princeton, NJ: Princeton University Press, 1958); Jean Saraillh, *La España ilustrada de la segunda mitad del siglo XVIII*, trans. Antonio Alatorre (Mexico City: Fondo de Cultura Económica, 1957); and Francisco Javier Puerto Sarmiento, *La ilusión quebrada: Botánica, sanidad y política científica en la España ilustrada* (Barcelona: El Serbal/SCIC, 1988).

4. These categories are derived from Lorraine Daston's discussion of early modern classifications of the natural, nonnatural, preternatural, and supernatural in "The Nature of Nature in Early Modern Europe," *Configurations* 6, no. 2 (1998), 149–72, esp. 154–58.

5. Neil Kenny, *Curiosity in Early Modern Europe: Word Histories* (Wiesbaden, Germany: Harrassowitz, 1998); Peter Harrison, "Curiosity, Forbidden Knowledge, and the Reformation of Natural Philosophy in Early Modern England," *Isis* 92 (2001), 265–90; and Barbara M. Benedict, *Curiosity: A Cultural History of Early Modern Inquiry* (Chicago: University of Chicago Press, 2001).

6. Thomas da Costa Kaufmann, "From Mastery of the World to Mastery of Nature: The *Kunstkammer*, Politics, and Science," in *The Mastery of Nature: Aspects of Art, Science, and Humanism in the Renaissance* (Princeton, NJ: Princeton University Press, 1993), 186; and Lorraine Daston and Katharine Park, *Wonders and the Order of Nature, 1150–1750* (New York: Zone Books, 2001), 305–6.

7. Daston and Park, *Wonders*, chap. 3.

8. Ibid., 306.

9. Ibid., chap. 8, esp. 305–16.

10. See, for example, Krystof Pomian, "Collections in Venetia in the Heyday of Curiosity" and "The Age of Curiosity," in *Collectors and Curiosities:*

Paris and Venice, 1500–1800, ed. Krystof Pomian, trans. Elizabeth Wiles-Porier (Cambridge: Polity Press, 1990); and da Costa Kaufmann, *The Mastery of Nature*. For the insatiability of curiosity, see Daston and Park, *Wonders and the Order of Nature*, 307–8.

11. Pomian, *Collectors and Curiosities*, 75 and "The Age of Curiosity." Although scholars have pointed to the difficulties of categorizing such varied collections, most agree that they were characterized by heterogeneity, a predilection for the singular and the rare, and a fine disregard for both utilitarianism and Aristotelian oppositions between art and nature. See Daston and Park, *Wonders and the Order of Nature*, 265–76.

12. Paula Findlen, *Possessing Nature: Museums, Collecting, and Scientific Culture in Early Moden Italy* (Berkeley: University of California Press, 1994), 400–401; Daston and Park, *Wonders and the Order of Nature*, chap. 9.

13. The historiography of curiosity cabinets is vast; I will cite here some of the more prominent English-language works: Pomian, *Collectors and Curiosities*; Findlen, *Possessing Nature*; Joy Kenseth, *The Age of the Marvelous* (Hanover, NH: Hood Museum of Art, Dartmouth College, 1991); John Elsner and Roger Cardinal (eds.), *The Cultures of Collecting* (Cambridge, MA: Harvard University Press, 1994); Oliver Impey and Arthur MacGregor (eds.), *The Origins of Museums: The Cabinets of Curiosities in Sixteenth- and Seventeenth-Century Europe* (Oxford: Clarendon Press, 1985); Lawrence Wechsler, *Mr. Wilson's Cabinet of Wonder* (New York: Pantheon Books, 1995); Susan M. Pearce, *On Collecting: An Investigation into Collecting in the European Tradition* (New York: Routledge, 1995); and four works by Thomas da Costa Kaufmannn: "From Treasury to Museum: The Collections of the Austrian Habsburgs," in Elsner and Cardinal, *The Cultures of Collecting*, 137–54; "From Mastery of the World to Mastery of Nature"; "Remarks on the Collections of Rudolf II: The *Kunstkammer* as a Form of Representation," *Art Journal* 38 (1978), 22–28; and *The School of Prague: Painting at the Court of Rudolf II* (Chicago: University of Chicago Press, 1988).

14. These concepts and arguments come largely from scholars of the "new museology" and collecting, who have used Foucault's conceptions of "epistemes" and European ideas about nature and taxonomy and applied them to the development of museums. See Michel Foucault, *The Order of Things: An Archaeology of Knowledge* (New York: Pantheon Books, 1971). For an excellent overview of recent works in museum studies, see Randolph Starn, "A Historian's Brief Guide to New Museum Studies," *American Historical Review* 110, no. 1 (2005), 68–98. See also Evelyn Hooper-Greenhill, *Museums and the Shaping of Knowledge* (New York: Routledge, 1992); Peter Vergo (ed.), *The New Museology* (London: Reaktion Books, 1989); Pearce, *On Collecting*, chap. 6; Paula Findlen, "Epilogue," in *Possessing Nature*; and Kaufmann, "From Mastery of the World to Mastery of Nature." The same assumptions run through a number of other works, including Kenseth (ed.), *Age of the Marvelous*; Barbara Maria Stafford, *Artful Science: Enlightenment, Education, and the Eclipse of Visual Education* (Cambridge, MA: MIT Press, 1994); Laura Auricchio, "Pahin de la Blancherie's Commercial Cabinet of Curiosity (1779–1787)," *Eighteenth-Cen-*

tury Studies 36, no. 1 (2002), 47–61; and Arthur MacGregor, "The Ashmolean as a Museum of Natural History, 1683–1860," *Journal of the History of Collections* 13 (2001), 125–44.

15. Bettina Dietz and Thomas Nutz, "Collections Curieuses: The Aesthetics of Curiosity and Elite Lifestyle in Eighteenth-Century Paris," *Eighteenth-Century Life* 29, no. 3 (Fall 2005), 44–75.

16. Ibid., 45–46. See also Pearce, *On Collecting*, 123.

17. I would like to thank Daniela Bleichmar for astutely pointing this out to me. For Jefferson's Indian Hall, see Joyce Henri Robinson, "An American Cabinet of Curiosities: Thomas Jefferson's 'Indian Hall' at Monticello," in *Acts of Possession: Collecting in America*, ed. Leah Dilworth (New Brunswick, NJ: Rutgers University Press, 2003), 19–24. For discussion of Linneaus's private collection, see Lisbet Koerner, "Carl Linneaus in His Time and Place," in *Cultures of Natural History*, ed. N. Jardine, J. A. Secord, and E. C. Spary (Cambridge: Cambridge University Press, 1996), 153–54.

18. According to Kaufmann, Bacon's utopian vision of a scientific community inspired dramatic changes in the nature of collecting, and its "specific utilitarian end . . . explicitly does not belong to the culture of curiosity." Kaufmann, *Mastery of Nature*, 185. See also Marjorie Swann, *Curiosites and Texts: The Culture of Collecting in Early Modern England* (Philadelphia: University of Pennslvania Press, 2001), chap. 2. See also William Ashworth, "Natural History and the Emblematic World View," in *Reappraisals of the Scientific Revolution*, ed. David C. Lindberg and Robert S. Westman (Cambridge, UK: Cambridge University Press, 1990), 303–32 and Paula Findlen, "Francis Bacon and the Reform of Natural History in the Seventeenth Century," in *History and the Disciplines: The Reclassification of Knowledge in Early Modern Europe*, ed. Donald R. Kelley (Rochester, NY: University of Rochester Press, 1997). Although scholars like da Costa Kaufmannn recognize that Bacon's work and thought were specific to seventeenth-century England, Kaufmann argues that "his utilitarian considerations have been found to bear a broader cultural significance with regard to developments on the continent as in Britain." *Mastery of Nature*, 186–87.

19. See Antonio Barrera: *Experiencing Nature*; "Empire and Knowledge: Reporting from the New World," *Colonial Latin American Review* 15, no. 1 (2006), 39–54; and "Local Herbs, Global Medicines: Commerce, Knowledge, and Commodities in Spanish America," in *Merchants and Marvels: Commerce, Science, and Art in Early Modern Europe*, ed. Pamela H. Smith and Paula Findlen (New York: Routledge, 2001), 163–81. See also Jorge Cañizares-Esguerra, "Iberian Science in the Renaissance: Ignored How Much Longer?" *Perspectives on Science* 12, no. 1 (2004), 91. I also see Spanish antecedents to Baconian science in the program for economic botany established by the crown in Paula De Vos, "The Science of Spices: Empiricism and Economic Botany in the Early Spanish Empire," *Journal of World History* 17, no. 4 (2006), 399–427.

20. Cañizares-Esguerra, "Iberian Science in the Renaissance," 91. Kaufmann (in *Mastery of Nature*, 187) even admits that Bacon "articulated many of the tendencies in collecting that were evolving on the continent, and which he may indeed have observed before they came into play in England."

21. Barrera-Osorio, *Experiencing Nature*, 85. See also David Goodman, *Power and Penury: Government, Technology, and Science in Philip II's Spain* (Cambridge: Cambridge University Press, 1988) and Enrique Martínez Ruiz (ed.), *Felipe II, la ciencia y la técnica* (Madrid: Actas Editorial, 1999) for a discussion of Philip II's interest in science.

22. For the collections of Philip II, Charles V, Philip III, and Philip IV, see Cabello Carro, *Coleccionismo americano indigena* and Miguel Moran and Fernando Checa, *El coleccionismo en España: De la cámara de maravillas a la galería de pinturas* (Madrid: Ediciones Cátedra, 1985). For the collection of Americana in Europe outside of Spain, see Christian Feest, "Mexico and South America in the European Wunderkammer," in Impey and MacGregor, *Origins of Museums*, 237–44; D. Heikamp and F. Anders, *Mexico and the Medici* (Florence: Editrice Edam, 1972); Findlen, *Possessing Nature*; Anthony Alan Shelton, "Cabinets of Transgression: Renaissance Collections and the Incorporation of the New World," in Elsner and Cardinal, *The Cultures of Collecting*, 177–203; Carina Johnson, "Negotiating the Exotic: Aztec and Ottoman Culture in Habsburg Europe, 1500–1590" (PhD diss., University of California, Berkeley, 2000).

23. Barrera-Osorio, *Experiencing Nature*, 85. See also Ronald Lightbrown, "Some Notes on Spanish Baroque Collectors," in Impey and MacGregor, *Origins of Museums*, 136–46.

24. Barrera-Osorio, *Experiencing Nature*, 120–27. See also Marcy Norton, "Tasting Empire: Chocolate and the European Internalization of Mesoamerican Aesthetics," *American Historical Review* 111, no. 3 (June 2006), 660–91 on the commodification of nature in the early Spanish empire.

25. For a history of the natural history cabinet, see Agustín J. Barreiro, *El Museo Nacional de Ciencias Naturales 1771–1935* (Aranjuez, Spain: Doce Calles, 1992). For a catalogue of the Archive's documents, see Maria Ángeles Calatayud Arinero, *Catálogo de documentos del Real Gabinete de Historia Natural (1787–1815)* (Madrid: CSIC, Museo Nacional de Ciencias Naturales, 1987).

26. AGI/S, Estado 53, no. 70, 1a, fol. 1r.

27. I am particularly indebted to the essay by Susan Deans-Smith, "Creating the Colonial Subject: Casta Paintings, Collectors, and Critics in Eighteenth-Century Mexico and Spain," *Colonial Latin American Review* 14, no. 2 (2005), 169–204, which lays out much of this early eighteenth-century history of royal orders for collecting curiosities.

28. Order of 23 July 1712. Francisco Esteve Barba, "Papeles varios de interés americano de la Colección Borbón Lorenzana de la Biblioteca Pública de Toledo," *Revista de Indias* 77–78 (1959), 321–71, cited in Cabello Carro, *Coleccionismo americano indígena*, 60.

29. Ibid., 61.

30. AGI/S, Estado, 81, no. 28, fol. 1r and Estado, 52, no. 117, fol. 1r.

31. There are interesting connections to be made between the collection of curiosities, and particularly of ethnographica, with arguments about material culture that I do not have the space to explore fully here. Several scholars making this connection use Pierre Bordieu's *Distinction: A Social Critique of the*

Judgement of Taste, trans. Richard Nice (Cambridge, MA: Harvard University Press, 1984). These ideas are also related to growing consumerism in early modern Europe. See, for example, Chandra Mukerji, *From Graven Images: Patterns of Modern Materialism* (New York: Columbia University Press, 1983) and John Brewer and Roy Porter (eds.), *Consumption and the World of Goods* (London and New York: Routledge, 1993). For anthropological studies of the relationship between ethnography, material culture, and empire, see the works of Nicholas Thomas.

32. For Dávila's collection, see María Angeles Calatayud Arinero, *Pedro Franco Dávila: Primer director del Real Gabinete de Historia Natural fundado por Carlos III* (Madrid: CSIC, Museo Nacional de Ciencias Naturales, 1988). For the history of how the museum obtained Dávila's collection, see Barreiro, *El Museo Nacional de Ciencias Naturales*, chap. 1.

33. Barreiro, *El Museo Nacional de Ciencias Naturales*, 62. Dietz and Nutz name Dávila's collection as one of their examples of an eighteenth-century Parisian natural history collection that included significant amounts of curiosities ("Collections Curieuses," 50). Dávila, who was born in Ecuador, moved to Paris in 1740 where he lived until he moved to Madrid in 1771 to take over as director of the museum.

34. Order of Gálvez, 1776, in Cabello Carro, *Coleccionismo americano indígena*, 62. This order is also cited in Deans-Smith, "Creating the Colonial Subject," 179.

35. Ibid.

36. Cabello Carro, *Coleccionismo americano indígena*, 61.

37. Although I have taken the eighteenth-century definitions of "curiosity" to heart, in discerning what was curious and what was useful in the hundreds of natural history shipments' documentation, I have had to use my own judgment. Undoubtedly, my twenty-first-century bias has come into play, particularly in the divisions I made between what I thought was useful and what I deemed curious, but I did follow some general rules. I designated all plants as useful. Although no doubt some were sent because of their particular size, rarity, or beauty, it was impossible to know which ones, and since plants could be cultivated and thus reproduced, I considered them useful. If a medicinal use was designated for an animal or animal part, I omitted it from consideration as a curiosity. And although many of the minerals listed here were chosen explicitly for their usefulness, I included them because they constituted a curiosity as a "representative example." Since minerals cannot be cultivated elsewhere, I counted these as singular examples in the Noachian tradition of making a representative. I counted all artificialia as curiosities unless, as was the case in one shipment, there were so many examples of one item—hundreds of dish sets or thousands of silk stockings, for example— that it was clear that these were being shipped to Spain in order to be sold, not displayed. As for the categorization of the specimens here, I have also tried to stay true to the types of categories specific in the royal orders for what constituted a curiosity, and the various subcategories that they used as well.

38. *Picure* and *cuchicuchi*: AGI/S, Indiferente 1549, Maracaibo, 1777, "No-

ticia de las cosas particulares de Isttoria Natural." *Tunata*: AGI/S, Indiferente 1549, Santa Fe, 1768.

39. Manila birds: AGI/S, Indiferente 1549, Manila, 1778, "Lista general de lo que para el Gabinete Histórico de S.M. llevan la Fragata Atrea y Urca Santa Ynes." Paraguay parrot: AGI/S, Estado, 81, no. 7.

40. Iguana: AGI/S Indiferente 1549, Manila, 1780, "Lista de los cajones que remite al Rey Nuestro Senior para su Real Gabinete el Gobernador de Philipinas." Alligator: AGI/S Indiferente 1549, Quayaquil, 1777, "Razón de los animales quadrupeos, bolatiles, y reptiles."

41. AGI/S Indiferente 1549, Quayaquil, 1777, "Razón de los animales quadrupeos, bolatiles, y reptiles."

42. Birds from Cartagena: AGI/S Lima, 798, Cartagena, 1789, "Expediente sobre la remission a Madrid de aves, animals, y esqueletos que ha enviado para S.M. el Virrey de Santa Fe." Birds from Manila: AGI/S Indiferente 1549, Manila, 1778, "Lista general de lo que para el Gabinete historico de S.M. llevan la Fratgata Astea y Urca Santa Ynes." Ducks from Guayaquil: AGI/S Indiferente 1549, Quayaquil, 1777, "Razón de los animales quadrupeos, bolatiles, y reptiles."

43. AGI/S Indiferente 1549, Quayaquil, 1777, "Razón de los animales quadrupeos, bolatiles, y reptiles."

44. AGI/S Indiferente 1549, Manila 1780, "Lista de los cajones que remite al Rey Nuestro Senior para su Real Gabinete el Gobernador de Philipinas."

45. AGI/S Lima 798, 1789, "Expediente sobre la remission a Md. de aves, animals, y esqueletos que ha enviado pa S.M. el Virrey de Sta. Fe."

46. AGI/S Indiferente 1549, Quayaquil, 1777, "Razón de los animales quadrupeos, bolatiles, y reptiles."

47. AGI/S Lima, 798, Cartagena, 1789, "Expediente sobre la remission a Madrid de Aves, Animals, y Esqueletos que ha enviado para S.M. el Virrey de Santa Fe."

48. Shells: AGI/S Indiferente 1549, S/L, 1777, "Nomina de las conchas, caracoles, y otras cosas curiosas de la naturaleza." For the very interesting history of shell collecting and conchology, see Henry Coomans, "Conchology before Linnaeus," in Impey and MacGregor, *Origins of Museums*, 188–92; and S. Peter Dance, *Shell Collecting: An Illustrated History* (London: Faber, 1966).

49. AGI/S, Indiferente 1549, Santa Fe, 1768.

50. See Table 14.4 for references to various items.

51. Santa Fe: AGI/S, Estado, 52, no. 117; Buenos Aires: AGI/S, Estado, 81, no. 28.

52. AGI/S, Estado, 81, no. 28.

53. Ibid.

54. Pita thread and pre-Columbian textile fragments: AGI/S Indifernte 1549, Maracaibo, 1777, "Noticia de las coasas particulares de Isttoria Natural." The fact that they wanted samples of thread that made stockings/hose could indicate a commercial intent—although as Chandra Mukerji has argued, there is sometimes little difference between capital accumulation and luxury spending; the same might be said here, that at times there is a fine line, and I have used my judgment to distinguish the two.

55. Wooden hats: AGI/S Indiferente 1549 Manila, 1778, "Lista general de lo que para el Gabinete histórico de S.M. llevan la Fragata Astrea y Urca Santa Ynes." Straw hats: AGI/S Indiferente 1549, Manila, 1780, "Lista de los cajones que remite al Rey Nuestro Senior para su Real Gabinete el Gobernador de Philipines." Purse: AGI/S Indiferente 1549, Manila, 1777, "Relación de lo que conduce Don Pedro Vares."

56. Feather rug: AGI/S, Indiferente 1549, Peru, 1772. Cups, bottles, glasses, tobacco case, and inkstand: Maracaibo, 1777.

57. Light machine: AGI/S Indiferente 1549, Peru, 1772. Walking sticks: AGI/S Indiferente 1549 Manila, 1778, "Lista general de lo que para el Gabinete histórico de S.M. llevan la Fragata Astrea y Urca Santa Ynes."

58. AGI/S Indiferente 1549, Manila, 1780, "Lista de los cajones que remite al Rey Nuestro Senior para su Real Gabinete el Gobernador de Philipines."

59. Ibid.

60. Coconuts: AGI/S Indiferente 1549, Maracaibo, 1777, "Noticia de las coasas particulares de Isttoria Natural." Cross: Santa Fe, n.d., cited in Angel L. Viloria and B. Urbani, "Curiosidades para el Rey: Relación de objetos enviados en el siglo XVIII al Real Gabinete de Historia Natural de Madrid desde el Nuevo Mundo," *Llull: Revista de la Sociedad Española de Historia de las Ciencias y de las Técnicas* 25, no. 52 (2002), 213. Tamarind seed: Santo Domingo, 1777, "Curiosidades para el Rey," 207, 210.

61. In using the terms "monster" and "monstrous," I am translating directly from the Spanish *monstruo* or *monstruoso* in the documents.

62. Daston, "The Nature of Nature," 162–63. Of course, it could be argued that there are only four examples of "monsters" among thousands of specimens, but the shipment of the horse in particular indicates to me that monsters were still of considerable interest. In addition, Aguirre states (in Barreiro, *El Museo Nacional de Ciencias Naturales*, 28) that there was a considerable interest in monsters throughout the museum's early history, as evident in the 1784 publication of a catalogue of the museum's "animals and monsters" by Juan Bautista Brú titled *Tomo I de la colección de animales y monstruos del Real Gabinete de Historia Natural de Madrid*. It was followed by a second volume published in 1786. See Barreiro, *El Museo Nacional de Ciencias Naturales*, 71–73.

63. Daston and Park, chap. 5, esp. 201–14. This chapter is a revision of an earlier paper. See also Lorraine Daston and Katharine Park, "Hermaphrodites in Renaissance France," *Critical Matrix* 1 (1985), 1–19.

64. AGI/S, Indiferente 1550 (Nicaragua, 1782).

65. The following account comes from "Expediente sobre la remisión a V. Md. de aves, animales, y esqueletos que ha enviado para S.M. el Virrey de Sta. Fe," AGI/S, Lima 798 (Cartagena, 1789). Letter from José de Astigarragas to Antonio Valdés (Santa Marta, 2 April 1789). Although the horse was considered hermaphrodite, its sender referred to it in the feminine, thus I have done the same here.

66. Pomian, *Collectors and Curiosities*.

67. The term is taken from da Costa Kaufmannn's "From Mastery of the World to Mastery of Nature," 176–84, though the theme is also present in his other works.

68. This idea was promoted early on in a now-classic work on collecting: Julius von Schlosser, *Die Kunst- und Wunderkammer der Spattrenaissance: Ein Beitrag zur Geschite des Sammelwesens* (Leipzig: Klinkhardt and Biermann, 1908). The political importance of the curiosity cabinet is also discussed in the works of da Costa Kaufmannn, Findlen, and Pomian as well as Gerald l'E. Turner, "The Cabinet of Experimental Philosophy," in Impey and MacGregor, *Origins of Museums*, and the work of Susan Stewart.

69. Maya Jasanoff, "Collectors of Empire: Objects, Conquests, and Imperial Self-Fashioning," *Past and Present* 184 (August 2004), 110, 112.

70. See ibid. and also Janet Owen, "Collecting Artifacts, Acquiring Empire: Exploring the Relationship between Enlightenment and Darwinist Collecting and Late-Nineteenth-Century British Imperialism," *Journal of the History of Collections* 18, no. 1 (2006), 9–25, who argues that artifacts collected in the eighteenth century acquired further significance in the nineteenth century as concrete material objects used to justify British imperialism by showing the "primitive" nature of the artifacts' cultures of origin.

71. Jorge Cañizares-Esguerra, "Eighteenth-Century Spanish Political Economy: Epistemology and Decline," in *Eighteenth-Century Thought*, ed. James G. Buickerwood (New York: AMS Press, 2003), 1:295–314.

72. The quote is taken from Deans-Smith, "Creating the Colonial Subject," 176.

73. For more information on this "dispute of the New World," see Jorge Cañizares-Esguerra, *How to Write the History of the New World: Histories, Epistemologies, and Identities in the Eighteenth-Century Atlantic World* (Stanford, CA: Stanford University Press, 2001); and Antonello Gerbi, *The Dispute of the New World: The History of a Polemic, 1750–1900*, trans. Jeremy Moyle (Pittsburgh: University of Pittsburgh Press, 1973).

74. Deans-Smith, "Creating the Colonial Subject," 178.

Chapter Fifteen

1. On Mutis and the expedition he led, see Marcelo Frías Núñez, *Tras El Dorado vegetal: José Celestino Mutis y la Real Expedición Botanica del Nuevo Reino de Granada (1783–1808)* (Seville: Diputación Provincial de Sevilla, 1994); A. Federico Gredilla, *Biografía de José Celestino Mutis* (1911; Bogotá: Plaza & Janés, 1982); Gonzalo Hernández de Alba, *Quinas amargas, el sabio Mutis y la discusión naturalista del siglo XVIII* (Bogotá: Academia de Historia de Bogotá, 1991); Enrique Pérez Arbeláez, *José Celestino Mutis y la Real Expedición Botánica del Nuevo Reino de Granada*, 2nd ed. (Bogotá: Instituto Colombiano de Cultura Hispánica, 1983); Benjamín Villegas (ed.), *Mutis y la Real Expedición Botánica del Nuevo Reyno de Granada*, 2 vols. (Madrid and Barcelona: Villegas Editores/Lunwerg Editores, 1992).

2. Daniela Bleichmar, "Painting as Exploration: Visualizing Nature in Eighteenth-Century Colonial Science," *Colonial Latin American Review* 15, no. 1 (June 2006), 81–104.

3. On the importance of naming plants honorifically at the time, see Londa Schiebinger, *Plants and Empire: Colonial Bioprospecting in the Atlantic World* (Cambridge, MA, and London: Harvard University Press, 2004), 194–225 and Mauricio Nieto Olarte, *Remedios para el imperio: Historia natural y la apropiación del Nuevo Mundo* (Bogotá: Instituto Colombiano de Antropología e Historia, 2000), chap. 2.

4. My interpretation of this painting is supported by another portrait depicting Mutis in an enclosed study, sitting with a sprig of *mutisia* in his hand and in front of a table on which rests a microscope. On scientific portraiture, see Ludmilla Jordanova, *Defining Features: Scientific and Medical Portraits* (London: Reaktion, 2000).

5. Cédula real (Aranjuez, Spain, 8 April 1777), Archivo del Museo de Ciencias Naturales (AMCN), Madrid, item 13 in María de los Ángeles Calatayud Arinero, *Catálogo de las expediciones y viajes científicos españoles a América y Filipinas (siglos XVIII y XIX)* (Madrid: CSIC, 1984). The New Spain expedition received almost identically phrased instructions, *Real Orden* (El Pardo, 20 March 1787), Archivo del Real Jardín Botánico, Madrid (hereafter, ARJBM) V, 1, 1, 17. All translations are mine.

6. Londa Schiebinger vividly describes the high-stakes world of eighteenth-century economic botany in *Plants and Empire*, 1–104.

7. Lisbet Koerner, *Linnaeus: Nature and Nation* (Cambridge, MA: Harvard University Press, 1999).

8. Yves Laissus (ed.), *Les naturalistes français en Amérique du Sud XVIe–XIXe siècles* (Paris: Comité des Travaux Historiques et Scientifiques, 1995); David Philip Miller and Peter Hans Reill (eds.), *Visions of Empire: Voyages, Botany and Representations of Nature* (Cambridge: Cambridge University Press, 1996); Schiebinger, *Plants and Empire*, 23–104; Londa Schiebinger and Claudia Swan (eds.), *Colonial Botany: Science, Commerce, and Politics in the Early Modern World* (Philadelphia: University of Pennsylvania, 2005).

9. On the Spanish case, see Carmen Añón Feliú, *Real Jardín Botánico de Madrid, sus orígenes 1755–1781* (Madrid: Real Jardín Botánico, 1987); María de los Ángeles Calatayud Arinero, *Pedro Franco Dávila: Primer director del Real Gabinete de Historia Natural fundado por Carlos III* (Madrid: CSIC, Museo Nacional de Ciencias Naturales, 1988); and Francisco Javier Puerto Sarmiento, *La ilusión quebrada: Botánica, sanidad y política científica en la España ilustrada* (Madrid: CSIC, 1988) and *Ciencia de cámara: Casimiro Gómez Ortega (1741–1818), el científico cortesano* (Madrid: CSIC, 1992). See also Lucile H. Brockway, *Science and Colonial Expansion: The Role of the British Royal Botanical Gardens* (London: Academic Press, 1979); John Gascoigne, *Science in the Service of Empire: Joseph Banks, the British State and the Uses of Science in the Age of Revolution* (Cambridge: Cambridge University Press, 1998); Donald P. McCracken, *Gardens of Empire: Botanical Institutions of the Victorian British Empire* (London: Cassell and Leicester University Press, 1997); Chandra Mukerji, *Territorial Ambitions and the Gardens of Versailles* (Cambridge: Cambridge University Press, 1997); Susan Scott Parrish, *American Curiosity: Cultures of Natural History in the Colonial British Atlantic World* (Chapel Hill:

University of North Carolina Press, 2006); and Emma Spary, *Utopia's Garden: French Natural History from Old Regime to Revolution* (Chicago and London: University of Chicago Press, 2000).

10. Jorge Cañizares-Esguerra, "Eighteenth-Century Spanish Political Economy: Epistemology and Decline," in *Eighteenth-Century Thought*, vol. 1 (2003), 295–314.

11. Pedro Rodríguez de Campomanes, *Reflexiones sobre el comercio español a Indias* (1762; Madrid: Instituto de Estudios Fiscales, 1988).

12. Pedro Rodríguez de Campomanes, "Del comercio del tabaco, azúcar y cacao," *Reflexiones*, 67–88.

13. Casimiro Gómez Ortega, "Oración gratulatoria al tomar posesión de su plaza de académico supernumerario (Real Academia de la Historia)" (1770), cited in Puerto Sarmiento, *Ciencia de cámara*, 54.

14. Casimiro Gómez Ortega, *Instrucción sobre el modo más seguro y económico de transportar plantas vivas por mar y tierra a los países más distantes ilustrada con láminas: Añádese el método de desecar las plantas para formar herbarios* (1779; Madrid: Biblioteca de Clásicos de la Farmacia Española, 1992), 22.

15. Casimiro Gómez Ortega, *Historia natural de la malagueta o pimienta de Tavasco, y noticia de los usos, virtudes y exención de derechos de esta saludable y gustosa especia, con la lámina de su árbol* (Madrid, 1780), 1–2. The mention of "free trade" refers to a series of trade reforms ennacted by the Bourbons, collectively known as *comercio libre*, on which see Marcelo Bitar Letayf, *Economistas españoles del siglo XVIII: Sus ideas sobre la libertad del comercio con Indias* (Madrid: Ed. Cultura Hispánica, 1968).

16. Casimiro Gómez Ortega to José de Gálvez (Madrid, 23 February 1777), quoted in Puerto Sarmiento, *Ciencia de cámara*, 154–56.

17. Antonio José Cavanilles, "Materiales para la historia de la botánica," *Anales de Historia Natural* 2, no. 4 (Madrid: Imprenta Real, 1800; facsimile edition, Aranjuez, Spain: Doce Calles, 1993), 24. On Cavanilles, see Antonio González Bueno, *Antonio José Cavanilles (1745–1804): La pasión por la ciencia* (Madrid: Fundación Jorge Juan, 2002).

18. Koerner, *Linnaeus*, 152.

19. José Celestino Mutis to King Charles III (20 June 1764), ARJBM III, 2, 6, 10. Reproduced in José Celestino Mutis, *Archivo epistolar del sabio naturalista Don José C. Mutis*, 2nd ed., 4 vols., ed. Guillermo Hernández de Alba (Bogotá: Instituto Colombiano de Cultura Hispánica, 1983), 1:31–43.

20. The number comes from Antonio Lafuente and Nuria Valverde, "Linnaean Botany and Spanish Imperial Biopolitics," in Schiebinger and Swan, *Colonial Botany*, 136. The only English-language monographs on the eighteenth-century Spanish expeditions are two pioneering studies: Iris H. W. Engstrand, *Spanish Scientists in the New World: The Eighteenth-Century Expeditions* (Seattle: University of Washington Press, 1981); and Arthur Robert Steele, *Flowers for the King: The Expedition of Ruiz and Pavon and the Flora of Peru* (Durham, NC: Duke University Press, 1964). See also Daniela Bleichmar, "The Visual Culture of Natural History: Botanical Illustrations and Expeditions in the Eighteenth-

Century Spanish Atlantic" (PhD diss., History Department, Princeton University, 2005); Jorge Cañizares-Esguerra, "Spanish America: From Baroque to Modern Colonial Science," in *The Cambridge History of Science*, vol. 4, *Eighteenth-Century Science*, ed. Roy Porter (Cambridge: Cambridge University Press, 2003), 718–38; Paula S. De Vos, "Research, Development, and Empire: State Support of Science in the Later Spanish Empire," *Colonial Latin American Review* 15, no. 1 (June 2006), 55–79; Rob Iliffe, "Science and Voyages of Discovery," in Porter, *The Cambridge History of Science*, 4:618–48; Agustí Nieto-Galán, "The Images of Science in Modern Spain," in *The Sciences in the European Periphery during the Enlightenment*, ed. Kostas Gavroglu (Dordrecht, Netherlands: Kluwer Academic Publishing, 1999), 73–94; Mauricio Nieto Olarte, "Remedies for the Empire: The Eighteenth-Century Spanish Botanical Expeditions to the New World" (PhD thesis, Department of History of Science and Technology, Imperial College, London, 1993); and Marie Louise Pratt, *Imperial Eyes: Travel Writing and Transculturation* (London and New York: Routledge, 1992). From the vast Spanish-language literature, see Alejandro Díez Torres, et al. (eds.), *La ciencia española en ultramar* (Aranjuez, Spain: Doce Calles, 1991); Antonio González Bueno (ed.), *La expedición botánica al virreinato del Perú (1777–1788)* (Madrid: Lunwerg, 1988); Antonio Lafuente, Arturo Elena, and María L. Ortega (eds.), *Mundialización de la ciencia y la cultura nacional* (Aranjuez, Spain: Doce Calles, 1993); Antonio Lafuente and José Sala Catalá (eds.), *Ciencia colonial en América* (Madrid: Alianza Editorial, 1992); Félix Muñoz Garmendia (ed.), *La botánica al servicio de la Corona: La expedición de Ruiz, Pavón y Dombey al virreinato del Perú* (Madrid: Lunwerg Editores, 2003) and *La botánica ilustrada, Antonio José Cavanilles (1745–1804) jardines botánicos y expediciones científicas* (Madrid: Caja Madrid Obra Social, and Barcelona: Lunwerg Editores, 2004); Nieto Galán, *Remedios para el imperio*; Juan Pimentel, *La física de la monarquía: Ciencia y política en el pensamiento colonial de Alejandro Malaspina (1754–1810)* (Aranjuez, Spain: Doce Calles, 1998); José Sala Catalá, *Ciencia y técnica en la metropolización de América* (Aranjuez, Spain: Doce Calles, 1994); María Pilar de San Pío Aladrén (ed.), *El águila y el nopal: La expedición de Sessé y Mociño a Nueva España (1787–1803)* (Madrid: Lunwerg, 2000); B. Sánchez, Miguel Ángel Puig-Samper, and J. de la Sota (eds.), *La real expedición botánica a Nueva España 1787–1803* (Madrid: Real Jardín Botánico, 1987); José Luis Peset (ed.), *La ciencia moderna y el Nuevo Mundo* (Madrid: CSIC, 1985); José Luis Peset (ed.), *Ciencia, vida y espacio en Iberoamérica*, 3 vols. (Madrid: CSIC, 1989); and José Luis Peset, *La expedición Malaspina, 1789–1794*, 9 vols. (Madrid: Lunwerg Editores, 1987–1996).

21. James E. McClellan and François Regourd, "The Colonial Machine: French Science and Colonization in the Ancien Régime," in *Nature and Empire: Science and the Colonial Enterprise*, ed. Roy MacLeod, *Osiris* 15 (2000), 31–50.

22. These questionnaires had their origin in the sixteenth-century *Relaciones geográficas*, on which see Antonio Barrera-Osorio, *Experiencing Nature: The Spanish American Empire and the Early Scientific Revolution* (Austin: University of Texas Press, 2006); Barbara Mundy, *The Mapping of New Spain: Indigenous*

Cartography and the Maps of the Relaciones Geográficas (Chicago: University of Chicago Press, 1996); and Howard F. Cline, "The *Relaciones Geográficas* of the Spanish Indies, 1577–1648," in *Handbook of Middle American Indians*, general ed. Robert Wauchope, vol. 12, *Guide to the Ethno-historical Sources*, ed. Howard F. Cline (Austin: University of Texas Press, 1964–1976), 183–242. On the eighteenth-century *relaciones*, see De Vos, "Research, Development, and Empire."

23. Hipólito Ruiz López, *Quinología, o tratado del árbol de la quina o cascarilla, con su descripción y la de otras especies de quinos nuevamente descubiertas en el Perú* (Madrid, 1792); *Disertaciones sobre la raíz de la ratánhia de la calaguala y de la china y acerca de la yerba llamada canchalagua* (Madrid, 1796; facsimile edition, Madrid: Biblioteca de Clásicos de la Farmacia Española, 1992); "Memoria sobre las virtudes y usos de la planta llamada en América bejuco de la estrella," *Variedades de Ciencias, Literaturas y Artes* 3 (Madrid, 1805), 59–62. This last memoir was translated as "Memoir on the Virtues and Uses of the Plant Called in Peru the Star-Reed (*bejuco de la estrella*)," in Aymler Bourke Lambert, *An Illustration of the Genus Cinchona* (London, 1821).

24. On tea, see Mutis, *Archivo epistolar*, 3:61–62, 65–68. On cinnamon, see ibid., 1:347–48; 3:41, 61, 73. On nutmeg, see ARJBM III, 1, 2, 18. On the cinchona controversies, see Gonzalo Hernández de Alba, *Quinas amargas, el sabio Mutis y la discusión naturalista del siglo XVIII* (Bogotá: Academia de Historia de Bogotá, 1991). The Spanish cinchona trade is the subject of a dissertation in progress by Matthew Crawford (UCSD).

25. Hipólito Ruiz and José Pavón, *Prodromus flora peruvianae et chilensis* (Madrid, 1794); and *Flora peruviana et chilensis*, 3 vols. (Madrid, 1798–1802; facsimile edition, Madrid: CSIC, 1995).

26. *Real orden* (El Pardo, 20 March 1787), ARJBM V, 1, 1, 17.

27. Robert Steele, *Flowers for the King*, 85–86, 137, 140, 156. These shipments did not always reach their destination: a 1780 one was lost when the English intercepted the ship that carried it; the 1784 cargo disappeared when the vessel that conveyed it shipwrecked. The total number of images received in Spain is less significant for my argument than the volume that was produced and shipped from the colonies, which attests to the importance of image making.

28. Casimiro Gómez Ortega to José de Gálvez (24 November 1780), in Hipólito Ruiz, *Relación histórica del viage, que hizo a los reinos del Perú y Chile el botánico D. Hipólito Ruiz en el año 1777 hasta el de 1788, en cuya época regresó a Madrid*, 2nd ed., 2 vols., ed. Jaime Jaramillo Arango (Madrid: Real Academica de Ciencias Exactas Físicas y Naturales, 1952), 445.

29. Casimiro Gómez Ortega, "Instrucción que deberán observar los dibujantes que pasan al Perú de orden de S.M. para servir con el ejercicio de su profesión en la expedición botánica" (approved 9 April 1777), AMCN, item 7 in Calatayud Arinero, *Catálogo de las expediciones y viajes*, reproduced in Ruiz, *Relación histórica del viage*, 416–18.

30. Such handling of coloring was common practice at the time, as described regarding the Cook exhibition in H. Walter Lack and V. Ibáñez, "Recording Colour in Late Eighteenth-Century Botanical Drawings: Sydney Parkinson,

Ferdinand Bauer and Thaddäus Haenke," *Curtis's Botanical Magazine* 14, part 2 (1997), 87–100.

31. Jorge Escobedo, Visitador Superintendente General del Perú (29 June 1788), quoted in Ruiz, *Relación histórica del viage*, 469–70.

32. Mutis, *Archivo epistolar*, 2:65.

33. José Celestino Mutis, *Diario de observaciones de José Celestino Mutis (1760–1790)*, ed. Guillermo Hernández de Alba, 2nd ed., 2 vols. (Bogota: Instituto Colombiano de Cultura Hispánica, 1983), 1:154–56.

34. I discuss the acquisition of visual expertise in "Training the Naturalist's Eye in the Eighteenth Century: Perfect Global Visions and Local Blind Spots," in *Skilled Visions: Between Apprenticeship and Standards*, ed. Cristina Grasseni (New York and Oxford: Berghahn Books, 2007), 166–90.

35. Paula Findlen describes a similar role for books in sixteenth-century natural history in "The Formation of a Scientific Community: Natural History in Sixteenth-Century Italy," in *Natural Particulars: Nature and the Disciplines in Renaissance Europe*, ed. Anthony Grafton and Nancy Siraisi (Cambridge, MA: MIT Press, 1999), 369–400.

36. On the uses of books in botanical travel, see Daniela Bleichmar, "Exploration in Print: Books and Botanical Travel From Spain to the Americas in the Late Eighteenth Century," *Huntington Library Quarterly* 70, no. 1 (March 2007), 129–151.

37. Vicente Cervantes to José Celestino Mutis (Mexico City, 27 December 1788), ARJBM III, 1, 1, 83, fol. 6r–v., reproduced in *Archivo epistolar*, 3:219–23. Sessé praised Echeverría's talents and sent especially fine samples of his work to Spain; Martín de Sessé to Casimiro Gómez Ortega (Mexico City, 27 June 1788), 2v–3r, ARJBM V, 1, 1, 23.

38. Luis Née to José Celestino Mutis (Guayaquil, ca. 22 October 1790), ARJBM, III, 1, 1, 230, fol. 2r; *Archivo epistolar*, 4:74–76.

39. The drawings, specimens, and correspondence are held today in the archive of the Linnaean Society of London, and include a set of thirty-two very fine drawings (Ms. BL 1178).

40. Casimiro Gómez Ortega to José de Gálvez (24 November 1780), in Ruiz, *Relación histórica del viage*, 445.

41. Two classic works on this subject are Bernard Smith, *European Vision and the South Pacific*, 2nd ed. (New Haven, CT, and London: Yale University Press, 1985); and Barbara Maria Stafford, *Voyage into Substance: Art, Science, Nature and the Illustrated Travel Account, 1760–1830* (Cambridge, MA: MIT Press, 1984).

42. José Celestino Mutis to Juan José de Villaluenga, president of Quito Audiencia (10 July 1786), ARJBM III, 2, 2, 196 and 197; reproduced in *Archivo epistolar*, 1:316.

Afterword

1. J. H. Parry, *The Age of Reconnaissance* (New York: World Publishing Company, 1963).

2. Juan Gil and Consuelo Varela, *Temas colombinos* (Seville: Escuela de Estudios Hispanoamericanos, 1986), 1–48; Juan Gil and Consuelo Varela (eds.), *Cartas de particulares a Colón y relaciones coetáneas* (Madrid: Alianza Editorial, 1984), 152–76.

3. Martín de la Cruz, *Libellus de medicinalibus indorum herbis: Manuscrito azteca de 1552 según traducción latina de Juan Badiano* (Mexico City: Fondo de Cultura Económica, 1991).

4. Nicolás Bautista Monardes, *Historia medicinal de las cosas que se traen de nuestras Indias occidentales que sirven en medicina* (1574; Seville: Padilla Libros, 1988).

Index

Printed and bound by CPI Group (UK) Ltd, Croydon, CR0 4YY

16/04/2025

14658401-0005